MONO
STATISTICS AND ,

Gene.... .._....

D.R. Cox, D.V. Hinkley, D.B. Rubin and B.W. Silverman

(Full details concerning this series are available from the Publishers.)

Analysis of Quantal Response Data

B. J. T. MORGAN

Professor of Applied Statistics
Institute of Mathematics and Statistics
University of Kent
Canterbury

CRC Press
Taylor & Francis Group
Boca Raton London New York

CRC Press is an imprint of the
Taylor & Francis Group, an **informa** business
A CHAPMAN & HALL BOOK

First published 1992 by Chapman & Hall

Published 2019 by CRC Press
Taylor & Francis Group
6000 Broken Sound Parkway NW, Suite 300 Boca
Raton, FL 33487-2742

© 1992 B. J. T. Morgan
CRC Press is an imprint of Taylor & Francis Group, an Informa business
First edition 1992

First issued in paperback 2019

No claim to original U.S. Government works

ISBN-13: 978-0-367-45024-3 (pbk)
ISBN-13: 978-0-412-31750-7 (hbk)

Visit the Taylor & Francis Web site at
http://www.taylorandfrancis.com

and the CRC Press Web site at
http://www.crcpress.com

A catalogue record for this book is available from the British Library

Library of Congress Cataloging-in-Publication data available
Morgan, Byron J. T., 1946–
    Analysis of quantal response data / B. J. T. Morgan. – 1st ed.
      p. cm. – (Monographs on statistics and applied probability; 46)
    Includes bibliographical references and indexes.
    ISBN 0–412–31750–8
    1. Biometry. 2. Probits. 3. Medical statistics. 4. Quantum statistics. I. Title. II. Series.
QH323.5.M67 1992   92–19056
574'.01'5195–dc20   CIP

Typeset in 10/12 Times by Thomson Press (India) Ltd, New Delhi

'All substances are poisons; there is none which is not a poison. The right dose differentiates a poison and a remedy.'

Paracelsus (1493–1541)

Contents

Preface

This book has grown out of a lecture course on biometry given to M.Sc. students in statistics at the University of Kent. The standard reference for the course was the book *Probit Analysis* by Professor D. J. Finney. It is now 20 years since the appearence of the 3rd edition of *Probit Analysis* and there have been many developments in statistics of relevance for the analysis of quantal response data during this time, in design, sequential methods, non-parametric procedures, over-dispersion, robust methods, Bayesian approaches, extended models, influence and diagnostics, synergy and many other areas. The single most important development is probably the introduction of generalized linear models, allied to specialist computer packages for fitting these models. Most computer packages now provide a menu of relevant procedures for quantal assay data. Additionally the whole computing scene has changed dramatically, with the move towards powerful personal computers and workstations. The aim of this book has been to describe the new developments for the analysis of quantal response data, and to emphasize the links between the various different areas.

Several extra-mural courses have been given on the text material. The first of these was at the Royal Melbourne Institute of Technology, given jointly with Professor R. G. Jarrett. The last was at Duphar, in Weesp in The Netherlands, and in between two residential courses were given at the University of Kent. The interest shown by the course participants was one motivation for writing this book.

Quantal response data are quite often used to illustrate statistical techniques, and readers of the book will find that they will encounter many different areas of statistics. The book may be read by people with a range of different backgrounds. It is designed to be read as a coherent text or as a source of reference. Numerate scientists should be able to follow many of the arguments. However, for a full understanding a mathematics background is necessary. Much of the material should be accessible to third year mathematics and statistics undergraduates in British universities who have had foundation and second-level courses in statistics in their first two years of study. The book should be ideal for study at the postgraduate level by students of statistics and biometry. There are 267 Exercises to help with the use of the book as a course text.

The first four Appendices help to make the book complete, and the fifth summarizes useful computing facilities. For illustration, a number of GLIM macros appear in the text, and a small number of examples are given in BASIC and MINITAB. However, prior knowledge of these packages/languages is not a prerequisite for understanding the material of the book.

Well over 50 data sets are presented. Several of these now have **classic** status, in that they are, sometimes uncritically, regularly used to illustrate new procedures. Some of the examples have arisen from consulting experience with the Division of Animal Health at CSIRO, Melbourne, with Pfizer Central Research, Sandwich, Kent, and with Shell Research, Sittingbourne, Kent.

I am grateful to many individuals for their help and comments while this book has been written. At the Biometry Division of Pfizer, Kent, a range of problems were raised and discussed by P. Colman, R. Hews, T. Lewis, H. Ross-Parker and R. White. I have been particularly fortunate in supervising two postgraduate students working in relevant areas. The material of Chapters 5 and 6 owes a clear debt to the Ph.D. thesis of Simon Pack, who worked on a CASE award with Wellcome Research Laboratories, Beckenham, Kent, while Chapter 4 has likewise benefited from the M.Sc. dissertation of Paul Goedhart.

Deborah Ashby read the entire book as referee, and both Martin Ridout and David Smith read particular chapters. To these three I am most grateful for many helpful corrections and comments. Prominent amongst the others whom I should thank are: Beverley Balkau, John Fenlon, Janneke Hoestra, Hans Jansen and Richard Jarrett. Michael Bremner provided computing advice and help with the troff system. Encouragement was provided by Professors B. M. Bennett and A. A. Rayner, and the late David Williams, and useful advice by Sir David Cox. Parts of the book were typed by Lilian Bond and Arija Crux but the lion's share of the labour was carried out by Mavis Swain, who surpassed even her legendary typing skills with great humour and patience.

Finally I thank my wife, Janet, and children, Chloë and Leo, for their tolerant acceptance of my regular weekend absences over the four-year period when the book was written.

Byron J. T. Morgan
Canterbury

Glossary and notation

Except where noted below, a standard notation is used throughout the book. A small number of notational clashes have been adopted between different chapters if that improved comprehension or corresponded to standard usage.

corpus luteum: glandular tissue in the ovary, which forms after rupture of the follicle at ovulation. It secretes progesterone.

dominant lethal test: experiment in which experimental units are male animals, and each male is mated to one or more females.

fecundability: probability of conception per menstrual cycle.

implant: used here to denote egg implanted in womb following fertilization.

isolate: a pure culture of an organism.

micromelia: abnormally small size of the arms or legs.

minimum inhibitory concentration: lowest concentration (of an anti-infective agent) at which a particular organism's growth is inhibited.

phocomelia: congenital absence of the upper arm and/or upper leg (e.g. as side-effect of thalidomide).

E[ ]: expectation.

V( ): variance.

Pr( ): probability.

L: likelihood.

l: log-likelihood.

D: deviance.

X^2: Pearson goodness-of-fit statistic.

$N(\mu, \sigma^2)$: normal distribution, mean μ, variance σ^2.

$\Phi(x)$: standard normal c.d.f.

$\phi(x)$: standard normal p.d.f.

$\Phi(x, y)$: bivariate standard normal c.d.f.

ED_{100p} $(LD_{100p},\ EC_{100p}) = \theta_p$ (for $p \neq 0.5$); $ED_{50} = \theta$ (but see also beta-distributions, below).

$\hat{\theta}_R$: Reed–Muench estimator of θ; $\hat{\theta}_D$: Dragstedt–Behrens estimator of θ.

E_M, E_B, E_{DB}, estimators of θ from the up-and-down experiment.

k: number of doses, cases or treatments.

m: number of signal presentations (Chapter 3)/number of sampling times (Chapter 5)/number of litters in a group (Chapter 6).

$\{d_i,\ 1 \leqslant i \leqslant k\}$: doses.

[More generally $\{x_i\}$, or $\{z_i\}$, when doses are not involved, or a transformation is used.]

$\Delta_i = (d_{i+1} - d_i)$.

n_i individuals are treated at dose d_i and r_i respond.

In the time interval $(0,\ t_j)$, r_{ij} respond to dose d_i; $n_{ij} = r_{ij} - r_{i,j-1}$ (Chapter 5).

Of n_{ij} insects exposed to a_i units of A and b_j units of B, r_{ij} die (Chapter 3).

$P_i = P(d_i) =$ probability of response to dose d_i/probability that x_j respond out of n_j in the mixture model of equation (6.12).

$\mathbf{P}_j$: transition matrix for movement between states between times t_{j-1} and t_j (Chapter 5).

$p_i = r_i/n_i$ (Chapter 2).

$P(t|d) =$ probability of response to dose d by time t.

$p_{ij} =$ probability of response to dose d_i in time-interval (t_{j-1}, t_j).

$s_i = r_i - n_i P_i$.

$$\tilde{s}_i = \frac{r_i - n_i \hat{P}_i}{\sqrt{n_i \hat{P}_i (1 - \hat{P}_i)}}$$

t: used to denote time/iterate number, as in $\zeta^{(t)}$.

Bin (n, p): binomial distribution, index n and probability p.

$B(\alpha, \beta)$: beta function: $\Gamma(\alpha)\Gamma(\beta)/\Gamma(\alpha + \beta)$.

$I_{k,\varsigma}(x)$: incomplete gamma integral.

μ: mean of a random variable, especially (Chapter 6) for a beta random variable.

$\hat{\mu}$: Spearman–Kärber estimate of θ (estimate of mean of tolerance distribution).

$\hat{\mu}_\alpha$: $\alpha\%$ trimmed Spearman–Kärber estimate.

μ_j: jth non-central moment of tolerance distribution.

$\mu_{(j)}$: jth central moment of tolerance distribution.

h: number of hits in multi-hit model.

$h(\mu)$, $h(\mu; \lambda)$: link functions.

$h_v(q)$: quantits.

$R(\alpha)$: Mills ratio.

$F(x)$: cumulative distribution function (c.d.f.).

$\hat{F}(x)$: empirical c.d.f.

$\tilde{F}(x)$: estimate of c.d.f. by linear interpolation from ABERS estimate.

$\text{tr}(\mathbf{A})$: trace of matrix $\mathbf{A}$.

$\mathbf{I}(v, \zeta)$, $\mathbf{J}(v, \zeta)$; Fisher information/expected information matrices.

$\psi(u)$: kernel function (Chapter 7).

$\phi_{T,F}(x)$: influence curve.

$SC_{T,m}(x)$: Tukey's sensitivity curve.

$IC_{T,F,D}(d, y)$: influence curve based on dose mesh D.

$IC_{T,F}(d, y)$: influence curve.

(α, β): location-scale pair of parameters/standard parameterization for beta distribution.

(μ, θ): alternative parameterization for beta distribution (but note also use of θ for ED_{50}).

$\rho = \theta/(1 + \theta)$, and also more generally as correlation between littermates/also ρ is used to denote relative potency.

m_i: number of litters in ith treatment group.

n_{ij}: size of jth litter in ith treatment group.

x_{ij}: number responding out of n_{ij} (Chapter 6).

p_i: probability of response for all litters in ith treatment group (Chapter 6).

[NB: subscript i is sometimes dropped for simplification.]

$\mathbf{X} = \{x_{ij}\}$, the design matrix (except Chapter 6).

Index of data sets

CHAPTER 1

Data, preliminary analyses and mechanistic models

1.1 Introduction

The names Hiroshima, Chernobyl and Thalidomide are synonymous with twentieth century tragedies. Many years after the exploding of atomic bombs over Hiroshima and Nagasaki, the effects of exposure to nuclear radiation on the survivors are quantifiable in terms of increased incidence of leukaemia, as described in the paper by Armitage and Doll (1962). An example involving chromosome aberration in survivors of Hiroshima is considered in Chapter 6. The drug **Thalidomide** had been prescribed as a safe hypnotic drug, but the winter of 1961 saw the horrifying reports of its use resulting in babies born with deformities of **phocomelia**, or **micromelia**. As Beedie and Davies (1981) ominously wrote, 'It had not been tested in animals for teratogenicity, but thousands of babies born to mothers who had taken the drug during pregnancy provided the missing data.'

Common to these two illustrations is the exposure of human beings to substances that are either unnatural, or provided at unnaturally high levels. The response of individuals, adults or embryos, is binary: they are either affected by the time they are inspected, or they are not. In more general terms, discrete responses may take a variety of forms, such as reduction of pain, alleviation of breathing problems, improvement in acne, remission from leukaemia, and so on. Data which quantify the effect of exposure of individuals to substances such as new drugs, or to radiation, are often described as discrete, or quantal. Responses need not just be binary, and later we shall see examples of quantal data which may result in three or more possible outcomes.

This book is concerned with the analysis of quantal response data, sometimes called dose-response data, or quantal assay data. Such data may arise in a wide variety of different areas as we shall see, and may be collected from a properly designed scientific experiment, or result from observational studies. Thus, for example, girls of different ages may be classified by whether or not they have started menstruation; patches of woollen fabrics may be assessed for the degree of 'prickle' they elicit in human subjects; viruses ingested by insects may or may not kill them; widely used food additives may be tested for their undesirable side-effects. An example of this was cited in an article in *The Independent* newspaper of 23 September 1987: 'Several preservatives may cause asthmatic reactions in susceptible people. And one, *methyl paraben* (E218), is the main volatile component in the vaginal secretions of beagles – it may cause socially embarrassing behaviour in dogs. E218 is used in beer and coffee and many other foods.'

The main emphasis in the examples which illustrate the methodology of the book will be on the evaluation and testing of substances, mainly drugs, for use in humans. Frequently the effect investigated is whether or not there is a positive outcome from using the drugs resulting in efficacy studies; however, it is also of vital importance to consider the possible harmful side-effects of otherwise potentially beneficial treatments. Thus, for example, patients suffering from the spine-fusing disease, Ankylosing Spondylitis, may be treated by radiation therapy, but leukaemia may result as an undesired side-effect (Court Brown and Doll, 1957). The radiation used in mammography has been estimated as likely to cause just one excess cancer per 10^6 million women screened (Whitehouse, 1985; see also Breslow and Day, 1980, p. 62). The Babylonian Code of Hammurabi, of 2200 BC, ordained that if a patient died, the treating physician should lose his hands, and this is regarded as the first example, indeed a somewhat extreme one, of the need for regulation of procedures for treating human beings. In modern times, pharmacopoeias have been devised throughout the world, presenting standards for drug purity. The first statute to control drug quality in America was passed in 1848, while as recently as 1968 the Medicines Act of Great Britain produced new safeguards for the development, production and use of new drugs.

Because a number of the examples in this book are drawn from toxicology, it is worthwhile outlining important aspects of toxicology before we start, and this is done in the next section. An excellent introduction to the statistical aspects of the full range of drug development and testing is given by Salsburg (1990).

1.2 Aspects of toxicology

The activity of chemical substances can sometimes be gauged from their physico-chemical properties, and the Quantitative Structure–Activity Relationship (QSAR) procedures described by Bergman and Gittins (1985) are designed to search for new active substances using physical structure and electrochemical property correlates with established substances of known performance.

New chemicals may also be tested *in vitro*. Thus for example the Ames test for mutagenicity positively identified 157 out of a series of 175 known carcinogens (McCann *et al.*, 1975). Ultimately, however, tests *in vivo* are necessary. The revolutionary oral anti-fungal drug fluconazole was not found to be especially effective *in vitro*: the 'modest *in vitro* profile understates the excellent *in vivo* activity of fluconazole demonstrated in animal models of fungal infections and in clinical trials' (Marriott and Richardson, 1987). See also Exercise 1.26. In using non-human animals as models for humans the basic assumption is always that the model is appropriate. With the possible exception of arsenic, all known chemical carcinogens in humans are carcinogenic in some, but not all species of animals used in laboratories, so the model has to be chosen with care (Klaassen, 1986). For further discussion on the extrapolation from animals to humans, see Mantel and Bryan (1961), Cornfield (1977) and Park and Snee (1983).

Carageenan, which is a seaweed extract, is used in products such as ice cream and biscuits, yet it causes changes resembling ulcerative colitis in the bowels of guinea pigs, rabbits and mice. Inevitably effects such as these are the result of doses given at far higher levels than those commonly encountered in foods, and to relatively small groups of animals. This is a standard toxicological procedure, and is necessary in order to reduce cost and unnecessary suffering in experimental animals. The difficult problem is then to extrapolate from a known dangerous dose in animals to a virtually safe one for human consumption, and we discuss this fundamental problem in sections 1.6 and 4.6.

Different toxins may be administered in different ways, for example

Table 1.1

	Weight (g)	Dosage (mg/kg)	Dose (mg/animal)	Surface area (cm²)	Dosage (mg/cm²)
Mouse	20	100	2	46	0.043
Rat	200	100	20	325	0.061
Guinea pig	400	100	40	565	0.071
Rabbit	1500	100	150	1270	0.118
Cat	2000	100	200	1380	0.145
Monkey	4000	100	400	2980	0.134
Dog	12000	100	1200	5770	0.207
Human	70000	100	7000	18000	0.388

through ingestion, by contact with the skin, or by intravenous injection, and their effect can be radically affected by the size of the animal tested. Thus it is quite usual for dosages to be given in mg/kg of body weight, for example, or mg/cm² of body area. Table 1.1, taken from Klaassen (1986), shows how a constant dosage measured in mg/kg translates into different overall doses per animal, for a variety of species, and different dosages in terms of mg/cm². It is difficult to appreciate what a dosage measured in mg/kg actually becomes when scaled up to life-size, and Table 1.2, also taken from Klaassen (1986), provides the required interpretation, together with a crude toxicity rating to describe the different lethal doses. The distinction between dose and dosage that is drawn here will be maintained throughout the book.

Before new drugs can be tested in the standard progression of clinical trials on human subjects, they may be screened on a variety

Table 1.2

Toxicity rating or class	Probable lethal oral dose for humans	
	Dosage	For average adults
1. Practically nontoxic	> 15 g/kg	More than 1 quart
2. Slightly toxic	5–15 g/kg	Between pint and quart
3. Moderately toxic	0.5–5 g/kg	Between ounce and pint
4. Very toxic	50–500 mg/kg	Between teaspoonful and ounce
5. Extremely toxic	5–50 mg/kg	Between 7 drops and teaspoonful
6. Supertoxic	< 5 mg/kg	A taste (fewer than 7 drops)

of animals, with tests designed for a corresponding range of different effects. These include acute and chronic toxicity, with experiments in the latter case possibly running for a number of years, and usually performed on rats. Rabbits are the preferred animal for tests for eye and skin irritation, while the guinea pig is usually used for tests for skin sensitization, when this seems appropriate. Tests for possible teratological effects usually involve rats and rabbits, and substances are administered to males and/or females, before mating, and, for females, during gestation, and during lactation. Observations include the pregnancy rate and the viability of progeny, and study may continue for several generations. Mutation effects can be sought through a number of *in vivo* and *in vitro* procedures. The dominant lethal test, which we encounter again in Chapter 6, involves giving a male animal (usually a rodent) a single dose of the compound prior to mating with one or two females. The females are then killed before term, and numbers of live embryos and *corpora lutea* recorded for analysis. The extrapolation from animals to humans takes us through what has been referred to as the 'species barrier'.

We see that substances may be administered in a variety of ways, and by single or repeated doses. Substances which are toxic by one route of application may not prove toxic by another: the skin may prove to be an effective barrier to poisons; the liver may detoxify a substance given orally, which may be far more toxic if inhaled, for example. While a compound itself may not be toxic, a metabolite of it might be. Clearly tests must try to reflect the intended use of substances. If they are likely to find their way into water, they need to be assessed for possible effects on fish, crustacea and so forth. Aquatic experiments may differ from those on mammals in that exposure to the toxic agent may be continuous.

Many of the features described in this section will be encountered in the examples which now follow.

1.3 Examples

We shall now present a number of examples to illustrate the wide range of problems to be considered, and to provide instances of the different types of experiment described in the last section. In all cases response is quantal, and in most cases there is a single covariate, such as mean age group or dose level, which is deemed likely to affect the response. In some cases there are several covariates, which may, singly or in conjunction, influence the response. We shall see also the kinds of questions that arise and require answers in an appropriate statistical setting. The remainder of this chapter also serves as an introduction to the rest of the book.

Example 1.1

An experiment to assay an anti-pneumococcus serum (dose measured in cc).

Irwin (1937) analysed the data of Table 1.3. Groups of mice were given a serum inoculation, at various doses, prior to being infected with pneumococci.

We see that as the dose of serum is increased, the proportion of mice protected increases. The relationship between dose and resulting proportion is frequently simplified by transformation in each case. Here we have logarithms of doses, and the commonly used transformation, logit $(p) = \log_e \{p/(1 - p)\}$ of proportions. Of interest here is the serum level to set for routine anti-pneumococcus inoculation.

Table 1.3 *Effect of anti-pneumococcus serum on mice*

$10.158 + log_2$ (serum dose)	No. of mice protected	No. of mice in experiment	Proportion protected (p)	Logit (p)
1	0	40	0.000	–
2	2	40	0.050	– 2.944
3	14	40	0.350	– 0.619
4	19	40	0.475	– 0.100
5	30	40	0.750	1.099

Example 1.2   Age of menarche in 3918 Warsaw girls

This example differs from most of the others presented in this chapter in that the data arise from an **observational** study rather than an experimental one. However, we can see the qualitative similarity between the data of Tables 1.3 and 1.4, and we shall see later how they may be analysed by the same methods. Nevertheless there remains an important distinction between the two different types of study, and we shall at times find it necessary to emphasize this distinction.

Table 1.4 *Age of menarche in Polish girls*

Mean age of group (years)	No. having menstruated	No. of girls	Proportion having menstruated (p)	Logit (p)
9.21	0	376	0.000	–
10.21	0	200	0.000	–
10.58	0	93	0.000	–
10.83	2	120	0.017	– 4.076
11.08	2	90	0.022	– 3.784
11.33	5	88	0.057	– 2.809
11.58	10	105	0.095	– 2.251
11.83	17	111	0.153	– 1.710
12.08	16	100	0.160	– 1.658
12.33	29	93	0.312	– 0.792
12.58	39	100	0.390	– 0.447
12.83	51	108	0.472	– 0.111
13.08	47	99	0.475	– 0.101
13.33	67	106	0.632	0.541
13.58	81	105	0.771	1.216
13.83	88	117	0.752	1.110
14.08	79	98	0.806	1.425
14.33	90	97	0.928	2.554
14.58	113	120	0.942	2.781
14.83	95	102	0.931	2.608
15.08	117	122	0.959	3.153
15.33	107	111	0.964	3.283
15.58	92	94	0.979	3.829
15.83	112	114	0.982	4.025
17.58	1049	1049	1.000	–

These data were presented by Milicer and Szczotka (1966) and record, for a sample of 3918 Warsaw girls taken in 1963, whether or not they had reached menarche (onset of menstruation). This is probably the best known of a number of studies of age of menarche. Other studies include those by Burrell *et al.* (1961) and Milicer (1968). Data resulting from the second of these papers are presented in Exercise 2.23. Interestingly, differences are detectable between individuals of different race and of different socio-economic status. From a purely statistical point of view, in the **experimental** context, data sets as large as these are less frequently encountered than much smaller sets, such as that of Table 1.3, and may allow discrimination between competing simple probability models which usually are indistinguishable.

Example 1.3   The effect of insecticide on flour-beetles

Hewlett (1974) observed the effect of insecticide sprayed onto flour-beetles at four different concentrations. The data given in Table 1.5 differ from those of Table 1.3 in that insects are used, application is topical, by spraying, different **sexes** are distinguished and also the observations are made at a number of times, rather than just one. The data of the last two rows present the responses for the entire length of the experiment, or endpoint mortalities as they are called, and so are qualitatively similar to the data of Table 1.3.

When presented with such data we might look for sex differences, both in terms of overall response and speed of response. When summarizing overall responses rates, or when comparing these between sexes, we might question the extent to which precision and

Table 1.5 *Numbers of male (M) and female (F) flour-beetles (Tribolium castaneum) dying in successive time intervals following spraying with insecticide (Pyrethrins B) in Risella 17 oil. The beetles were fed during the experiment in an attempt to eliminate natural mortality. Data from Hewlett (1974)*

Time interval (days)	Concentration (mg/cm² deposit)							
	0.20		0.32		0.50		0.80	
	M	F	M	F	M	F	M	F
0–1	3	0	7	1	5	0	4	2
1–2	11	2	10	5	8	4	10	7
2–3	10	4	11	11	11	6	8	15
3–4	7	8	16	10	15	6	14	9
4–5	4	9	3	5	4	3	8	3
5–6	3	3	2	1	2	1	2	4
6–7	2	0	1	0	1	1	1	1
7–8	1	0	0	1	1	4	0	1
8–9	0	0	0	0	0	0	0	0
9–10	0	0	0	0	0	0	1	1
10–11	0	0	0	0	0	0	0	0
11–12	1	0	0	0	0	1	0	0
12–13	1	0	0	0	0	1	0	0
No. survivors	101	126	19	47	7	17	2	4
No. treated	144	152	69	81	54	44	50	47

power have been increased by collecting data over time. We consider this point in detail in Chapter 5.

The beetles involved here are *Tribolium castaneum*, the rust-red flour-beetle. They are small insects, 3–4 mm long, infesting flour, and eating this or broken grain (Hewlett, P. S., personal communication). The fact that the insects were sprayed means that different beetles receive different doses, for a given concentration. The analysis in Chapter 5 ignores this feature, but it is discussed in section 3.9.

Example 1.4   Recovery of insects

An important feature of aerosol fly sprays is whether they knock flies down, and not necessarily whether the flies are actually killed in the process – sometimes flies recover from 'knock-down', as the data of Table 1.6 show. How might we compare the results of the two experiments? We discuss a mechanistic model for such data in Chapter 5.

Example 1.5   Experiments to investigate the effect of arboviruses on chicken eggs

Jarrett *et al.* (1981) analysed experiments carried out to investigate the effects of arboviruses injected into chicken embryos. The aim was to quantify the potency of arboviruses, with a view ultimately to assessing how these might affect lamb foetuses. Two examples of the resulting data are given in Table 1.7.

In this example there are three possible responses, and, as was implicit also in the last two examples, we are interested in comparisons between sets of data. Data of this kind frequently result from making observations over time, as in the last two examples, but the time information is suppressed in this case. Thus in Table 1.7 eggs were classified 18 days after injection of the virus; non-specific deaths in the first few days were excluded, each group of eggs having been originally of size 20. An illustration of time-dependent data for this kind of experiment is given in Table 1.8. Eggs were candled, i.e. held up to the light, each day to see whether the embryo was dead or alive.

In many investigations responses may be due to different causes. Presented with pairs of different woollen fabrics, with only one of each pair being 'prickly', subjects who cannot discriminate between

Table 1.6 *For two experiments, A and B, the data below give the numbers of houseflies (Musca domestica) airborne at several times after the initial dose of spray was administered: a fixed amount of spray was released into a wind tunnel in which the flies were allowed to fly freely. Data from Pack (1986a)*

	Experiment					
	A			B		
Time (minutes)	Concentration ($\mu g/l$)			Concentration ($\mu g/l$)		
	0.3	1.0	2.0	0.3	1.0	2.0
1	18	12	9	19	19	10
5	15	0	0	10	0	0
10	12	0	0	12	0	0
20	15	2	0	13	0	0
60	18	4	0	18	13	0
180	18	16	17	20	22	10
group size	18	16	22	20	22	20

Table 1.7 *The effect of two arboviruses on chicken embryos*

				Alive	
Virus	Inoculum titre (PFU/egg)	No. of eggs	Dead	Deformed	Not deformed
Facey's	3	17	3	1	13
Paddock	18	19	4	1	14
	30	19	8	2	9
	90	20	17	1	2
Tinaroo	3	19	1	0	18
	20	19	2	0	17
	2400	15	4	9	2
	88000	19	9	10	0
Control		18	1	0	17

the fabrics by touch may correctly identify the prickly item by chance. In other cases the correct response can result from a clear perception of prickle on the part of the subjects. Death may result from a cause other than the application of a poison. Even onset of menstruation may, in some cases, be incorrectly ascribed to bleeding due to

Table 1.8 *Time course of an experiment to investigate the effect of an arbovirus on chicken embryos. The data give the cumulative number dead out of 20, except for log dose 0.65, when an egg was dropped on day 8*

							Day							
Log dose	1	2	3	4	5	6	7	8	9	10	11	12	13	14
0.65	0	0	1	1	3	3	3	3	4	4	4	4	4	4
2.50	2	2	2	2	2	2	2	3	3	3	3	3	4	4
4.32	2	2	2	2	2	2	2	4	4	6	7	9	9	11
6.23	0	1	1	1	2	2	3	6	7	10	11	12	12	14
Control	0	0	1	1	1	1	1	1	1	1	1	1	1	1

pathological causes. In Example 1.3, beetles were fed in order to minimize natural mortality. In cases where natural response is possible, it is advisable for control groups to be employed, as in Table 1.8. Further illustrations are given in Example 1.7. Ways of dealing with natural response as in a control group are considered in section 3.2.

Example 1.6 Hypersensitivity reactions to a drug

The data in Table 1.9 are taken from a much larger study into the possible side-effects of a drug. Differing experimental protocols at different sites resulted in experiments of appreciably variable lengths

Table 1.9 *Hypersensitivity reactions to a drug, administered at four sites, A, B, C or D*

Site				Time on drug (days)	Presence of a reaction (1 = reaction)	Sex (2 = female)	Dose (mg)
A	B	C	D				
1	0	0	0	11	1	2	250
1	0	0	0	22	0	2	250
1	0	0	0	20	0	1	250
1	0	0	0	7	1	1	100
0	0	0	1	78	0	2	250
0	0	0	1	27	0	2	50
0	1	0	0	399	0	1	150
0	0	1	0	55	0	1	125

being run before the studies were terminated. Here, as in the last two examples, times are recorded in addition to whether a response took place. Of primary importance to the pharmaceutical company involved was whether there was evidence of hypersensitivity reactions being related to the dose level used.

Example 1.7   Foetal death in a control population

New drugs need to be tested carefully for any possible effects on pregnant animals. The data in Table 1.10 are taken from Haseman and Soares (1976) and just describe control groups from dominant lethal assays, mentioned in section 1.2. In this experiment a drug's ability to cause damage to reproductive genetic material, sufficient to kill the fertilized egg or developing embryo, is tested by dosing a male mouse and mating it to one or more females. A significant

Table 1.10a *Sample No. 1 of Haseman and Soares (1976)*

Litter size														
	0	1	2	3	4	5	6	7	8	9	10	11	12	13
1	2													
2	2													
3	3													
4	5	1	1											
5	2	2												
6	2	2												
7	2	2	2	1										
8	6	1		1	1									
9	2	3	1											
10	2	4	2		2									
11	19	11	3	3										
12	33	24	11	5	4	4							1	
13	39	27	12	6	5	2			1					
14	34	30	14	6	6		1							
15	38	22	18	4	2	1								
16	13	16	14	4	3	1								
17	8	4	3	3	2	1		1						
18		4	2	1										
19	2	1												
20														1

Table 1.10b *Sample No. 3 of Haseman and Soares (1976)*

Litter size	Observed frequency distribution of foetal death in mice									
	Number of dead foetuses									
	0	1	2	3	4	5	6	7	8	9
1	7									
2	7									
3	6									
4	5	2	1							
5	8	2	1		1	1				
6	8									
7	4	4	2	1						
8	7	7	1							
9	8	9	7	1	1					
10	22	17	2		1			1	1	
11	30	18	9	1	2		1		1	
12	54	27	12	2	1		2			
13	46	30	8	4	1	1		1		
14	43	21	13	3	1					1
15	22	22	5	2	1					
16	6	6	3		1	1				
17										
18	3		2	1						

increase in foetal deaths is then indicative of a mutagenic effect. We need to consider how we might describe such data sets in a relatively simple manner, and how we might make comparisons with similar data sets corresponding to treated animals. This is the topic of Chapter 6, where the basic assumption of a binomial distribution for responses is relaxed to accommodate **extrabinomial variation**, which usually arises when different litters of animals are involved.

Example 1.8 Signal detection experiments

A common experiment in psychology involves presenting subjects with stimuli which may either just be noise (N), or may involve a signal superimposed upon noise (SN). The subjects indicate whether or not they thought the signal was present, sometimes qualifying their responses with a measure of confidence. The performance of subjects may be monitored under a variety of adverse environmental conditions, and it is then of interest to measure the extent to which

Table 1.11 *Data resulting from a signal detection experiment on three different subjects. Each subject responds 'Yes' if the stimulus was thought to be present, 'No' if it was not, etc.*

			Responses	
Subject	Stimulus	No	Not sure	Yes
1	N	30	10	15
	SN	9	7	35
2	N	25	17	5
	SN	2	18	30
3	N	18	7	3
	SN	2	10	16

performance may change as conditions change. Such experiments may model behaviour such as the vigilance of radar screen monitors in submarines. The data of Table 1.11, taken from Grey and Morgan (1972), provide an illustration.

Again we want to summarize the data, and make comparisons between subjects. Relevant analysis is provided in section 3.5. An alternative form of signal detection experiment arises when subjects are informed that a stimulus is present on just one of m occasions, for $m > 2$, and the subjects have to select the occasion they think corresponds to the signal presentation. This is called an m-alternative forced-choice experiment, and will also be discussed again in Chapter 3.

Example 1.9 The Australian bovine tuberculosis eradication campaign

In work aimed at the eradication of bovine tuberculosis in Australia, suitably treated bovine tissue is placed on culture plates and examined for the growth of colonies of *Mycobacterium bovis*. Material for culture is decontaminated prior to inoculation onto culture media and the data in Table 1.12 describe colony counts when two different decontaminants (HPC and oxalic acid) are applied, in varying concentrations. While there are obvious similarities between this experiment and, say, that of Example 1.1, there is no universal upper limit to a colony count, and the data of Table 1.12

Table 1.12 Colony count data, taken from Trajstman (1989). The value marked by a * was omitted from all analyses

[HPC] % weight/volume — Decontaminant: HPC. No. M. Bovis colonies at stationarity

% weight/volume											Sample mean	Sample variance
0.75	2	4	8	9	10	1	0	5	14	7	6.0	19.6
0.375	11	12	13	12	11	13	17	16	21	2	12.8	24.4
0.1875	16	6	20	23	23	39	18	23	33	21	22.2	80.6
0.09375	33	46	42	18	35	20	19	29	41	36	31.9	102.3
0.075	30	30	27	53	51	39	31	36	38	22	35.7	100.0
0.0075	53	62	38	54	54	38	46	58	54	57	51.4	66.5
0.00075	3*	42	45	49	32	39	40	34	45	51	41.9	40.6

[Oxalic acid] % weight/volume — Decontaminant: oxalic acid. No. M. Bovis colonies at stationarity

% weight/volume											Sample mean	Sample variance
5	14	15	6	13	4	1	9	6	12	13	9.3	23.1
0.5	27	33	31	30	26	41	33	40	31	20	31.2	39.1
0.05	33	26	32	24	30	52	28	28	26	22	30.1	70.8
0.005	36	54	31	37	50	73	44	50	37	–	45.8	164.4

Control experiment where no decontaminant is used. No. M. Bovis colonies at stationarity

									Sample mean	Sample variance
52	80	55	50	58	50	43	53	54	51.8	110.8
44	51	34	37	46	56	64	67	40		

will require different probability models, to be described in sections 3.3 and 6.5.2. See also Exercise 1.22.

Example 1.10   Serological data

The results of a serological survey carried out in Zaire into the extent of malarial infection in individuals aged greater than 6 months are given in Table 1.13. In this example the percentage sero-positive is bounded above by a factor reflecting the overall incidence of malaria. We consider modelling these data in Chapter 3.

Table 1.13 *Data from Bongono (Zaire) showing the proportions of individuals in different age groups with antibodies present, as assessed by a particular serological test. Data from Marsden (1987)*

Mean age group (years)	No. of individuals examined	No. sero-positive	Percentage sero-positive
1.0	60	2	3.3
2.0	63	3	4.8
3.0	53	3	5.7
4.0	48	3	6.3
5.0	31	1	3.3
7.3	182	18	9.9
11.9	140	14	10.0
17.1	138	20	14.5
22.0	84	20	23.8
27.5	77	19	24.7
32.0	58	19	32.8
36.8	75	24	32.0
41.6	30	7	23.3
49.7	62	25	40.3
60.8	74	44	59.5

Example 1.11   Mixtures of drugs

The data of Table 1.14 result from an experiment designed to investigate how two insecticides (*A* and *B*) may act in combination. Of interest here is whether insecticides interact to produce enhanced performance (synergy), or a reduction in performance (antagonism). An analysis of these data is provided by Giltinan *et al.* (1988) and we discuss their findings in section 3.7.

Table 1.14 *The results of a study to investigate the contact insecticidal activity of mixtures of two insecticides, A and B. The target insect was the tobacco budworm, Heliothis virescens. Treatment was administered by means of direct application of one microlitre for each dosage to the body of each insect. Mortality was measured 96 hours after treatment (Data from Giltinan et al., 1988)*

Mixture	Amount of A(ppm)	Amount of B(ppm)	Number of dead insects	Number of insects tested
B	0	30.00	26	30
B	0	15.00	19	30
B	0	7.50	7	30
B	0	3.75	5	30
A25:B75	6.50	19.50	23	30
A25:B75	3.25	9.75	11	30
A25:B75	1.625	4.875	3	30
A25:B75	0.812	2.438	0	30
A50:B50	13.00	13.00	15	30
A50:B50	6.50	6.50	5	30
A50:B50	3.25	3.25	4	29
A50:B50	1.625	1.625	0	29
A75:B25	19.50	6.50	20	30
A75:B25	9.75	3.25	13	30
A75:B25	4.875	1.625	6	29
A75:B25	2.438	0.813	0	30
A	30.00	0	23	30
A	15.00	0	21	30
A	7.50	0	13	30
A	3.75	0	5	30

1.4 Preliminary graphical representations

An obvious first approach to the kind of data illustrated so far is to plot proportions affected against dose, or log dose, or time, or whatever appears appropriate. This is done in Figures 1.1–1.3 for the data in Tables 1.3–1.5, respectively. The value of doing this is illustrated in Figure 1.3, for example: we can appreciate that males appear to be more susceptible than females and, furthermore, that when they respond they appear to do so more quickly than females. We shall quantify these differences by using mixture models, from the area of survival analysis, in Chapter 5.

One may well consider fitting a straight line to points such as those of Figure 1.1. However, it is preferable to transform the

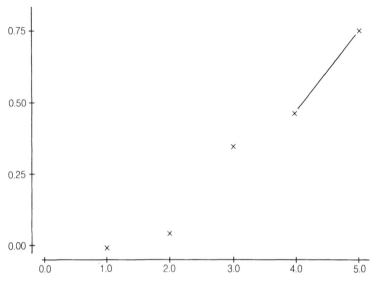

Figure 1.1 *A plot of the proportions protected versus* $\log_2$ *(dose) for the data of Table 1.3. The reason for connecting the two proportions shown is given in section 1.6.*

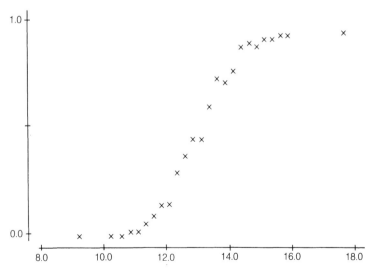

Figure 1.2 *A plot of the proportions of Table 1.4 versus mean age of groups.*

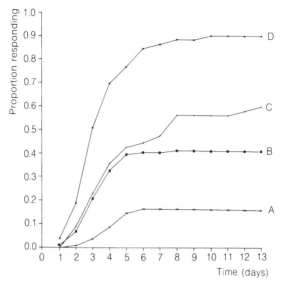

Figure 1.3a *A plot of the proportions of female beetles responding versus time, from Table 1.5, reproduced from Pack (1986a).*

Key

$0.2 \, \text{mg/cm}^2$	A
$0.32 \, \text{mg/cm}^2$	B
$0.50 \, \text{mg/cm}^2$	C
$0.80 \, \text{mg/cm}^2$	D

proportions first. In many cases the plot corresponding to that of Figure 1.1 has a more sigmoid appearance, as is true of the points of Figure 1.2. If we plot the logits of the proportions versus age for the data of Table 1.4, we obtain the more linear plot of Figure 1.4. Finite logits do not exist for proportions of 0 or 1. Corresponding doses are indicated by arrows on the graph. Special graph paper may be used if the plotting is to be done by hand.

For Figure 1.4 the least squares linear regression line is,

$$\text{logit} \, \{p(x)\} = -20.8 + 1.58x \tag{1.1}$$

where $p(x)$ denotes the proportion that have reached menarche by age x. The product–moment correlation between logit $\{p(x)\}$ and x has value 0.992, and so one might feel that the data are well described by equation (1.1). However, the proportions of Figure 1.2 result from

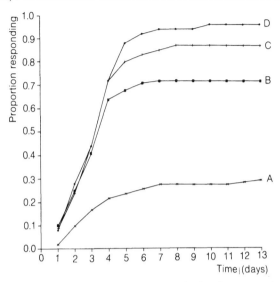

Figure 1.3b *A plot of the proportions of male beetles responding versus time, from Table 1.5, reproduced from Pack (1986a).*

Key

0.2 mg/cm²	A
0.32 mg/cm²	B
0.50 mg/cm²	C
0.80 mg/cm²	D

binomial distributions, and will have unequal variances. A weighted regression, weighting inversely with respect to the variance of logit (p) (Exercise 1.2) gives the regression line:

$$\text{logit}\{p(x)\} = -20.0 + 1.54x \qquad (1.2)$$

revealing little difference from equation (1.1) – but see also Exercise 1.3. The fitted line of equation (1.1) is called a minimum chi-square line, while that of equation (1.2) results from the technique of minimum logit chi-square. These methods are compared in section 2.6, after a full discussion of the maximum-likelihood estimation procedure for standard quantal response data.

Before fitting lines to data we can also consider whether a preliminary transformation of the explanatory variate, x, might improve the fit. A logarithmic transformation is often used routinely

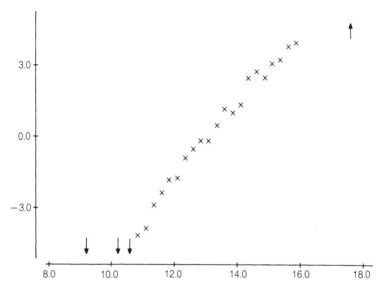

Figure 1.4 *A plot of the logits of the proportions of Table 1.4 versus age, excluding points for which the logit is not finite.*

for this purpose, though in some cases it has impaired the fit, rather than improved it; we shall discuss this further in detail in Chapter 4. In many exploratory investigations of new substances, their potency may be uncertain before the experiment. In such a case a wide range of dose levels is therefore sensible, and a natural device is to space the doses equally on a log scale, and then also for convenience to present results, plots and analyses in terms of that scale.

If we set $x = 0$ in equation (1.2) we see that birth equates to onset of mentruation in a very small, but non-zero, fraction of newborn female children. This should not worry us unduly, since it involves extrapolation well outside the age range over which data were collected. Over the age range of the collected data the model may provide a succinct description of the data. However, we can see that if the model had been formulated in terms of log (age), this problem would not have arisen. In fact, as we shall see in Chapter 4, the logarithmic transformation improves the fit of the model for these data. Extrapolation is the subject of section 1.6. The use of logarithms can also appear naturally from various mechanistic models which we shall now describe.

1.5   Mechanistic models

When we consider the data of Table 1.4, on age of onset of menstruation, it is natural to suppose that age of menarche, in a homogeneous population of girls, has some distribution, with cumulative distribution function $F(\alpha + \beta x)$, where α, β are a location and scale pair of parameters. In such a case, the probability, P, of, for example, 79 girls out of 98 with mean age of 14.08 years having reached menarche can be approximated by the binomial form

$$P \approx \binom{98}{79} F(\alpha + 14.08\beta)^{79} \{1 - F(\alpha + 14.08\beta)\}^{19}$$

The exact form for P takes into account the interval nature of the data – see Exercise 1.13. An obvious contender for the form of $F(\ )$ is the normal distribution. An example of this is cited by Biometrika tables and refers to the detonation of explosives at varying distances from cardboard discs. In each experiment the proportion of discs perforated is noted (Exercise 1.14).

This same model is also used for standard quantal assay data, as in Table 1.3, and it may be justified by the classical, or threshold model for quantal response data. In this model it is assumed that each individual in the relevant population has a dose tolerance, or threshold, T say, to a particular substance. If the dose administered, d, is greater than T then the individual responds. Otherwise it does not. If the tolerances are distributed throughout the population with distribution $F(\alpha + \beta x)$, say, as in the above illustration, then the probability of individual response to dose d is simply:

$$\Pr(T \leqslant d) = F(\alpha + \beta d)$$

In practice this model may also be used when d is a dosage, rather than a dose (see Exercise 1.15 for further discussion). In this book we shall refer to tolerance/threshold models/distributions. A small fraction of individuals may have high tolerances, giving rise to a positively skewed tolerance distribution. A logarithmic dose transformation might then be advantageous if the model to be fitted assumes a symmetric tolerance distribution.

Historically the favoured form for $F(\ )$ has been normal, resulting in what is called **probit** analysis. The greater simplicity of the cumulative distribution function of the similar logistic distribution has resulted in emphasis now being placed on use of the logistic distribution and the resulting **logit** analysis, a particular example of logistic regression discussed in section 2.8. However, a computational advantage of the probit model arises if a dose d is observed with error, and we return to this errors-in-variables situation in Chapter 3.

Probit or logit models may be adopted from the pragmatic viewpoint of simply requiring an appropriate description of the data, and this has already been done of course in the fitted model of equation (1.2), without any reference to threshold models. There are cases where the threshold model is not appropriate; in cancer formation, for example, tumours may result from a change to a single cell initially and similarly death may follow from infection by a single virus particle. However, threshold models usually provide a useful way of thinking about the data. We shall consider an extension of the threshold model in Chapter 5, when we model times to response. We may note here that there are also other areas where ideas of thresholds have been found to be useful for analysing data; for example, Anderson and Aitkin (1985), and Exercise 1.24.

In psychology, a similar model provides the justification for signal detection theory which is used to describe data such as those of Example 1.8 (Exercises 1.12 and 2.8). This theory has been given a general setting by McCullagh (1980), as a way of analysing contingency table data with ordered categories; here too the terminology of probit and logit (and other) models is used. The resulting models are discussed in section 3.5, for the analysis of multiple response data, as in Table 1.7.

In general terms we may describe models as **mechanistic** or **descriptive** (in the latter case, Ripley, 1987, prefers the term **convenient**). The latter type of model does not rely on a specification of a mechanism, and simply aims to summarize the data, and provide a framework for inference. Thus we may regard a simple linear regression model as descriptive. The end-product of a mechanistic modelling exercise may be a descriptive model, whose parameters play no role other than fitting the model to the data, and we shall encounter several examples of this.

Puri and Senturia (1972) proposed an elaborate mechanistic model for the way in which insects might attempt to shed insecticide, through a random sequence of losses of random amounts. This model was then used to fit data by supposing that the individual insect hazard rate at any time was a function of the amount of insecticide remaining by that time. We consider this model in detail in

Chapters 4 and 5. One-hit and multi-hit models are also described in Chapters 4 and 5. Originally devised as models of carcinogenicity, these models have quite recently been employed for describing quite general quantal response data – see Rai and Van Ryzin (1981). The basic premise is that individuals exposed to a substance can be likened to a target bombarded with arrows, at an intensity determined by the dose level adopted. In the **one-hit** model it is supposed that a single arrow on target is sufficient to elicit a response. The **multi-hit** model is more stringent in requiring several hits. The **multi-stage** model supposes that various stages have to be completed, either in series or in parallel before a response is obtained. For a comprehensive review, see Kalbfleisch et al. (1983). If a toxic response results from at least h hits from arrows arriving in a Poisson process at rate ζd for some fixed (say unit) time, where d corresponds to the dose level, then (see Exercise 1.17) the probability of response is given by:

$$P(d; h, \zeta) = \int_0^d \frac{\zeta^h x^{h-1} e^{-\zeta x} dx}{\Gamma(h)} \quad \text{for} \quad 0 < d < \infty$$

and we can see therefore that this model is equivalent to a threshold model with a gamma-distributed tolerance distribution. This is in fact a limiting case of an extended threshold model presented by Prentice (1976b), based on the $\log F$ distribution, with cumulative distribution function,

$$F(z) = \int_{-\infty}^z \frac{e^{wm_1}(1 + e^w)^{-(m_1 + m_2)}}{B(m_1, m_2)} dw, \quad -\infty < z < \infty$$

The logit model is 'extended' by this model, as it forms the special case of $m_1 = m_2 = 1$. There is full discussion of this and other extended models in Chapter 4.

The Puri–Senturia model can also result in a descriptive model with a positively skewed tolerance distribution (Chapter 5), and this finding can help to explain why a preliminary logarithmic dose transformation followed by analysis based on a model which incorporates a symmetric tolerance distribution may often be a useful procedure (for more discussion, see Chapter 4).

At the other end of the dose-response spectrum from a supposed 'target' is the administration of the substance under test. A fundamental defect of many mechanistic models is that they take the administered dose as the effective dose, when in fact the dose level may be inaccurately measured, much sprayed dose may miss the insect, and so forth. Van Ryzin and Rai (1987) have produced a mechanistic model of this process, based on a compartmental-model assumption (Exercise 2.33). A simpler assumption is to suppose that the effective dose has a Poisson distribution, with mean value the supposed administered dose. This is the approach adopted by Williams (1965) in a mechanistic model for microbial infections based on the stochastic linear birth and death process. This model, considered also by Morgan and Watts (1980), is different from others considered in this book in that the dose refers to a suspension of bacteria, which by their development can effectively change the dose during the course of the infection (Exercise 1.18). The work of section 3.9 considers how errors in dose measurement may be described and built into the model. Ridout and Fenlon (1991) consider microbial control of insect pests. In this case the pathogen is delivered through food, which for a given concentration of pathogen may even result in some insects receiving no pathogen.

Mechanistic models have been found to be particularly useful in justifying various forms of low dose extrapolation procedures, and we describe how in the next section.

1.6 Interpolation and extrapolation

A partial answer to the question: 'What are models for?' has already been provided in the last section: models can simplify, summarize and provide a basis for inference – as in making comparisons; for example, the topic of section 3.7. Mechanistic models may provide much more, since the parameters of the model might correspond to definite aspects of the supposed mechanism. By simplifying the data, models may be thought of as smoothing out variation, and so fitted models may be used for interpolation. Another major use of models in general is in prediction. This is abundantly clear in areas such as time-series analysis, but it is also the objective of low dose extrapolation, mentioned in section 1.2.

As is discussed in detail in section 2.7, we are frequently interested in estimating dose levels, dosage levels or concentrations which correspond to an average given percentage, $100p$, of individuals responding. Traditionally such levels are denoted by expressions such as ED_{100p} or LD_{100p}. Here E denotes 'effective', and L denotes 'lethal'. In some studies it may be an EC_{100p} level which is estimated, corresponding to a particular **concentration**. An example is provided by Ridout and Fenlon (1991) who also discuss how to convert concentrations to doses. By analogy, a single summary of the age-of-menarche data displayed in Figure 1.2 would be the age by which we would expect 50% of girls to have reached menarche, the median of the threshold distribution.

The values in Table 1.15 (taken from Klaassen, 1986) show how an ED_{50} may be useful in providing a simple indication of the relative toxicity of a range of different chemicals (response here is death in treated animals). A review article is given by Zbinden and Flury-Roversi (1981). There is more discussion of ED_{100p} estimation, with particular reference to interval estimation, in section 2.7. While such values are readily estimated from models, it may be argued that a non-parametric procedure could be more relevant. Thus in Figure 1.1, for example, we could simply use the indicated linear interpolation if we wanted to estimate that dose resulting in a 50% response rate. This simple procedure is the simplest illustration of a trimmed Spearman–Kärber estimate, and we discuss this and other non-parametric methods in Chapter 7.

As an alternative to using the ED_{50} for evaluating the toxicity of substances, the British Toxicology Society (1984) proposed a fixed dose procedure. The approach here is to use four dosages, namely 5, 50, 500 and 2000 mg/kg and test 10 animals at that dosage which

Table 1.15

Agent	ED_{50} (mg/kg)
Ethyl alcohol	10 000
Sodium chloride	4 000
Ferrous sulphate	1 500
Morphine sulphate	900
Phenobarbital sodium	150
Picrotoxin	5
Strychnine sulphate	2
Nicotine	1
d-Tubocurarine	0.5
Hemicholinium-3	0.2
Tetrodotoxin	0.10
Dioxin (TCDD)	0.001
Botulinum toxin	0.00001

is judged *a priori* to produce a toxic but not lethal effect. Other dosages may or may not then be used, depending on the observed response. This procedure has been evaluated by van den Heuvel *et al.* (1990) and Whitehead and Curnow (1991). It has been found to be more conservative in general than simply using the ED_{50}, but it requires fewer animals to be tested on average.

While Finney (1971, p. 210) emphasizes 'the evils of extrapolation', extrapolation is frequently required in the context of quantal response data, for reasons explained in section 1.2. A review is given by Armitage (1982). This is clearly an area where different models can give quite different predictions. While a model based on the logistic distribution, say, may provide a satisfactory fit to data, as judged perhaps by a chi-square goodness-of-fit test, significant improvements may result from fitting more complex models such as that of Prentice, described above. Improvements of fit may typically be anticipated in the tail areas, and so be particularly important for extrapolation. In the evaluation of virtually safe doses of possible carcinogens, response rates as low as 10^{-6} or 10^{-8} may be involved (Rai and Van Ryzin, 1981). This is an area to which we shall return at various points throughout the book. In practice a variety of procedures is used in such extreme cases of extrapolation (Exercise 1.19).

1.7 Discussion

The material of this chapter provides an introduction to the work of later chapters. The logit transformation was employed in section 1.4, but many others may also have been used, e.g. the probit transformation, with

$$F(z) = \frac{1}{\sqrt{2\pi}} \int_{-\infty}^{z} e^{-\frac{1}{2}x^2} dx$$

which has slightly lighter tails than the logistic form, the sine curve, with

$$F(z) = \{1 + \sin(z)\} \quad \text{for } -\frac{\pi}{2} \leqslant z \leqslant \frac{\pi}{2}$$

and the Cauchy cumulative distribution function,

$$F(z) = \frac{1}{2} + \frac{1}{\pi} \tan^{-1}(z), \quad \text{for } -\infty < z < \infty$$

sometimes called Urban's curve. Appropriate graphs are given in Ashton (1972, p. 12). For most data sets it is impossible to distinguish between these different models (Cox and Snell, 1989, p. 22). An advantage of the logit model is its computational simplicity. It is also widely used in other areas, such as in the analysis of case-control data (Rosenbaum, 1991) and for projections of population size (Leach, 1981). For data sets containing as many individuals as Table 1.4, for example, it can prove possible to prefer, say, the probit model to the logit model. We return to this point in Chapter 4, where we also encounter a wide range of more complex models for quantal response data.

The binomial distribution was encountered in section 1.5, and it clearly forms the fundamental basis for much statistical analysis of quantal response data. We shall see later when it is replaced by alternatives such as Poisson and trinomial. A particularly important development is needed when 'over-dispersion', relative say to binomial or Poisson distributions, is present. In such cases variances are inflated, due to heterogeneity, relative to the simpler distributional forms. This is the topic of Chapter 6.

A common feature of the examples of section 1.3 is that the data are not obtained in a sequential manner. In some situations it is possible to employ a sequential design, and conduct new experiments at dose-levels suggested by responses to earlier experiments. This can result in a concentration of resources on the regions of doses of most overall interest, and we consider sequential methods as well as the general topic of how to design experiments in Chapter 8.

A fundamental feature of much bioassay, implicit in Examples 1.3, 1.4 and 1.9 amongst others, is the need for comparisons. This is a topic which recurs throughout the book, in section 3.7, and Chapters 6 and 7.

We have seen that analysis may be confounded by the doses which actually have an effect sometimes being different from those which nominally are administered. A further complication which may arise is if the substances tested are impure, or otherwise result in a mixture of substances being used. This is a subject which is dealt with at length in Plackett and Hewlett (1979). In some cases dose-response curves may be non-monotonic, in contrast to most of the examples above, and this may be due to the effect of mixtures of substances (e.g. Finney, 1971, p. 266), or possibly to high dose toxicity of substances which produce positive responses to lower doses (e.g.

Simpson and Margolin, 1986 and Exercise 1.16). These features, of response to mixtures, and of high-dose toxicity are considered respectively in section 3.7 and Chapter 4.

Note finally that much of the material that follows can be thought of as building blocks and these may be combined as appropriate for any particular problem, in ways which are not covered in this book.

The main texts in the area are those by: Aldrich and Nelson (1984), Ashton (1972), Cox and Snell (1989), Cramer (1991), Dobson (1990), Finney (1971, 1978), Govindarajulu (1988), Hubert (1992), Plackett and Hewlett (1979) and Salsburg (1986). Relevant material is also to be found in Agresti (1984), Bishop *et al.* (1975) and Haberman (1974). A useful case-study in logistic regression is given by Kay and Little (1986), and issue no. 4 of volume 42 of *Statistica Neerlandica* of 1988 is devoted to aspects of logistic regression.

The model-fitting required in the analysis of quantal response data nearly always involves numerical procedures for fitting non-linear models, and so inevitably computers are involved. In some cases, specialist computer packages and algorithms are available (e.g. Finney, 1976; Smith 1983; Russell *et al.* 1977; Morgan and Pack, 1988; Morgan *et al.* 1989). In more complex applications recourse is made to non-linear optimization routines (as in Rai and Van Ryzin, 1981, for example). The facilities of packages for generalized linear models, available for instance in GLIM, SAS and GENSTAT 5, are often particularly convenient. Also the MLP package (Ross, 1987) has routines for the analysis of quantal response data. At times throughout the book illustrations are made through computer code, with examples in GLIM, MINITAB and BASIC. A review of relevant programs and packages is given in Appendix E.

1.8 Exercises and complements

The exercises vary in complexity. It is particularly desirable to attempt the five exercises marked with a †. Exercises marked with a * are generally more difficult or speculative.

1.1 Identify at least three examples of quantal response data which do not arise from toxicology or efficacy studies.

1.2 Use a Taylor series expansion to verify the following first-order approximation, for sufficiently small σ and well-behaved function

$\psi(\ )$

$$V\{\psi(X)\} \approx \{\psi'(\mu)\}^2\sigma^2, \quad \text{where } E[X] = \mu, \, V(X) = \sigma^2$$

Deduce that if a random variable R has the binomial $\text{Bin}(n, p)$ distribution, and is observed to be $R = r$, then we may estimate

$$V\left\{\log\left(\frac{R}{n - R}\right)\right\} \approx \frac{1}{r} + \frac{1}{(n - r)}$$

1.3[†]   The following data sets were part of a collection put together by Copenhaver and Mielke (1977), and analysed also by Morgan (1985). Each set is given in the form: dose, dosage, concentration (or transformation of this), number responding, number in experiment. Details of dose transformations are given in Copenhaver and Mielke (1977, Table 1) and the raw data are given in Egger (1979), who excluded data set 14 on account of response being partly due to natural mortality. (note that this topic will be discussed in section 3.2.) The same numbering of data sets as in Copenhaver and Mielke (1977) has been adopted. In which cases do you think a preliminary log transformation of the dose scale has taken place? In each case, plot the proportions, and the logits of the proportions, and fit weighted and unweighted regressions on the logits. Comment on similarities and differences, with regard to estimates of error, as well as point estimates.

Set 2	2	10	30	Set 3	1	0	6
	3	14	30		2	0	6
	5	20	30		4	1	6
	9	23	30		8	0	6
					16	2	6
					32	4	6
					64	4	6
					128	6	6
					256	5	6
Set 4	0.41	6	50	Set 7	0	2	30
	0.58	16	48		1	8	30
	0.71	24	46		2	15	30
	0.89	42	49		3	23	30
	1.01	44	50		4	27	30

Set 9	1.6907	2	29	Set 10	0.71	16	49
	1.7242	7	30		1.0	18	48
	1.7552	9	28		1.31	34	48
	1.7842	14	27		1.48	47	49
	1.8113	23	30		1.61	47	50
	1.8369	29	31		1.7	48	48
	1.8610	29	30				
	1.8839	29	29				
Set 11	0.4	7	47	Set 14	1.57	34	132
	0.71	22	46		2.17	40	51
	1.0	27	46		2.49	114	127
	1.18	38	48		2.66	115	117
	1.31	43	46		2.79	125	125
	1.4	48	50				
Set 18	1	0	40	Set 22	0.4472	0	10
	2	2	40		0.55	2	10
	3	14	40		0.6232	3	10
	4	19	40		0.7	4	10
	5	30	40		0.7482	7	10
					0.8	8	10

Performing these calculations by hand can be quite time-consuming, even with the aid of special graph paper. Access to a computer package such as MINITAB can result in an enormous simplification. The following MINITAB commands may be used to answer this question for data in a file named: 'QUANTAL' and in the same format as above.

```
READ 'QUANTAL' INTO COLUMNS C1,C2,C3
LET C4=1/C3
LET C5=C2-C3
LET C6=1/C5
LET C7=C4+C6
LET C8=1/C7
NOTE: C8 CONTAINS THE WEIGHTS
LET C9=C3/C5
LET C10=LOG(C9)
NOTE: C10 CONTAINS THE LOGITS
PLOT C10 VS C1
REGRESS C10, 1 COVARIATE, ON C1
REGRESS C10, WEIGHTS IN C8, 1 COVARIATE, ON C1
```

Problems will be encountered with 0% or 100% responses: MINITAB overcomes these by failing to return a logit in that case.

and the above program can then still be run. A further collection of sets of quantal response data, in this case describing toxic response, is given in Rai and Van Ryzin (1979).

1.4 The data below, describing the mortality of adult flour-beetles (*Tribolium confusum*), after 5 hours' exposure to gaseous carbon disulphide (CS_2), are reproduced in Prentice (1976b).

Dose (CS_2 mg/l)	49.06	52.99	56.91	60.84	64.76	68.69	72.61	76.54
Number of beetles in experiment	59	60	62	56	63	59	62	60
Number of insects killed	6	13	18	28	52	53	61	60

Investigate whether a preliminary log transformation of the dose simplifies a plot of the logits. We shall discuss the analysis of these data further in Chapter 4. Taken from a paper by Bliss (1935), they form one of the standard data sets which are used to illustrate any new model for analysing quantal response data. This is probably due to their use as an illustration by Prentice (1976b), which was an early paper presenting new models.

1.5 The following data come from an experiment conducted at the Stanford School of Medicine to measure the toxicity of guthion on mice:

Dosage (mg/kg of body weight)	Number of mice treated	Number of mice killed
4	30	1
5	46	3
6	46	13
7	46	23
8	46	29
10	46	44

For these data, perform the same investigation as in Exercise 1.4.

$1.6^{\dagger}$ A group of 16 pregnant female rats was fed a control diet during pregnancy and lactation. The diet of a second group of 16 pregnant females was treated with a chemical. For each litter the number n of pups alive at 4 days and the number r of pups that survived the 21-day lactation period were recorded. The resulting data were as follows:

Control	r	13	12	9	9	8	8	12	11	9	9	8	11	4	5	7	7
Group	n	13	12	9	9	8	8	13	12	10	10	9	13	5	7	10	10
Treated	r	12	11	10	9	10	9	9	8	8	4	7	4	5	3	3	0
Group	n	12	11	10	9	11	10	10	9	9	5	9	7	10	6	10	7

Provide a brief critical discussion of these data, and a simple test of whether the treatment affects survival. (Data from Weil, 1970, analysed in Williams, 1975.) We continue discussion of these data in section 6.2.

1.7 Kooijman (1981) presented the data below which are the cumulative mortality counts for *Daphnia magna* in water containing cadmium chloride. Plot the data and provide an analysis and a description of your conclusions. This is part of a larger data set, given in full in Table 5.1.

	Concentration ($\mu g/l$)					
Day	0.0	3.2	5.6	10.0	18.0	32.0
11	0	1	2	5	27	50
14	0	1	3	8	36	50
16	0	1	3	10	36	50
18	0	1	5	10	38	50
21	1	2	7	12	42	50
Group size	50	50	49	50	53	50

1.8* Carter and Hubert (1981b, 1984) consider ways of analysing data of the following kind (taken from Carter and Hubert, 1981b) in which groups of 20 trout fry are exposed to different concentrations of copper sulphate. Observations were made over a period of 48 hours, and the experiment was repeated each week for 5 weeks. Presented are the cumulative numbers of dead fish. The stock of fish was homogeneous with respect to age at the start of the 5-week period, so that the ages of the fish used each week increased. At the end of the experiment, all fish still alive were killed, and then all fish were dried. The mean dry weight of the 20 fish in each tank was recorded and is also given in the table. These data illustrate the complexities of experimental design which can occur. Consider how you might illustrate and analyse these data.

Week	Concentration ($\mu g/l$)	Time (hours)					Mean dry weight (g)
		8	14	24	36	48	
	270	0	0	5	5	5	0.6195
	410	0	2	6	6	6	0.5305
1	610	0	10	15	18	18	0.5970
	940	3	13	20	20	20	0.6385
	1450	9	20	20	20	20	0.6645
	270	0	1	4	4	4	0.5685
	410	1	2	6	6	6	0.6040
2	610	1	9	19	20	20	0.6325
	940	2	14	20	20	20	0.6845
	1450	4	17	20	20	20	0.7230
	270	0	0	0	0	2	0.6695
	410	0	1	3	5	7	0.6405
3	610	0	3	19	19	19	0.7290
	940	0	11	19	20	20	0.7700
	1450	2	27	20	20	20	0.5655
	270	0	0	0	1	2	0.7820
	410	0	1	3	4	5	0.8120
4	610	2	9	19	19	19	0.8215
	940	2	14	20	20	20	0.8690
	1450	7	19	20	20	20	0.8395
	270	0	1	4	4	4	0.8615
	410	0	0	3	5	5	0.9045
5	610	0	8	18	20	20	1.0280
	940	1	13	20	20	20	1.0445
	1450	6	17	20	20	20	1.0455

$1.9^{\dagger}$ Investigate, for the menarche data of Example 1.2, whether you think a preliminary log transformation of the age scale will improve the fit.

1.10 Consider how you would illustrate the data of Table 1.11.

1.11 Does the binomial distribution provide an adequate model for the data of Table 1.10?

1.12 (Anderson, 1985a, b) Devise experiments to compare the 'prickliness' of different types of woollen fabrics.

*1.13** The data of Table 1.4, on age of menarche in Polish girls, is described by a model which makes use of the group mean ages. Consider how you would analyse the data if you were given:

(a) the age of each girl separately;
(b) the endpoints of each age-group.

The latter case corresponds to what is termed, 'interval-censoring' in survival analysis, and can be seen to occur also in Table 1.5, for example. We return to this topic in Chapter 5. Müller and Schmitt (1990) omitted the last age class. Why do you think this was done? For males the corresponding event to menarche is spermarche, the estimation of which is the subject of the paper by Jørgensen *et al.* (1991).

1.14 Biometrika Tables (Vol. I – see Pearson and Hartley, 1970) describe quantal response data as arising in '... biological assay and in tests on the detonation of explosives'. The illustration given was of an experiment in which a given weight of explosive was detonated at different distances from a standard disk of cardboard. The experiment was repeated 16 times for each distance and on each occasion it was recorded whether or not the disk was perforated. Analyse the resulting data shown below.

No. of disks perforated	Distance in feet
0	53
9	49
9	45
12	41
16	37

*1.15** Consider how a known distribution of weights of individuals may be used to relate probabilities of response to a dose and to a dosage. Cf. Exercise 3.39. Finney (1971, p. 181) suggests accommodating different body weights of individuals by an analysis which relates response explicitly to both dose and weight. Smits and Vlak (1988) examined the response of the beet armyworm, *Spodoptera exigua*, to different concentrations of a virus. Data on the intake of individual larvae was used to convert LC_{50}s to LD_{50}s.

*1.16** Simpson and Margolin (1986) consider the data shown below, which present the numbers of revertant colonies in plates containing *Salmonella* bacteria of strain TA98, given different doses of a substance called Acid Red 114 (Ames *et al.*, 1975).

| | | | *Dose ($\mu g/ml$)* | | | |
Replicate	0	100	333	1000	3333	10000
	22	60	98	60	22	23
1	23	59	78	82	44	21
	54	50	59	33	25	35
	19	15	26	39	33	10
2	17	25	17	44	26	8
	16	24	31	30	23	–
	23	27	28	41	28	16
3	22	23	37	37	21	19
	14	21	35	43	30	13

Illustrate the data and consider how to analyse them.

1.17 In the multi-hit model (Maynard Smith, 1971, p. 87) a toxic response results from at least h hits from arrows fired at a target in a Poisson process of rate ζd in an interval of time which may be taken to be the unit interval. Here d denotes the dose level, and h and ζ are parameters. Show that the probability of response is then given by:

$$P(d; h, \zeta) = \int_0^d \frac{\zeta^h x^{h-1} e^{-\zeta x} dx}{\Gamma(h)}, \quad \text{for } 0 < d < \infty$$

Why is this equivalent to assuming a gamma tolerance distribution?

*1.18** Suppose that an individual is infected by a single bacterium, whose development within the host can be described by a supercritical linear birth and death process, with birth rate λ, and death rate μ, and $\lambda > \mu$ (Bailey, 1964, p. 194). Show that the probability that there will be i bacteria in the host by time t is given by:

$$p_i(t) = (1 - \alpha)(1 - \beta)\beta^{i-1} \quad \text{for } i \geqslant 1$$

$$p_o(t) = \alpha$$

where

$$\alpha = \mu(1 - e^{(\mu - \lambda)t})(\lambda - \mu e^{(\mu - \lambda)t})^{-1}$$

$$\beta = \lambda(1 - e^{(\mu - \lambda)t})(\lambda - \mu e^{(\mu - \lambda)t})^{-1}$$

Suppose further that the incubation period, T, of the disease caused by the infecting bacteria, and which commenced with the initial infection, ends, with the appearance of symptoms, after a (large) threshold number of bacteria n is reached. Deduce that the distribution of T is given by:

$$P(T \leqslant t) \approx (1 - \alpha)\beta^{n-1}(1 - \mu/\lambda)^{-1}$$

This and other possible models for the progress of microbial infections are considered in Morgan and Watts (1980). See also Armitage (1959).

1.19* One result of the mechanistic **multistage** model for carcinogenesis is the prediction that the proportion of animals with tumours after a fixed period of continuous dosage at dose level d is given (approximately) by:

$$P(d) = 1 - \exp\left(-\sum_{i=0}^{m} \zeta_i d^i\right)$$

for constants m (the number of stages) and $\{\zeta_i\}$. Investigate the form taken for $P(d)$ as $d \downarrow 0$, and discuss how you would use this model for low dose extrapolation. For further discussion see Crump *et al.* (1977) and Armitage (1982). A time-dependent version of $P(d)$ will be considered in section 5.4.2.

1.20 Discuss the potential problems of performing experiments involving very large numbers of experimental animals in order to estimate extreme doses such as ED_{01} with reasonable precision – the so-called 'mega-mouse' experiments. See Schneiderman *et al.* (1975) and Staffa and Mehlman (1979) for discussion.

1.21 Line transect sampling is a technique used in ecology for the estimation of mobile animal populations. In this technique an observer traverses a route, recording observations of the animals, birds or whatever that are being censused. A possible defect with this procedure lies in animals that are present being missed. Suggest an experiment to investigate how the missing of objects might vary with distance from the transect, and consider how resulting data may be analysed using quantal assay procedures. (See Otto and Pollock, 1990, for one solution to this problem.)

*1.22** Fruit infected by fruit-fly larvae may be treated by exposure to cold, to try to kill the infecting larvae. Consider how you might analyse the artificial data below, presented by Wadley (1949):

No. of days of exposure to cold:	1	2	3	4	5	6
No. of flies emerging from a specified quantity of fruit:	715	396	158	55	32	4

Discussion of these data continues in section 3.3.

1.23 Discuss whether you think variation in group size tested, with respect to dose may be important. (Example 1.5, Exercises 1.4 and 1.5, and section 6.6.5).

*1.24** (Ashford and Sowden, 1970) The data below describe the incidence of **two** symptoms in a sample of **working** coalminers. The

	Breathlessness	*Yes*		*No*		
	Wheeze	*Yes*	*No*	*Yes*	*No*	*Total*
Age	20–24	9	7	95	1841	1952
group	25–29	23	9	105	1654	1791
(years)	30–34	54	19	177	1863	2113
	35–39	121	48	257	2357	2783
	40–44	169	54	273	1778	2274
	45–49	269	88	324	1712	2393
	50–54	404	117	245	1324	2090
	55–59	406	152	225	967	1750
	60–64	372	106	132	526	1136
	Total	1827	600	1833	14022	18282

data were collected as part of the National Coal Board's Pneumo-
coniosis Field Research. They are discussed in detail in McCullagh and
Nelder (1989) and section 3.8.

Consider how the threshold model of section 1.5 may be extended
to describe these data. Discuss the implications of the fact that the
miners were working at the time of examination. Suggest two further
practical examples resulting in multivariate data of this kind.

1.25[+] (Healy, 1988, p. 85) The data below classify women by age,
smoking habit, and whether or not the menopause has been
reached. Plot the data and discuss.

	Non-smokers		Smokers	
Age (years)	Total	Number menopausal	Total	Number menopausal
45-	67	1	37	1
47-	44	5	29	5
49-	66	15	36	13
51-	73	33	43	26
53	52	37	28	25

1.26[+] (Marriott and Richardson, 1987) The data below compare
the performance of two anti-fungal drugs *in vivo* and *in vitro*. Discuss
the findings. *In vivo*: mice were infected intravenously with *Candida
albicans* before treatment. ED_{50}s follow from fitting a logit model.
The estimation of the standard errors is explained in Chapter 2.

Compound	Route of administration	$ED_{50}(mg/kg)$ (estimated standard error)	
		2 days post infection	5 days post infection
Fluconazole	oral	0.08 (0.02)	0.7 (0.5)
	intravenous	0.06 (0.04)	1.2 (0.2)
Ketoconazole	oral	9.5 (1.2)	> 20

In vitro activity against pathogenic yeasts and dermatophytes on a defined tissue culture-based medium, measured by minimum inhibitory concentration (MIC).

Organism	No. of isolates	Geometric mean MIC (*mug./mcurlyl.*)	
		Fluconazole	*Ketoconazole*
Candida albicans	159	0.39	0.008
Candida glabrata	3	1.9	<0.01
Microsporum canis	4	9.4	0.14
Trichophyton mentagrophytes	21	>100	0.64

1.27 What factors should be considered at the design stage of experiments like that resulting in the data of Table 1.5? Comment on the distribution of insects over concentrations in this experiment.

CHAPTER 2

Maximum-likelihood fitting of simple models

2.1 The likelihood surface and non-linear optimization

Instead of the simple approaches to model-fitting given in Chapter 1, the method of maximum-likelihood is usually employed. In this chapter we describe the operation of this method, concentrating on the logit model as an illustration. More complex models are encountered in later chapters, and so there is some discussion of alternative numerical procedures that may be used for the maximum likelihood fitting of such models.

Suppose that k doses of a substance are tested in an experiment, and that the ith dose, d_i is given to n_i individuals, of whom r_i respond. If P_i denotes the probability that any individual responds at the ith dose, then assuming independence between individuals within doses, and between doses, we obtain the likelihood as the following product of binomial terms:

$$L(\alpha, \beta) = \prod_{i=1}^{k} \binom{n_i}{r_i} P_i^{r_i} (1 - P_i)^{n_i - r_i}$$

We may model the probabilities of response by assuming a location and scale model, with

$$P_i = F(\alpha + \beta d_i)$$

for the location/scale pair of parameters (α, β) and (usually) cumulative distribution function $F(\ )$ of some continuous random variable. The likelihood, L, then becomes a function, $L(\alpha, \beta)$, of the two parameters, α and β. The **logit** (sometimes also called **logistic**) model results if we take $F(x)$ as the cumulative distribution function of the standard logistic distribution, with mean 0 and variance $\pi^2/3$:

$$F(x) = \frac{1}{1 + e^{-x}}, \; -\infty < x < \infty$$

When we set $\partial L/\partial \alpha = \partial L/\partial \beta = 0$ we obtain two non-linear equations to solve for the maximum-likelihood estimates of α and β. It is necessary in this case to use numerical optimization in order to obtain the maximum-likelihood estimates, $\hat{\alpha}$ and $\hat{\beta}$.

Prior knowledge may suggest bounds for the slope parameter β, which may be useful in initiating the iteration for numerical optimization. Thus Burges and Thomson (1971) and Hughes *et al.* (1984) suggest that slopes greater than 2 are rare in the particular context of the models they fit to insect pathology assays. Mantel and Bryan (1961) anticipate much larger slopes from systemic poisons than in studies of therapeutic effects of antibiotics. The use of a lower bound for β has been suggested to produce a conservative procedure for estimating dose levels which will result in very low incidence of responses, a subject which is discussed further in section 4.6.

Figure 2.1 provides an isometric projection and contour map of the log-likelihood surface, $l(\alpha, \beta) = \log L(\alpha, \beta)$ which results when the logit model is applied to the data in Table 2.1. In this application the maximum-likelihood parameter estimates (with estimates of

Table 2.1 *Busvine (1938) examined the toxicity of ethylene oxide, applied to the grain beetle, Calandra granaria. Insects were classified one hour after exposure to the chemical*

Log_{10} dose of chemical (mg/100 ml)	Number of insects affected	Number of insects
0.394	23	30
0.391	30	30
0.362	29	31
0.322	22	30
0.314	23	26
0.260	7	27
0.225	12	31
0.199	17	30
0.167	10	31
0.033	0	24

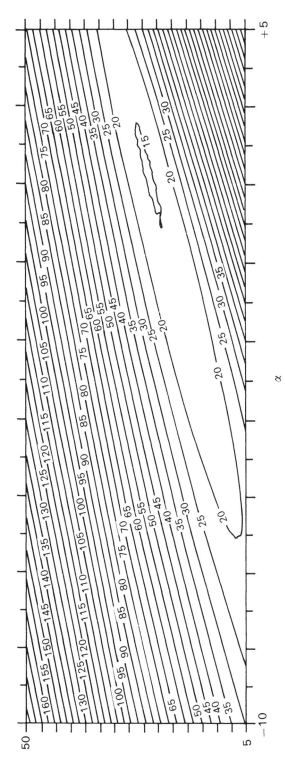

Figure 2.1 The log-likelihood surface for the logit model applied to the data in Table 2.1. Here we use the (α, β) parameterization; see also Exercise 2.1.

standard error in parentheses) are: $\hat{\alpha} = -4.52$ (0.4), $\hat{\beta} = 18.15$ (1.5), with the estimated correlation, $\text{Corr}(\hat{\alpha}, \hat{\beta}) = -0.93$. We discuss the construction of these errors and correlation later. The size of the correlation results because the doses $\{x_i\}$ have not been 'centred' – see Marquardt (1980) – and is mirrored in the elongated contours of negative slope. See also Exercise 2.1. In this example the logit model does not provide a good fit to the data. For discussion, see Exercise 2.28.

Fitting the logit or a similar model by maximum likelihood is therefore a straightforward problem of numerical optimization in a small number of variables. Necessary and sufficient conditions for the existence of a unique optimum for the logit and probit models are given in Silvapulle (1981) – see Exercise 2.27. In section 2.2 we shall employ the method-of-scoring approach to obtain the maximum likelihood estimates of (α, β). Later in the book we shall consider a range of more complicated models. For fitting these by maximum likelihood, a wide variety of iterative numerical techniques exist (e.g. Adby and Dempster, 1974) and computer routines are provided by computer packages and libraries such as: NAG, IMSL, BMDP, SAS, MLP, GLIM and GENSTAT. (See Appendix E for examples.) The simplex method of Nelder and Mead (1965), available as NAG routine E04CCF, is extremely useful as it does not require any information other than the expression for the log-likelihood, and we shall see several examples of its use later in the book. Some numerical procedures use finite differences to approximate derivatives, and this is done in procedure $3R$ in BMDP for example. In some cases, computer algebra packages may be used to evaluate expressions for first- and second-order derivatives of the log-likelihood, and so facilitate the use of second-order optimization procedures, such as the method-of-scoring described below.

Before starting any optimization we should consider whether variables are centred (see, for example, Marquardt, 1980, and Exercise 2.1), and whether a reparameterization of the problem would be convenient. Ross (1975, 1990) has advocated the use of stable parameters in maximization problems. In the context of logit or probit modelling an example of a stable parameter is the ED_{50}, which is usually precisely estimated and has an estimator with low absolute correlation with other parameter estimators – see here for example the illustration of Figure 2.3. Note also the paper by Cox and Reid (1987), and Exercise 2.20. Another feature of model parameterization concerns parameters which are bounded; this can be especially problematical if a method like the simplex procedure is used, as successive trial simplices can take parameter estimates out of range. An illustration results from the mixture modelling of Lwin and Martin (1989), which is discussed later in Example 3.3. In that example, parameter constraints were: $0 < \zeta_i < 1$, $1 \leqslant i \leqslant r$,

$$\sum_{i=1}^{r} \zeta_i = 1$$

$$-\infty < \alpha_i < \infty, \quad 1 \leqslant i \leqslant r$$

$$0 < \beta_i < \infty, \quad 1 \leqslant i \leqslant r$$

The reparameterization they adopted was:

$$\zeta_i = e^{\tau_i} \Big/ \left\{ 1 + \sum_{j=1}^{r-1} e^{\tau_j} \right\}, \quad 1 \leqslant i \leqslant r-1$$

$$\zeta_r = 1 \Big/ \left\{ 1 + \sum_{j=1}^{r-1} e^{\tau_j} \right\}$$

$$\beta_i = e^{\beta_i^*}$$

permitting unconstrained optimization with respect to $\{\tau_i, \beta_i^*, \alpha_i\}$. Alternatively, a number of NAG routines allow parameters to be bounded – for instance, E04VCF.

We shall now demonstrate how the method-of-scoring procedure operates for the logit model. There are several reasons for doing this, rather than simply recommending a computer optimization routine. For a start, with the method-of-scoring we know precisely how the method operates. If it fails to converge, we may then try an alternative starting point, making use of how the procedure behaved previously. Numerical techniques provided by computer packages tend to operate as black boxes, so that users are often not entirely in control, and sure that they know how the procedure is operating. Furthermore, statisticians require more than just parameter estimates corresponding to likelihood maxima, they also need measures of error (such as those provided above following the analysis of the data in Table 2.1), and to produce confidence intervals and regions for parameters. One way of estimating error is to use measures of curvature of the likelihood surface at the maximum, and we can employ the expectation of second-order derivatives of the likelihood surface to provide us with the information we require. These are needed for the method-of-scoring procedure, as we shall see below, but packaged procedures, designed for general mathematical use, may not provide the necessary material.

Finally, we use the method-of-scoring, because of the connections that result with linear regression modelling. This link between two areas of statistics which appear at first sight to be quite different is surprising. It enables us to use a computer program for standard weighted linear regression (as exists in the MINITAB package, for example – Ryan *et al.* 1985) for non-linear modelling, by iterating the weighted linear regression. At this stage we might wonder whether there is a connection between maximum-likelihood estimation and the regressions of section 1.4. The key paper in this regard is the one by Little (1968) and we shall discuss its conclusions in section 2.6. First of all we shall give in detail the method-of-scoring procedure for the logit model.

2.2 The method-of-scoring for the logit model

What follows is a particular instance of logistic regression. We deal with the general context of logistic regression in section 2.8, but we shall find it useful to start with the detail of a particular, and most useful, example.

From section 2.1, the log-likelihood $l(\alpha, \beta) = \log_e L(\alpha, \beta)$ can be written as:

$$l(\alpha, \beta) = \text{constant} + \sum_{i=1}^{k} r_i \log_e P_i + \sum_{i=1}^{k} (n_i - r_i) \log_e(1 - P_i)$$

For the logit model, $P_i^{-1} = 1 + e^{-(\alpha + \beta d_i)}$.

If now ζ is used to denote either α or β,

$$\frac{\partial l}{\partial \zeta} = \sum_{i=1}^{k} \frac{r_i}{P_i} \frac{\partial P_i}{\partial \zeta} - \sum_{i=1}^{k} \left(\frac{n_i - r_i}{1 - P_i}\right) \frac{\partial P_i}{\partial \zeta} = \sum_{i=1}^{k} \frac{\partial P_i}{\partial \zeta} \frac{n_i(p_i - P_i)}{P_i(1 - P_i)}$$

where we have written $p_i = r_i/n_i$ and so,

$$\frac{\partial l}{\partial \alpha} = \sum_{i=1}^{k} \frac{n_i(p_i - P_i)}{P_i(1 - P_i)} f(d_i)$$

$$\frac{\partial l}{\partial \beta} = \sum_{i=1}^{k} \frac{n_i x_i(p_i - P_i) f(d_i)}{P_i(1 - P_i)}$$

where we have written $f(x)$ for $d\,P(x)/dx$. In the logit model $f(d)$ is given by the logistic density function,

$$f(d) = \frac{e^{-(\alpha + \beta d)}}{(1 + e^{-(\alpha + \beta d)})^2}$$

Solving equations

$$\frac{\partial l}{\partial \alpha} = \frac{\partial l}{\partial \beta} = 0$$

is equivalent to model-fitting by weighted least squares – Exercise 2.25.

Note that, for a model that is not the logit model, Chambers (1973) remarks on possible inaccuracies in computing $f(d)/[P(d)\{1 - P(d)\}]$ if d is large or small. He recommends using a power series expansion of $f(d)/\{1 - P(d)\}$ if d is large and positive, or of $f(d)/P(d)$ if d is large and negative. When probit analysis is used, $f(d)/\{1 - P(d)\}$ is the Mills ratio (Rabinowitz, 1969, and Exercise 2.36). However, for the logit case, we see that $f(d) = P(d)(1 - P(d))$, and so

$$\frac{\partial l}{\partial \alpha} = \sum_{i=1}^{k} n_i(p_i - P_i) \quad \text{and} \quad \frac{\partial l}{\partial \beta} = \sum_{i=1}^{k} n_i d_i(p_i - P_i)$$

Now for a general model,

$$\frac{\partial^2 l}{\partial \zeta^2} = \sum_{i=1}^{k} \frac{\partial P_i}{\partial \zeta} \frac{n_i\left(\dfrac{-\partial P_i}{\partial \zeta}\right)}{P_i(1 - P_i)} + \text{terms with factors of } (p_i - P_i)$$

Furthermore, $E[p_i] = P_i$, since $p_i = r_i/n_i$, and r_i results from the binomial distribution, $\text{Bin}(n_i, P_i)$, with mean $n_i P_i$.
Hence,

$$E\left[\frac{\partial^2 l}{\partial \zeta^2}\right] = -\sum_{i=1}^{k} \frac{n_i\left(\dfrac{\partial P_i}{\partial \zeta}\right)^2}{P_i(1 - P_i)}$$

and similarly,

$$E\left[\frac{\partial^2 l}{\partial \alpha \partial \beta^2}\right] = -\sum_{i=1}^{k} \frac{n_i\left(\dfrac{\partial P_i}{\partial \alpha} \dfrac{\partial P_i}{\partial \beta}\right)}{P_i(1 - P_i)}$$

In order to obtain a solution to the non-linear equations resulting from setting $\partial l/\partial \alpha = \partial l/\partial \beta = 0$, we use the method-of-scoring which provides the iteration:

$$\begin{pmatrix} \alpha^{(t+1)} \\ \beta^{(t+1)} \end{pmatrix} = \begin{pmatrix} \alpha^{(t)} \\ \beta^{(t)} \end{pmatrix} + \mathbf{A}^{-1} \begin{bmatrix} \dfrac{\partial l}{\partial \alpha} \\ \dfrac{\partial l}{\partial \beta} \end{bmatrix} \quad \text{for } t \geqslant 0 \qquad (2.1)$$

where

$$\mathbf{A} = -E \begin{bmatrix} \dfrac{\partial^2 l}{\partial \alpha^2} & \dfrac{\partial^2 l}{\partial \alpha \partial \beta} \\ \dfrac{\partial^2 l}{\partial \alpha \partial \beta} & \dfrac{\partial^2 l}{\partial \beta^2} \end{bmatrix}$$

for suitable starting vector $(\alpha^{(0)}, \beta^{(0)})'$. See, for instance, Silvey (1975, p. 71). We refer to $\mathbf{A}$ as the expected Fisher information matrix. The Newton–Raphson method results from the iteration of equation (2.1) if we do not take expectations in $\mathbf{A}$, but instead just use minus the Hessian matrix of second-order derivatives. $\mathbf{A}$ without expectations is called the Fisher information matrix.

Here $\mathbf{A}$ and the gradient vector, $(\partial l/\partial \alpha, \partial l/\partial \beta)'$ are evaluated at the tth iterate values $(\alpha^{(t)}, \beta^{(t)})'$.

A BASIC program for the logit model is given in Example 2.1. A variety of termination rules may be specified. In the illustration of Example 2.1, iteration continues until $|\partial l/\partial \alpha| < 10^{-5}$ and $|\partial l/\partial \beta| < 10^{-5}$, though this simple rule has been criticised by Kennedy and Gentle (1980, p. 437) as possibly susceptible to round-off error. They suggest an alternative procedure, due to Marquardt (1963) which is as follows.

Denote the parameter vector by $\zeta = \{\zeta_j\}$; accept $\zeta^{(t+1)}$ as the estimate of ζ whenever

$$|\zeta_j^{(t)} - \zeta_j^{(t+1)}| \leqslant \varepsilon_1 (|\zeta_j^{(t)}| + \varepsilon_2), \quad \text{for all } j$$

over a given number of iterations; values of $\varepsilon_1 = 10^{-4}$ and $\varepsilon_2 = 10^{-3}$ are often used. An alternative approach is adopted by the BMDP procedure 3R, for non-linear regression, and in practice a range of different termination criteria should be used in conjunction (Exercise 2.3).

An alternative to iteration is simply to take a single step, and return $(\alpha^{(1)}, \beta^{(1)})$ as point estimates. In some cases this can provide a useful correction to a good starting point and we discuss such an approach in Appendix A and section 3.4.

There is no general guarantee that methods such as the method-of-scoring will converge. An illustration of non-convergence is provided by Ridout (1990), in the context of modelling dilution series data (Exercise 2.42). Guaranteed convergence follows from using the lower bound principle, in which the Hessian of the Newton–Raphson method is replaced by a global lower bound, which leads to a series of iterations which monotonically increase the likelihood (Böhning and Lindsay, 1988).

Example 2.1

A BASIC program for fitting the logit model to quantal assay data, using the method-of-scoring. In this case users select their own starting values – see line 140. Experience with this program suggests that it is quite robust with respect to choice of starting values. When there is only one dose with a response between 0% and 100%, the likelihood is maximized at $\beta = \infty$; cf. Exercise 2.30. As we can see from line 410, in this case iteration continues until $|\partial l/\partial \alpha| < 10^{-5}$.

```
10      DIM Q(20)
20      DIM N(20),R(20),P(20),X(20)
30      REM PROGRAM TO PERFORM LOGIT ANALYSIS, USING
40      REM THE METHOD OF SCORING
50      FILE *1: "DATA7"
60      INPUT K
70      PRINT " LOGIT ANALYSIS OF QUANTAL ASSAY DATA "
80      PRINT " THE DATA ARE : "
90      PRINT " X      R      N "
100     FOR I=1 TO K
110      INPUT *1: X(I),R(I),N(I)
120      PRINT X(I),R(I),N(I)
130     NEXT I
140     INPUT A,B
150     PRINT "ESTIMATES OF A,B, FOLLOWED BY THE RESPECTIVE DERIVATIVES :"
160      FOR J=1 TO 100
170      LET D1=0
180      LET D2=0
190      LET B1=0
200      LET B2=0
210      LET B4=0
220      FOR I=1 TO K
230       LET A1=EXP(-A)
240       LET P(I)=1/(1+A1*EXP(-B*X(I)))
250       LET Z=R(I)-N(I)*P(I)
260       LET D1=D1+Z
270       LET D2=D2+Z*X(I)
280       LET Q(I)=P(I)
290       LET P(I)=P(I)*(1-P(I))*N(I)
300       LET B1=B1+P(I)
310       LET B2=B2+P(I)*X(I)
```

(Cont.)

```
320        LET B4=B4+P(I)*X(I)*X(I)
330        NEXT I
340        LET W=B1*B4-B2*B2
350        LET C1=B4/W
360        LET C2=-B2/W
370        LET C4=B1/W
380        LET A=A+C1*D1+C2*D2
390        LET B=B+C2*D1+C4*D2
400        PRINT J,A,B,D1,D2
410        IF ABS(D1) < 1E-5 THEN 430
420        NEXT J
430        PRINT " DOSE   EXPECTED PROPORTION,   EXPECTED NO. "
440        LET C1=0
450        LET L=0
460        FOR I=1 TO K
470        PRINT X(I),Q(I),Q(I)*N(I)
480        LET C1=C1+(R(I)-Q(I)*N(I))*(R(I)-Q(I)*R(I))/(Q(I)*N(I))
490        LET L=L-R(I)*LOG(Q(I))-(N(I)-R(I))*LOG(1-Q(I))
500        NEXT I
510        PRINT "CHI-SQUARE GOODNESS-OF-FIT STATISTIC = ",C1
520        PRINT "VALUE OF THE LOG-LIKELIHOOD AT CONVERGENCE =",L
530        END
```

It was convenient, in the above, to use the method-of-scoring, rather than Newton–Raphson, as that simplified the matrix **A**, used in equation (2.1). However, it is readily verified that for the logit model the method-of-scoring and Newton–Raphson are in fact identical. See Exercise 2.4 for further discussion of this feature. The matrix inversion of equation (2.1) can prove time-consuming, and the computer package QUAD, developed by Morgan and Pack (1988) and Morgan *et al.* (1989), uses an efficient Gauss–Jordan sweep algorithm described by Clarke (1982), which is discussed further by Ridout and Cobby (1989).

2.3   The connection with iterated weighted regression; generalized linear models

From the work of the last section, we see that we can write

$$
\mathbf{A} = \begin{bmatrix} \sum_{i=1}^{k} n_i P_i(1 - P_i) & \sum_{i=1}^{k} n_i P_i(1 - P_i)d_i \\ \sum_{i=1}^{k} n_i d_i P_i(1 - P_i) & \sum_{i=1}^{k} n_i d_i^2 P_i(1 - P_i) \end{bmatrix}
$$

$$
= \begin{bmatrix} \sum_{i=1}^{k} n_i w_i & \sum_{i=1}^{k} n_i w_i d_i \\ \sum_{i=1}^{k} n_i w_i d_i & \sum_{i=1}^{k} n_i w_i d_i^2 \end{bmatrix}
$$

where $w_i = P_i(1 - P_i)$. This then enables us to rewrite equation (2.1) in the form,

$$\mathbf{A} \begin{pmatrix} \alpha^{(t+1)} \\ \beta^{(t+1)} \end{pmatrix} = \mathbf{A} \begin{pmatrix} \alpha^{(t)} \\ \beta^{(t)} \end{pmatrix} + \begin{bmatrix} \dfrac{\partial l}{\partial \alpha} \\ \dfrac{\partial l}{\partial \beta} \end{bmatrix}$$

which becomes

$$\begin{bmatrix} \sum\limits_{i=1}^{k} n_i w_i & \sum\limits_{i=1}^{k} n_i w_i d_i \\ \sum\limits_{i=1}^{k} n_i w_i d_i & \sum\limits_{i=1}^{k} n_i w_i d_i^2 \end{bmatrix} \begin{pmatrix} \alpha^{(t+1)} \\ \beta^{(t+1)} \end{pmatrix}$$

$$= \begin{bmatrix} \sum\limits_{i=1}^{k} n_i(w_i^{(t)}\alpha^{(t)} + w_i^{(t)}d_i\beta^{(t)} + p_i - P_i^{(t)}) \\ \sum\limits_{i=1}^{k} n_i d_i(w_i^{(t)}\alpha^{(t)} + w_i^{(t)}d_i\beta^{(t)} + p_i - P_i^{(t)}) \end{bmatrix}$$

$$= \begin{bmatrix} \sum\limits_{i=1}^{k} n_i w_i^{(t)} y_i^{(t)} \\ \sum\limits_{i=1}^{k} n_i w_i^{(t)} d_i y_i^{(t)} \end{bmatrix}, \quad \text{say,} \tag{2.2}$$

where we have everywhere indicated dependence on the tth iteration values $\alpha^{(t)}$ and $\beta^{(t)}$, and set

$$y_i^{(t)} = \{\alpha^{(t)} + d_i\beta^{(t)} + (p_i - P_i^{(t)})/w_i^{(t)}\}, \quad 1 \leqslant i \leqslant k$$

The $\{y_i\}$ are termed 'working logits' by analogy with the logits, $(\alpha + \beta d)$.

We now see from equation (2.2) that obtaining the $(t+1)$th iterates, $(\alpha^{(t+1)}, \beta^{(t+1)})'$ is equivalent to a weighted linear regression of the working logits $\{y_i\}$ on $\{d_i\}$ (Cook and Weisberg, 1982, p. 209), with weights $\{n_i w_i\}$, and so,

$$\alpha^{(t+1)} = \frac{\left(\sum\limits_{i=1}^{k} n_i w_i^{(t)} y_i^{(t)}\right)\left(\sum\limits_{i=1}^{k} n_i w_i^{(t)} d_i^2\right) - \left(\sum\limits_{i=1}^{k} n_i w_i^{(t)} d_i\right)\left(\sum\limits_{i=1}^{k} n_i w_i^{(t)} d_i y_i^{(t)}\right)}{\left(\sum\limits_{i=1}^{k} n_i w_i^{(t)}\right)\left(\sum\limits_{i=1}^{k} n_i w_i^{(t)} d_i^2\right) - \left(\sum\limits_{i=1}^{k} n_i w_i^{(t)} d_i\right)^2}$$

$$\beta^{(t+1)} = \frac{\left(\sum\limits_{i=1}^{k} n_i w_i^{(t)} y_i^{(t)}\right)\left(\sum\limits_{i=1}^{k} n_i w_i^{(t)} d_i\right) - \left(\sum\limits_{i=1}^{k} n_i w_i^{(t)} d_i y_i^{(t)}\right)\left(\sum\limits_{i=1}^{k} n_i w_i^{(t)}\right)}{\left(\sum\limits_{i=1}^{k} n_i w_i^{(t)} d_i\right)\left(\sum\limits_{i=1}^{k} n_i w_i^{(t)} d_i\right) - \left(\sum\limits_{i=1}^{k} n_i w_i^{(t)} d_i^2\right)\left(\sum\limits_{i=1}^{k} n_i w_i^{(t)}\right)}.$$

$$\tag{2.3}$$

Hence the method-of-scoring can be performed using a suitable computer program/package which allows us to carry out weighted regression, ideally with a looping facility to iterate the procedure, changing the weights each time.

Example 2.2

The following MINITAB program was kindly provided by J. M. Bremner. The data have been read into columns *C1*, *C2* and *C3*, and in this example just three iterations are carried out. J. R. Sedcole has produced a MINITAB program for probit analysis, and this is presented in Exercise 2.5.

```
NAME     C1      'DOSE'
NAME     C2      'N'
NAME     C3      'R'
NAME     C4      'LOGDOSE'
LET      'LOGDOSE'=LOGTEN('DOSE')
NAME     C5      'P'
NAME     C6      'Q'
NAME     C7      'PSTAR'
NAME     C8      'DEP'
NAME     C9      'WEIGHT'
NAME     C10     'STRES'
NOTE:    WE NOW USE SIMPLE LINEAR REGRESSION
NOTE:    TO OBTAIN A STARTING VALUE
LET      'PSTAR'=LOG('R'/('N'-'R'))
LET      C11=C7
REGRESS  'PSTAR' 1 'LOGDOSE' 'STRES' 'PSTAR'
STORE
LET      'P'=EXPONENTIAL('PSTAR')/(1.0+EXPONENTIAL('PSTAR'))
LET      'Q'=1.0-'P'
LET      'WEIGHT'='P'*'Q'*'N'
LET      'DEP'='PSTAR'+('Y'-'N'*'P')/('WEIGHT')
REGRESS  'DEP' 'WEIGHT' 1 'LOGDOSE' 'STRES' 'PSTAR'
END
EXECUTE  3 TIMES
NOTE:    'PSTAR' VALUES ARE THE FITTED VALUES AT EACH ITERATION
NOTE:    'STRES' ARE THE VALUES OF STANDARDISED RESIDUALS
PLOT     'STRES' VS 'PSTAR'
M PLOT   C11 VS 'LOGDOSE' AND 'PSTAR' VS 'LOGDOSE'
```

Error estimation is standard, with minus the inverse of the expected Hessian evaluated at the maximum-likelihood estimate, $(\hat{\alpha}, \hat{\beta})$, providing an asymptotic estimate of the dispersion matrix for the estimators $(\hat{\alpha}, \hat{\beta})$ [Silvey (1975, p. 78) and Chambers (1973)]. Sowden (1971) warns that asymptotic estimators of standard errors are subject to appreciable small sample bias; see also Exercise 2.34.

The above approach, of solving the likelihood equations $\partial l/\partial \alpha = \partial l/\partial \beta = 0$ by means of the method-of-scoring, and noting that each iteration results in a simple weighted linear regression applies also in the far more general context of generalized linear models (Appendix B).

Example 2.3

The following instructions enable the GLIM package for fitting generalized linear models to fit the logit model: data input is from a file with the name: [.QADATA]MENARCHE.DAT, containing the 25 rows of data of Table 1.4. Channel 13 was used for data input, which is why 13 appears after $INPUT.

```
$UNITS  25$
$DATA   X R N $
$DINPUT  13$
FILE NAME  ?  [.QADATA]MENARCHE.DAT
$ERROR  B N$
$LINK   G$
$YVAR R$
$FIT X$
$DIS  R$
```

in the $ERROR command, 'b' indicates that the binomial distribution is to be selected, with index given by the entries in vector n. The letter 'g' in the $LINK command indicates that the logistic link is to be used. The $FIT and $DIS commands provide the model-fitting followed by display of residuals. Parameter estimates, estimates of covariances and correlation for parameter estimators result from following $DIS by, respectively, E, V, C.

2.4 Hand calculation and using tables

The three examples of the last section provide three different ways of using the computer to fit the logit model. Individuals with access to computers need not consider how to perform the necessary calculations by hand. However, the method-of-scoring with just two parameters is quite feasible by hand; the iterated weighted regression aspect can be used to ease hand calculations if tables are available to provide the weights $\{w_i\}$ and the 'working logits' at each stage. The method proceeds as follows:

After t iterations we have estimates $\alpha^{(t)}$, $\beta^{(t)}$. These then provide us with our latest estimates of the logits: $\{l_i^{(t)} = \alpha^{(t)} + \beta^{(t)}d_i, 1 \leqslant i \leqslant k\}$.

The working logits are given by:

$$y_i^{(t)} = \{\alpha^{(t)} + d_i\beta^{(t)} + (p_i - P_i^{(t)})/w_i^{(t)}\}$$

where

$$p_i = r_i/n_i,$$

$$P_i^{(t)} = 1/(1 + e^{-l_i^{(t)}}) \quad \text{and}$$

$$w_i^{(t)} = P_i^{(t)}(1 - P_i^{(t)})$$

Hence the $l_i^{(t)}$ values provide $P_i^{(t)}$ and $w_i^{(t)}$ values, leading ultimately to $y_i^{(t)}$ values. Equations (2.3) then provide $\alpha^{(t+1)}$, $\beta^{(t+1)}$. Table 2 of Ashton (1972) provides the 'antilogits' of $P_i^{(t)}$ for given values of $l_i^{(t)}$ and Table 3 of Ashton (1972) provides weights $\{w_i^{(t)}\}$ for given antilogits $P_i^{(t)}$. Both of these tables are taken from Berkson (1953). The main advantage of this approach is that it avoids having to evaluate $P_i^{(t)}$ and $w_i^{(t)}$ at each iteration. A single two-way table of working-logits may be used, indexed by p_i and $\alpha^{(t)} + \beta^{(t)}d_i$. For tables of working-probits, see for example Hubert (1992, p. 89 and Table 3), Table 6 of Biometrika tables (Pearson and Hartley, 1970), and Finney (1971, Table 4).

Finally, we note that the stages in maximum-likelihood fitting of logit and probit models are also outlined in the Appendix to Hewlett and Plackett (1979), but in that case with particular reference to use on a programmable hand calculator.

2.5   The chi-square goodness-of-fit test; heterogeneity

Consider the example fitted below by a probit model, using maximum likelihood. The data are data set 4 of Exercise 1.3.

Log_2 (concentration)	No. of insects (n_i)	No. of insects killed (r_i)	Expected no. killed $(n_i\hat{P}_i)$	$(r_i - n_i\hat{P}_i)$	$\dfrac{(r_i - n_i\hat{P}_i)^2}{n_i\hat{P}_i(1 - \hat{P}_i)}$
0.41	50	6	6.3	−0.3	0.02
0.58	48	16	15.8	0.2	0.00
0.71	46	24	24.8	−0.8	0.06
0.89	49	42	39.3	2.7	0.94
1.01	50	44	45.6	−1.6	0.64

Simple inspection by eye of the $(n_i \hat{P}_i)$ expected values assures us that the fit of the probit model to the data is apparently quite good, but we need a statistical yardstick to quantify this measure of fit. This is conveniently provided by the standard chi-square goodness-of-fit test:

$$\text{Set } X^2 = \sum_{i=1}^{k} \frac{(r_i - n_i \hat{P}_i)^2}{n_i \hat{P}_i (1 - \hat{P}_i)} \qquad (2.4)$$

If the model is appropriate, then asymptotically,

$$X^2 \sim \chi^2_{k-2}$$

We can justify this result by observing that r_i is the realization of a binomial, $\text{Bin}(n_i, P_i)$ random variable, R_i, with mean $n_i P_i$ and variance $n_i P_i (1 - P_i)$. As $n_i \to \infty$, for each i, then by the central limit theorem, and the consistency of the estimates of mean and variance,

$$\frac{(R_i - n_i \hat{P}_i)}{\sqrt{n_i \hat{P}_i (1 - \hat{P}_i)}}$$

has a distribution which tends to $N(0, 1)$, and so the distribution of X^2 becomes chi-squared. Alternatively, we can approach the problem from the point of view of a $k \times 2$ contingency table, the ith row of which has entries, $\{r_i, (n_i - r_i)\}$, so that the columns correspond to dead and alive categories, respectively. When the model is fitted, requiring the estimation of two parameters, the expected values for the ith row are $\{n_i \hat{P}_i, n_i (1 - \hat{P}_i)\}$ and the usual Pearson X^2 goodness-of-fit test results in the statistic,

$$X^2 = \sum_{i=1}^{k} \left\{ \frac{(r_i - n_i \hat{P}_i)^2}{n_i \hat{P}_i} + \frac{\{n_i - r_i - n_i(1 - \hat{P}_i)\}^2}{n_i (1 - \hat{P}_i)} \right\}$$

being referred to the chi-square distribution on $(k - 2)$ degrees of freedom (since each row provides a single degree of freedom and we lose two degrees of freedom due to the two parameters estimated). We note finally that we now obtain equation (2.4), since

$$\frac{(r_i - n_i \hat{P}_i)^2}{n_i \hat{P}_i} + \frac{\{n_i - r_i - n_i(1 - \hat{P}_i)\}^2}{n_i (1 - \hat{P}_i)} = \frac{(r_i - n_i \hat{P}_i)^2}{n_i \hat{P}_i (1 - \hat{P}_i)}$$

For further discussion and approximations in the general $k \times 2$ contingency table context, see Kendall and Stuart (1979, p. 609). The chi-square goodness-of-fit test is a useful omnibus test, of general use, but as a result it lacks power. In Chapter 4 we consider ways of restricting the family of possible alternative models to the logit, and hence of providing more powerful tests of fit.

When X^2 indicates a significant lack of fit, this may be due to a poorly fitting model. For instance, we may need to replace the logit model by a more complicated form embracing asymmetry, such as the model of Exercise 2.13. Alternatively, the assumption of a binomial distribution may be in error, due perhaps to a correlation between responses of individuals, or a basic heterogeneity in the way that different individuals respond. If the latter is the case, Finney (1971, p. 72) suggested using $X^2/(k-2)$ as a **heterogeneity factor**, which could be used to scale up all variances. In addition, normal distributions, used for confidence interval construction, could be replaced by t_{k-2} distributions.

In Chapter 6 we shall see how to interpret the heterogeneity factor through a quasi-likelihood approach to model fitting. In general it may well be difficult to distinguish between the two reasons for lack of fit outlined above. It is unfortunate that certain statistical packages routinely scale variances by means of the heterogeneity factor. This is done in the SAS procedure PROBIT, for example. In Chapters 4 and 6 we shall see how to extend the basic models to accommodate both types of model failure described here.

2.6 Minimum chi-square estimation

An alternative approach to the method of maximum likelihood for parameter estimation is to choose the parameters to minimize the X^2 expression of equation (2.4). Little (1968) has shown that when we do this the estimate of $\hat{\beta}$ is less than that which results from maximum-likelihood estimation (Exercise 2.24). Further enlightening discussion is to be found in Mantel (1985). A number of authors have shown that the use of iterated weighted least squares, in an attempt to minimize X^2, is equivalent to maximum-likelihood in certain cases. (See for example the papers by Moore and Zeigler, 1967, and Charnes *et al.*, 1976, and Exercise 2.25.)

In Equation (1.1) we have an example of fitting the least-squares regression line:

$$y = \alpha + \beta x.$$

where $y = \text{logit}\ (r/n)$.

Estimates of α and β are obtained from minimizing the sum-of-squares:

$$S(\alpha, \beta) = \sum_{i=1}^{k} (y_i - \alpha - \beta x_i)^2 \tag{2.5}$$

and this is frequently also referred to as minimum chi-square since in the corresponding linear regression model with normally distributed errors, $S(\hat{\alpha}, \hat{\beta})$ is proportional to a chi-square random variable on $(k-2)$ degrees of freedom (Wetherill, 1981, p. 120). This procedure is designed for situations with $\{y_i\}$ of constant variance, which is not true here (but see Exercise 2.9). An alternative approach (used to give equation (1.2)) is that of weighted regression, where we minimize:

$$\tilde{S}(\alpha, \beta) = \sum_{i=1}^{k} w_i (y_i - \alpha - \beta x_i)^2$$

where $w_i^{-1} = V(y_i)$.

Setting $\partial \tilde{S}/\partial \alpha = 0$ gives: $\sum_{i=1}^{k} w_i(y_i - \alpha - \beta x_i) = 0$.

Setting $\partial \tilde{S}/\partial \beta = 0$ gives: $\sum_{i=1}^{k} w_i x_i(y_i - \alpha - \beta x_i) = 0$,

i.e.
$$\alpha \left(\sum_{i=1}^{k} w_i \right) + \beta \left(\sum_{i=1}^{k} w_i x_i \right) = \sum_{i=1}^{k} w_i y_i$$

$$\alpha \left(\sum_{i=1}^{k} w_i x_i \right) + \beta \left(\sum_{i=1}^{k} w_i x_i^2 \right) = \sum_{i=1}^{k} w_i x_i y_i \tag{2.6}$$

These two equations may now clearly be solved in the usual way to provide explicit estimates $\hat{\alpha}$ and $\hat{\beta}$, provided we have forms for

$$w_i^{-1} = V(y_i)$$

As we saw in Exercise 1.2, if R has the binomial $\text{Bin}(n, p)$ distribution, then to a first order approximation,

$$V \left\{ \log \left(\frac{R}{n-R} \right) \right\} = \frac{1}{np(1-p)}$$

which may be *estimated* by $n/\{r(n-r)\}$ and which we may therefore take as an approximation to the inverse of the weight to be used. The resulting procedure is called **minimum logit chi-square**. It can also be described as a modified-transformed chi-square procedure, in the general terminology defined by Hsiao (1985). A table of weights, with adjustments for natural mortality is given in Hewlett and Plackett (1979, p. 32); natural mortality is discussed in section 3.2.

A drawback with this approach is that if $r = 0$ or $r = n$, these weights are zero. A similar problem was encountered in the construction of Figure 1.4. Various approaches have been proposed and investigated, such as replacing $0/n$ by $1/2n$, and n/n by $(n - 1/2)/n$, due to Berkson (1944, 1955). Alternatively, all proportions r/n can be adjusted to $(r + \frac{1}{2})/(n + 1)$ (McCullagh and Nelder, 1989, p. 117, and Exercise 2.10). Apart from this problem, the minimum logit chi-square method has the attractive feature that it provides explicit estimates of α and β – no iterations are necessary. Simulation studies by Hamilton (1979) and James et al. (1984) have suggested that it is advantageous to smooth response proportions, if necessary, before minimum logit chi-square estimation, to obtain a set of proportions of individuals responding that increases monotonically with increasing dose. How this may be done is described in section 7.2.

We may also consider modifying the method of minimum logit chi-square, to update the estimates of p in the weights. Thus once the model is fitted using minimum logit chi-square then we have estimates $(\tilde{\alpha}, \tilde{\beta})$ say, and so we can estimate $P(x)$ by $(1 + e^{-(\tilde{\alpha} + \tilde{\beta}x)})^{-1}$, re-estimate the weights, and continue in this way until there is no change, to the accuracy of working, in $(\tilde{\alpha}, \tilde{\beta})$. This then removes the criticism that treatment of 0% and 100% responses is arbitrary, but it requires an iterative way of proceeding. A compromise would be to iterate just once (cf. Appendix A).

While both the method of maximum-likelihood and the minimum logit chi-square method have optimal asymptotic properties (Hsiao, 1985), care is needed in the interpretation of what asymptotic means in this context. As pointed out by Silverstone (1957), and emphasized by Mantel (1985), for minimum logit chi-square it is essential that group sizes tend to infinity. However, for maximum-likelihood it can suffice that the number of doses used tends to infinity. We return to this point in Chapter 7 when we consider the relative efficiency of non-parametric estimators of the ED_{50}.

Comparisons of the two methods for small sample sizes may be carried out using approximate analytical techniques (Amemiya, 1980) or by computer simulation (Cramer, 1964; Smith et al., 1984; Smith, et al., 1985). In some cases there is evidence that the method of maximum-likelihood produces estimators that are more biased than the method of minimum logit chi-square, but the result is not uniformly true. This discussion is continued in Exercise 2.24. Both QUAD (Morgan *et al.*, 1989) and GLIM start the method-of-scoring iteration for the maximum likelihood fit of the logit model from the minimum logit chi-square estimates (McCullagh and Nelder, 1989, pp. 41 and 117).

2.7 Estimating the dose for a given mortality

When a model such as the logit is fitted to quantal assay data the model may be used to summarize the data, through the pair of parameter estimates, $(\hat{\alpha}, \hat{\beta})$, and the model may form the basis for comparisons between different sets of data (section 2.9). In many cases it may be desirable to estimate ED_{100p} values, which are doses which correspond, under the model, to $100p\%$ mortality. A commonly used summary of a fitted model is the ED_{50}, the dose corresponding to 50% mortality. Of course a single value, such as an ED_{50}, is an imperfect summary of the data, since test substances with the same ED_{50} could have quite different responses for doses away from the ED_{50}. The ED_{50} has a useful interpretation as the median of the tolerance distribution. Many countries have dropped the requirement that the reporting of new drugs must include an investigation of the ED_{50}. The British Toxicology Society (1984) has proposed a fixed dose procedure as an alternative to the ED_{50} as a measure of toxicity, as discussed in Chapter 1. However, it remains of general interest, especially in observational rather than experimental studies. In routine bioassays to monitor potentially toxic effluents, the American Environmental Protection Agency requires point and interval estimates of the ED_{50} (Committee on Methods for Toxicity Tests with Aquatic Organisms, 1975). A number of non-parametric and sequential techniques have been specifically designed to estimate the ED_{50}, as we shall see in Chapters 7 and 8. It will prove convenient later to use θ to specify the ED_{50}, and θ_p to denote the ED_{100p} value.

Usually, unless an experiment has been designed to estimate more extreme dose levels, or has been poorly designed for the given test substance, percentage responses span the 50% value and ED_{50} estimation may be carried out with reasonable precision. Estimating by eye, from Figure 1.1, we might estimate the ED_{50} by $2^{4.2} = 18.38$. From Figure 1.2 an estimate by eye of the ED_{50} is about 13 years, which may also be verified from Figure 1.4, as the ED_{50} corresponds to a zero logit. In many cases a more extreme dose level, such as an ED_{90} or an ED_{95}, say, may be regarded as a more suitable summary of the data, as has already been discussed in Chapter 1. Additional relevant discussion is provided by Brown (1967).

Alternative models to the logit would give very similar estimates of the ED_{50} in most cases, but potentially could give quite different estimates of extreme dose levels and associated errors. This is a subject which has attracted much attention in recent years and it is dealt with in some detail in Chapter 4.

For the logit model, we write

$$\alpha + \beta \theta_p = \text{logit}(p) = \log_e \left(\frac{p}{1-p} \right)$$

and so we have

$$\theta_p = \frac{1}{\beta} \left\{ \log_e \left(\frac{p}{1-p} \right) - \alpha \right\} \tag{2.7}$$

In particular, $ED_{50} = -\alpha/\beta$.

For any quantal assay model with the probability of response to dose d_i given by $F(\alpha + \beta d_i)$, then $p = F(\alpha + \beta \theta_p)$ and so

$$\frac{\{F^{-1}(p) - \alpha\}}{\beta} = \theta_p.$$

Thus for the probit model, or indeed any symmetric c.d.f. $F(\ )$, we also have

$$ED_{50} = -\frac{\alpha}{\beta}$$

but for the Aranda-Ordaz (1981) asymmetric model, discussed further in Exercise 2.13 and Chapter 4, for which

$$F(\alpha + \beta d) = 1 - (1 + \lambda e^{\alpha + \beta d})^{-1/\lambda}, \text{ we have}$$

$$\theta_p = \frac{-\alpha}{\beta} + \frac{1}{\beta} \log_e \left(\frac{1}{\lambda(1-p)^\lambda} - \frac{1}{\lambda} \right)$$

and so

$$ED_{50} = \frac{-\alpha}{\beta} + \frac{1}{\beta} \log_e \left(\frac{2^{\lambda-1}}{\lambda} \right) \tag{2.8}$$

which $= -\alpha/\beta$ when $\lambda = 1$, when the logit model results as a special (symmetric) case. Unlike estimation by eye, maximum-likelihood fitting of a parametric model enables us to provide an error estimate for the estimator of an ED_{100p} and to produce appropriate confidence-intervals. For any given model, this can be done in a variety of ways, the three most common of which we shall now describe.

2.7.1 Using the delta method

This well-known procedure relies on asymptotic approximations and properties of maximum-likelihood estimators. It is based on a truncated Taylor series expansion, as described in Appendix A. An illustration is provided in Example 2.4, for the logit model.

Example 2.4 Using the delta-method for the logit model to estimate the ED_{100p} (Goedhart, 1985)

Here,

$$\alpha + \beta\theta_p = \log_e\left(\frac{p}{1-p}\right)$$

and we suppose that the maximum-likelihood estimators $(\hat{\alpha}, \hat{\beta})$ have the estimated dispersion matrix,

$$\hat{\Sigma} = \begin{pmatrix} V_{11} & V_{12} \\ V_{21} & V_{22} \end{pmatrix}$$

We estimate the θ_p by

$$\hat{\theta}_p = \frac{1}{\hat{\beta}}\left(\log_e\left(\frac{p}{1-p}\right) - \hat{\alpha}\right)$$

which corresponds to $g(\boldsymbol{\theta})$ in Appendix A, with

$$\hat{\boldsymbol{\theta}}' = (\hat{\alpha}, \hat{\beta})$$

So, making use of the Appendix A results, the asymptotic approximation to the distribution of the estimator $\hat{\theta}_p$ is $N(\theta_p, \sigma^2)$, where we estimate σ^2 by (Exercise 2.12)

$$\hat{\sigma}^2 = \frac{1}{\hat{\beta}^2}(V_{11} + 2\hat{\theta}_p V_{12} + \hat{\theta}_p^2 V_{22}) \qquad (2.9)$$

Results of this kind are used in designing experiments for optimal precision (e.g. Exercise 8.9). The delta-method $\gamma\%$ confidence interval is then given by

$$\hat{\theta}_p \pm \frac{z_{\gamma/2}}{\hat{\beta}}(V_{11} + 2\hat{\theta}_p V_{12} + \hat{\theta}_p^2 V_{22})^{1/2} \qquad (2.10)$$

where $z_{\gamma/2}$ is the $100(1 - \gamma/2)$ percentage point of the standard normal distribution. ☐

Similar confidence intervals can be derived for other models (Exercises 2.13 and 2.14). The delta-method confidence intervals are necessarily symmetric, which may be a disadvantage in small samples, especially with regard to the estimation of extreme dose values. The remaining two methods of confidence interval construction which we shall now consider do not possess this feature. Cox (1990) shows that the delta-method estimate of variance of the ED_{50} is the same as that which arises from asymptotic theory if the model is directly parameterized in terms of the ED_{50}.

2.7.2 Using Fieller's theorem

We start with a general result for ratios of random variables. Suppose that random variables $\hat{\alpha}$ and $\hat{\beta}$ have a bivariate normal distribution with mean vector $(\alpha, \beta)'$ and dispersion matrix $\begin{pmatrix} V_{11} & V_{12} \\ V_{21} & V_{22} \end{pmatrix}$.

If $\theta = -\alpha/\beta$, then a natural estimate of θ is $\hat{\theta} = -\hat{\alpha}/\hat{\beta}$, and this is also the maximum-likelihood estimate of θ. A general result for the distribution of sums of bivariate normal random variables tells us that

$$\theta\hat{\beta} + \hat{\alpha} \sim N(0, V_{11} + 2\theta V_{12} + \theta^2 V_{22})$$

Hence a confidence interval for θ, with confidence probability $(1 - \gamma)$, is given by the set of θ values satisfying:

$$\frac{(\hat{\alpha} + \hat{\beta}\theta)^2}{(V_{11} + 2\theta V_{12} + \theta^2 V_{22})} < z_{\gamma/2}^2$$

The limits satisfying this inequality can be written as,

$$\hat{\theta} + \left(\frac{c}{1-c}\right)\left(\hat{\theta} + \frac{V_{12}}{V_{22}}\right) \pm \frac{z_{\gamma/2}}{\hat{\beta}(1-c)}$$
$$\times \left\{V_{11} + 2\hat{\theta}V_{12} + \hat{\theta}^2 V_{22} - c\left(V_{11} - \frac{V_{12}^2}{V_{22}}\right)\right\}^{1/2} \qquad (2.11)$$

where $c = z_{\gamma/2}^2 V_{22}/\hat{\beta}^2$. This is the $(1 - \gamma)$ Fieller interval (Fieller, 1944, 1954).

For models with probability of response at dose d given by $F(\alpha + \beta d)$, we have from previous work,

$$\theta_p = \frac{F^{-1}(p) - \alpha}{\beta}$$

and so the Fieller intervals for the θ_p are given by equation (2.11), replacing $\hat{\theta}$ by $\{F^{-1}(p) - \hat{\alpha}\}/\hat{\beta}$. For the ED_{50} we write the estimate as $\hat{\theta}$. Comparison of the two intervals given by equations (2.10) and (2.11) shows that the intervals are similar wherever c is small (say $c < 0.05$), but otherwise, interval (2.11) becomes increasingly skewed as c increases. Zerbe (1978) has provided a connection between Fieller's theorem and the general linear model, and Cox (1990) provides the extension for the generalized linear model.

2.7.3 The likelihood-ratio interval

The likelihood ratio (LR) interval for an expected dose level, θ_p, consists of all values of μ, say, for which the null hypothesis: $H_0 : \theta_p = \mu$ is not rejected in favour of the alternative: $H_1 : \theta_p \neq \mu$, by means of a likelihood ratio test. We shall illustrate the construction of this interval by using the logit model and the ED_{50} as an example. In this model we express the logit of the probability of response to dose d_i as

$$\log_e\left(\frac{P_i}{1 - P_i}\right) = \alpha + \beta d_i$$

An alternative parameterization is

$$\log_e\left(\frac{P_i}{1 - P_i}\right) = \beta(d_i - \theta)$$

where $\theta = -\alpha/\beta$ is the ED_{50}. Of course, maximization of the likelihood with respect to (θ, β) will give directly a symmetric confidence interval for θ, assuming asymptotic normality. As mentioned above, this is equivalent to the delta-method interval when the (α, β) parameterization is used (Cox, 1990).

However, the log-likelihood can now be written as $l(\theta, \beta)$ and in order to obtain an LR interval for θ we need the **profile** log-likelihood of $\max_\beta l(\theta, \beta)$, plotted against θ (Aitkin *et al.*, 1989, p. 190; Cox and Reid, 1987). An appropriate slice of the likelihood then produces the LR confidence interval (θ_-, θ_+), as shown in Figure 2.2.

It is equivalent to take the same slice (at a depth of 1.92 from the maximum in Figure 2.2) of the two-parameter log-likelihood surface $l(\theta, \beta)$, and then project the resulting contour onto the θ axis. Illustrations of this are shown in Figures 2.3 and 2.4. See Exercise 2.15 for further discussion. We may note the non-elliptical shape of the contours, reflecting more skew in the sampling distribution when we adopt a different parameterization from (α, β) – see Finney (1971, p. 86) for discussion of this point. This suggests that it is preferable to use a likelihood ratio interval for θ, rather than the symmetric interval which results from the usual asymptotic normal approximation to the sampling distribution of the maximum-likelihood estimator of θ.

We can justify this procedure as follows. Suppose that under H_0, β is estimated using maximum likelihood, and then we denote the value of the log-likelihood by $l_{H_0}(\theta)$, say. Suppose that under H_1 the overall maximum of the log-likelihood, maximized with respect to β and θ, is l_{H_1}. We know that, asymptotically, $2\{l_{H_1} - l_{H_0}(\theta)\}$ has a chi-square distribution with 1 degree of freedom. It follows that the $(1 - \alpha)$ likelihood-ratio interval for the ED_{100p} value is given by the set of values θ satisfying,

$$D(\theta) = 2\{l_{H_1} - l_{H_0}(\mu_0)\} < z_\alpha^2$$

where z_α is the $100(1 - \alpha)$ percentage point of the standard normal distribution.

Morgan *et al.* (1989) describe how the computer package, QUAD, automatically evaluates this interval for the logit model: for any value of θ, $l(\theta, \beta)$ is maximized with respect to β, using the method-of-scoring. The endpoints of the required interval, θ_- and θ_+ are then determined by the secant method (Adby and Dempster, 1974, p. 23). The same procedure applies for any ED_{100p} value. Williams (1986b) has provided a set of GLIM commands which may be used (Exercise 2.29). As well as the LR interval, the QUAD package also produces the delta-method interval. In practice, for data sets of reasonable size, little difference can usually be observed between the two intervals, although of course one is contrained to be symmetric, while the other does not have that constraint. Differences tend to increase as p departs from $p = 0.5$. Formally, we need to compare alternative methods of confidence-interval construction in a simulation experiment, and that has been done by Williams (1986b). His conclusions are outlined in the next section.

2.7.4 Comparing the likelihood-ratio and Fieller intervals

The Fieller interval has been shown to be slightly conservative when the two random variables involved in the ratio in question have an exact bivariate normal distribution (Milliken, 1982). In the context of ED_{100p} estimation, bivariate normality is an asymptotic approximation which may well be poor for experiments involving small sample sizes. Williams (1986b) compared the LR and Fieller intervals and concluded that there is overwhelming evidence that the Fieller intervals are conservative – far more so than when bivariate normality is exact, rather than approximate. He found that even in 'large' experiments, involving 120 experimental subjects and six doses, the true confidence probability for the Fieller interval could exceed 95% by 1% or more. In smaller experiments this probability could exceed 98%. In contrast, the LR interval was found to fall below 95% in small experiments involving 20 to 30 subjects, but it rarely fell below 93%. However, Hoekstra (1990) presented alternative simulations, with a different spacing of dose levels, and found likelihood-ratio intervals to be conservative, so a general conclusion here appears to be elusive.

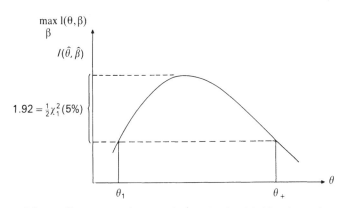

Figure 2.2 *An illustration of using the profile log-likelihood to obtain a 95% likelihood-ratio confidence interval for θ, the ED_{50}.*

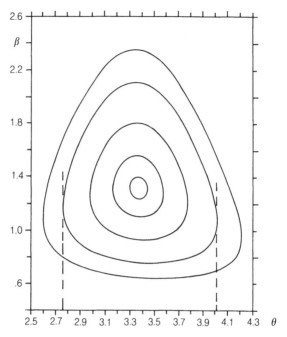

Figure 2.3 *Contours of constant likelihood for data from Reed and Muench (1938, p. 494), for which there were 9 dose levels and 54 experimental subjects in all. Here the log-likelihood is parameterized in terms of β and the ED_{50}, θ. Contours correspond to 20, 50, 90, 95 and 99 percentage points. Also shown is the 95% confidence interval for θ, formed by projecting the 95% contour onto the θ axis. Figure taken from Goedhart (1985).*

An additional problem with the Fieller approach is that with small sample sizes there are cases when the confidence-interval is indeterminate. An example is provided by Racine *et al.* (1986), which led them to suggest that the Bayesian approach, discussed later in Chapter 3, is superior in analysing quantal assay data. Several contributors to the discussion of that paper pointed out that the LR interval, rather than the Fieller interval, should have been used in comparison with the Bayes interval (Exercise 2.15). A parametric bootstrap approach to confidence-interval construction has been investigated, in a small study, by Goedhart (1985) – see Efron (1982). He concluded that the shape of bootstrap histograms of the ED_{50} and ED_{95} varied as the original data set changed. For data sets with small numbers of individuals per dose, problems of non-convergence

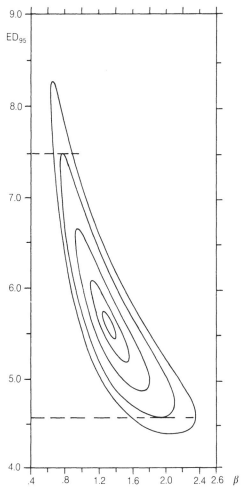

Figure 2.4 *For the same data as in Figure 2.3, the log-likelihood is now parameterized in terms of β and the* ED$_{95}$. *Contours correspond to 20, 50, 90, 95 and 99 percentage points. Figure taken from Goedhart (1985).*

may be encountered if bootstrap samples with one or no intermediate responses occur (e.g. Example A of Exercise 2.30). Cox (1990) and Sitter and Wu (1991) have compared Fieller and delta-method confidence intervals. Sitter and Wu (1991) find the Fieller interval to have better converge probability and suggest when the two intervals will differ.

2.8 Maximum-likelihood estimation for logistic regression

The logit model of section 2.2 provides a simple illustration of the more general case of logistic regression, in which more than one explanatory variable is involved. An example of data involving more than one explanatory variable, taken from Finney (1947), is shown in Table 2.2. Here the effect of rate and volume of air inspired on a transient vaso-constriction of the skin of the digits was investigated. We present these data solely to illustrate the fitting process. They are used also similarly by Aitkin *et al.* (1989, section 4.2).

In this example there are no repeated cases of the same volume and rate values, and so all responses are simply binary. If P denotes

Table 2.2. *Data on vaso-constriction of the skin of the digits, taken from Finney (1947). In common with other authors, we shall treat the data as 39 independent observations though in fact they result from 9, 8 and 22 repeated measures respectively on each of just 3 subjects (Pregibon, 1982c). The responses 1 and 0 denote, respectively, the occurrence and non-occurrence of vaso-constriction*

Volume	Rate	Response	Volume	Rate	Response
3.7	0.825	1	0.4	2.0	0
3.5	1.09	1	0.95	1.36	0
1.25	2.5	1	1.35	1.35	0
0.75	1.5	1	1.5	1.36	0
0.8	3.2	1	1.6	1.78	1
0.7	3.5	1	0.6	1.5	0
0.6	0.75	0	1.8	1.5	1
1.1	1.7	0	0.95	1.9	0
0.9	0.75	0	1.9	0.95	1
0.9	0.45	0	1.6	0·4	0
0.8	0.57	0	2.7	0.75	1
0.55	2.75	0	2.35	0.03	0
0.6	0.3	0	1.1	1.83	0
1.4	2.33	1	1.1	2.2	1
0.75	3.75	1	1.2	2.0	1
2.3	1.64	1	0.8	3.33	1
3.2	1.6	1	0.95	1.9	0
0.85	1.415	1	0.75	1.9	0
1.7	1.06	0	1.3	1.625	1
1.8	1.8	1	–	–	–

the probability that there is an occurrence of vaso-constriction for given rate and volume values, Pregibon (1981) fitted the following model by maximum likelihood:

$$\text{logit}(P) = \alpha + \beta_1 \log_e(\text{rate}) + \beta_2 \log_e(\text{volume}) \qquad (2.12)$$

and obtained the following estimates, given together with estimated standard errors:

$$\hat{\alpha} = -2.875 \ (1.319)$$
$$\hat{\beta}_1 = 5.179 \ (1.862)$$
$$\hat{\beta}_2 = 4.562 \ (1.835)$$

The residual deviance was 29.23 (Appendix B) and the corresponding chi-squared goodness-of-fit statistic was 34.15, both less than the degrees of freedom of 36 in this case; but this cannot be used reliably as a measure of fit, for reasons discussed by McCullagh and Nelder (1989, p. 122). See also Exercise 2.21. We shall return to this set of data when we consider measures of influence and regression diagnostics in section 3.4. A further analysis of these data is provided by Pregibon (1982c, 1983); Copas (1988) provides a particularly simple description in terms of a model for contamination (cf. section 3.9 and Exercise 3.40).

The operations involved in the maximum-likelihood fitting of a general logistic regression model are a straightforward extension of those detailed in section 2.2. The required procedure is now available in, for example, the BMDP package, as routine LR, or within the SAS package as procedures FUNCAT and LOGIST, and is described in detail in Hosmer and Lemeshow (1989).

We shall now provide this extension of the material of section 2.2, as it is needed for the discussion of influence in section 3.4, and also because it is attractive to see the general formulation of the iterative procedure.

The logit model of section 2.2 has a log-likelihood of the form:

$$l(\alpha, \beta) = \text{constant} + \sum_{i=1}^{k} r_i \log P_i + \sum_{i=1}^{k} (n_i - r_i) \log(1 - P_i)$$

where

$$P_i = 1/(1 + e^{-(\alpha + \beta x_i)}).$$

Hence

$$l(\alpha, \beta) = \text{constant} - \sum_{i=1}^{k} r_i \log(1 + e^{-(\alpha + \beta x_i)})$$

$$- \sum_{i=1}^{k} (n_i - r_i) \log(1 + e^{-(\alpha + \beta x_i)}) - \sum_{i=1}^{k} (n_i - r_i)(\alpha + \beta x_i)$$

i.e. $l(\alpha, \beta) = \text{constant} - \sum_{i=1}^{k} n_i \log(1 + e^{-(\alpha + \beta x_i)})$

$$- \sum_{i=1}^{k} (n_i - r_i)(\alpha + \beta x_i). \tag{2.13}$$

In general logistic regression, we replace the pair of parameters (α, β) by the vector $\boldsymbol{\beta}$, so that in the general context the parameter β_1 is now the intercept parameter, previously denoted by α, and the model of equation (2.12) is denoted by:

$$\text{logit}(P) = \beta_1 + \beta_2 \log_e(\text{rate}) + \beta_3 \log_e(\text{volume})$$

We shall adopt the convention that the logit of a vector is a vector whose elements are the logits of the elements of the original vector, in the original order.

If there are m explanatory variables, then the logistic regression model has the form,

$$\text{logit}(P) = X\boldsymbol{\beta}$$

where P is the column vector:

$$P' = (P_1, P_2, \ldots, P_k)$$

k is now the total number of different combinations of explanatory variables that is employed in the experiment, P_i is the probability of positive response to the ith combination of explanatory variables, and X is the $(k \times (m + 1))$ design matrix, $\{x_{ij}\}$, the ith row of which contains the values of the explanatory variables for P_i. It is customary to refer to the ith row of X as the ith case. Thus in the example of Table 2.2, $k = 39$, and

$$X = \begin{bmatrix} 1, & \log(3.70) & , & \log(0.825) \\ 1, & \log(3.50) & , & \log(1.090) \\ \cdot & \cdot & \cdot & \cdot \\ \cdot & \cdot & \cdot & \cdot \\ 1, & \log(0.75) & , & \log(1.900) \\ 1, & \log(1.30) & , & \log(1.625) \end{bmatrix}$$

We see therefore that the likelihood of equation (2.13) generalizes to give:

$$l(\boldsymbol{\beta}) = \text{constant} - \sum_{i=1}^{k} n_i \log(1 + e^{-\mathbf{x}_i'\boldsymbol{\beta}}) - \sum_{i=1}^{k} (n_i - r_i)\mathbf{x}_i'\boldsymbol{\beta} \quad (2.14)$$

where $\mathbf{x}_i'$ denotes the ith row of the matrix $\mathbf{X}$.

An equivalent form for equation (2.14) is readily verified to be:

$$l(\boldsymbol{\beta}) = \text{constant} - \sum_{i=1}^{k} n_i \log(1 + e^{\mathbf{x}_i'\boldsymbol{\beta}}) + \sum_{i=1}^{k} r_i\mathbf{x}_i'\boldsymbol{\beta}. \quad (2.15)$$

Differentiating equation (2.15) with respect to the elements of $\boldsymbol{\beta}$ gives:

$$\frac{\partial l}{\partial \beta_j} = \sum_{i=1}^{k} x_{ij}(r_i - n_i P_i), \quad 1 \leqslant j \leqslant m+1 \quad (2.16)$$

where

$$P_i = \frac{1}{(1 + e^{-\mathbf{x}_i'\boldsymbol{\beta}})}$$

and furthermore,

$$\frac{\partial^2 l}{\partial \beta_j \partial \beta_k} = - \sum_{i=1}^{k} x_{ij}x_{ik}n_i P_i(1 - P_i) \quad 1 \leqslant j, k \leqslant m+1 \quad (2.17)$$

We can see how equation (2.17) generalizes the results of section 2.2 by noting that

$$\frac{\partial P_i}{\partial \beta_j} = x_{ij}P_i(1 - P_i)$$

In vector notation, we can summarize the equations formed by setting equations (2.16) to zero, as:

$$\mathbf{X}'s = 0$$

where s is the column vector with ith element, $s_i = (r_i - n_i P_i)$. These equations are analogous to those which result in standard linear regression (Exercise 2.22). Furthermore, equations (2.17) allow us to write the method-of-scoring (or in this case equivalently the Newton–Raphson method) as:

$$\boldsymbol{\beta}^{t+1} = \boldsymbol{\beta}^t + (\mathbf{X}'\mathbf{V}\mathbf{X})^{-1}\mathbf{X}'s \quad (2.18)$$

where V is the diagonal matrix: $V = \text{diag}(n_i \hat{P}_i(1 - \hat{P}_i))$, and where both V and s are evaluated at the tth iterate, β^t.

If we now set $z^t = X\beta^t + V^{-1}s$, then equation (2.18) becomes:

$$\beta^{t+1} = (X'VX)^{-1}X'Vz^t \qquad (2.19)$$

We recognize the solution of equations (2.19) as the weighted least squares estimate of the parameter vector β in the linear regression model:

$$Y = X\beta + \varepsilon, \qquad (2.20)$$

where

$$\text{Var}(\varepsilon) = \sigma^2 V^{-2}$$

(e.g. Cook and Weisberg, 1982, p. 209), and so equation (2.19) generalizes the result of equation (2.2), that this standard iterative way of obtaining the maximum-likelihood estimate of β is equivalent to an iterated weighted linear regression, sometimes called IRLS (iterated reweighted least squares). We shall return to the work of this section in section 3.4. The paper by Kay and Little (1986) provides an interesting case-study of the use of logistic regression in a medical context and there is further mention of that work in section 3.4 also. The bias of maximum-likelihood estimators for probit analysis and logistic regression is discussed by Copas (1988), Sowden (1971) and Griffiths et al. (1987) (Exercise 2.34). The logit model has the advantage of simplicity in comparison with the probit and it also plays an important role in the analysis of case-control data (McCullagh and Nelder, 1989, pp. 111–114). More complex models will be considered in Chapter 4. Variable selection in logistic regression is discussed in Peduzzi et al. (1980), Slater (1981), Hosmer et al. (1989) and Robinson and Jewell (1991). Score tests in GLIM are presentd by Pregibon (1982b), for variable selection and for testing goodness-of-fit (in the latter case developing the work of Tsiatis, 1980) (Appendices B and D). Marx and Smith (1990) develop a method of principal component parameter estimation.

Note finally that there is an equivalence between fitting a logistic regression and fitting log-linear models, as described by McCullagh and Nelder (1989, section 6.4) and Fienberg (1981, Chapter 6). In the unlikely case that it would be more convenient in some applications to fit a log-linear model, then this provides an alternative approach. An example is discussed in Exercise 2.35 and its solution.

2.9 Making comparisons

Strictly, the term **bioassay** applies when a comparison is being carried out, as when a new substance is compared with an old one. The need for a comparison can be seen in many of the examples of Chapter 1, and comparisons are readily carried out by means of likelihood ratio tests, in GLIM, for example. Suppose, for illustration, that substance i has resulted in the model:

$$\Pr(\text{response to dose } d_j) = 1/(1 + e^{-(\alpha_i + \beta_i \log d_j)}) \quad i = 1, 2$$

Frequently one might expect $\beta_1 = \beta_2 = \beta$, which is commonly termed **parallelism**. If parallelism is found to be unacceptable, then in many cases the comparison of substances ends with that conclusion. However, if parallelism appears to be acceptable then a single value may be used to provide a comparison between the two substances. If substance i requires dose d_i to produce an expected proportion responding of p, for any $0 < p < 1$, and if the substances are measured on a log scale, as above, then we have

$$\alpha_1 + \beta \log d_1 = \alpha_2 + \beta \log d_2$$

and

$$\log(d_1/d_2) = (\alpha_2 - \alpha_1)/\beta = \text{constant}$$

The ratio, d_1/d_2, of **equally effective doses**, is termed the **relative potency**, ρ. If the relative potency of substance B, relative to substance A is ρ, then z units of B have the same expected effect as (ρz) units of A. We shall utilize the idea of relative potency in Chapter 3, when we consider the performance of mixtures of substances A and B.

Tests of parallelism, and of equality of intercepts, are readily carried out by a computer package such as GLIM. One illustration, resulting in a rejection of parallelism, is provided in the GLIM manual (Payne, 1986, p. 127). Another follows in Example 2.5.

Example 2.5   (Healy, 1988, p. 85)

The data of Exercise 1.25 classify a sample of women by age, smoking habit and whether or not the menopause has been reached. The GLIM analysis of the data, shifted so that the age origin is 40, can be described by either of the following sets of commands:

```
$UNITS  10                              $UNITS  10
$DATA  SMOK  AGE  N  R  $READ           $DATA  SMOK  AGE  N  R  $READ
1   5   67   1                          1   5   67   1
1   7   44   5                          1   7   44   5
1   9   66   15                         1   9   66   15
1  11   73   33                         1  11   73   33
1  13   52   37                         1  13   52   37
2   5   37   1                          2   5   37   1
2   7   29   5                          2   7   29   5
2   9   36   13                         2   9   36   13
2  11   43   26                         2  11   43   26
2  13   28   25                         2  13   28   25
$FACTOR  SMOK  2                        $FACTOR  SMOK  2
$ERROR  B  N                            $ERROR  B  N
$YVARIABLE  R                           $YVARIABLE  R
$FIT  AGE + SMOK                        $FIT  AGE * SMOK
$DISPLAY E R                            $FIT  -AGE.SMOK
$FIT  +AGE.SMOK                         $DISPLAY  E  R
```

The conclusion is that parallelism is acceptable, and that the age-shift for the same expected proportion of menopausal women is 1.23 years. Preisler (1989a) presents GLIM code for the corresponding situation which is complicated by the presence of natural mortality (section 3.2). A resistant approach to making comparisons is given in Green (1984) applied to the data of Exercise 2.38.

2.10   Testing for trend in proportions

An obvious question in connection with quantal assay data is whether the substance being tested has any effect in comparison with a placebo, and whether there is a dose-response. A simple illustration is provided by the data of Table 2.3. In carcinogenicity studies, demonstration of a dose-response can be used to suggest causality (Breslow and Day, 1980, p. 86).

To test for an increase with dose level in the proportion responding, Cochran (1954) and Armitage (1955) proposed the test statistic:

$$X^2 = \frac{\left(\sum\limits_{i=1}^{k} r_i d_i - \hat{p} \sum\limits_{i=1}^{k} n_i d_i \right)^2}{\hat{p}(1-\hat{p})\left\{ \sum\limits_{i=1}^{k} n_i d_i^2 - \left(\sum\limits_{i=1}^{k} n_i d_i \right)^2 \middle/ \sum\limits_{i=0}^{k} n_i \right\}}$$

where

$$\hat{p} = \frac{\sum\limits_{i=0}^{k} r_i}{\sum\limits_{i=0}^{k} n_i} \tag{2.21}$$

Table 2.3 (Tarone, 1982a) *Lung tumor incidence in female F344 rats resulting from the bioassay of nitrilotriacetic acid*

Dose level	No. of animals	No. of animals with tumour
0	15	0
1	49	3
2	46	7

If there is no difference between the $(k + 1)$ groups, this statistic has, asymptotically, a χ_1^2 distribution. Tarone (1985) shows that X^2 results from a score test of $\beta = 0$ (Exercise 2.39 and Appendix D). For the example of Table 2.3, $X^2 = 4.04$, which is significant at the 5% level. For further discussion, see Tarone (1982a), who considers how historical control information may also be incorporated. We return to this topic in section 6.7.4. See also Dietz (1985) for a non-parametric method for testing $\beta = 0$. The Williams test (Williams, 1971, 1972) was originally designed for comparing treatments with a control for non-quantal response. However, it may also be applied to the case of proportions – see Williams (1988b), who also considers tests for a range of alternative hypotheses. The paper by Shuster and Yang (1975) presents a non-parametric method for estimating the minimum dose level for a given level of response. A related topic is that of **dose-ranging**, which is a premarketing study to determine reasonable doses at which new drugs may be administered. This topic is discussed by Sheiner *et al.* (1989).

2.11 Discussion

So far the material of this chapter has been mainly in connection with the logit model. Historically, the emphasis was more on probit analysis, based on the normal cumulative distribution function, which has somewhat lighter tails than the logistic form. The logit model is more convenient in terms of analysis (cf. also Exercise 2.4) and is now more usually used. An interesting exception occurs when dose values are imperfectly observed (section 3.10). More generally, the logit model forms the basis for logistic regression. Kay and Little (1986) describe a case-study in the use of logistic regression which employs a variety of modern approaches, many of which we shall encounter in Chapters 3, 4 and 7. They also point out that the sampling basis for logit modelling of quantal assay data is different, in general, from that found in logistic regression. Further two-parameter models, based on the Cauchy, normal and extreme-value distributions, are discussed in Exercises 2.4 and 2.6. Three-parameter models are encountered in Exercises 2.13 and 2.14, and these and several others are discussed fully in Chapter 4.

We have in this chapter concentrated on fitting simple models to data using the method of maximum likelihood, and we have seen the need for numerical optimization of some kind. For many statisticians today that need is met by the GLIM package (Payne, 1986). Here and in later chapters we shall on occasion present GLIM code for model-fitting. The range of models which may be fitted in GLIM continues to be expanded, and useful references here are the papers by Green (1984) and Stirling (1984). We have assumed that multiple maxima do not occur. However, if sample sizes are small it is possible to obtain multiple maxima. An illustration is provided by Racine *et al.* (1986), and see also Williams (1986b) and the solution to Exercise 2.27. For the example in the first of these two papers, 20 animals were distributed evenly over four doses. It is not surprising that difficulties can arise with model-fitting and inference in such cases.

Some numerical analysis procedures may be better suited to a particular optimization problem than others, which may even fail to converge. For example, Chambers (1973) found the general optimization procedures of Fletcher (1970) converged more rapidly than iterated linear least squares in a number of logistic regression examples. For this reason he suggested that statisticians should have a variety of optimization routines as part of their armoury, comment-ing that 'explicit practical guarantees of performance cannot be had for nonlinear calculations'. As we shall see later, there are also examples for which the numerical complexity of likelihood formation renders maximum likelihood impractical for model-fitting.

The ubiquity of computers for fitting models such as the logit can give rise to a cavalier attitude towards such aspects as careful selection of starting values for an iterative procedure (Exercise 2.3). When computing was more costly, care in this respect was far more important (e.g. Cox, 1970, p. 89). Finney (1971) preceded the Newton–Raphson approach with an initial simplex search, while the package POLO, described by Russell *et al.* (1977), used the method of conjugate gradients. Morgan *et al.* (1989) found the minimum logit chi-square starting values worked extremely well on a routine basis in the QUAD computer package. This is also the starting value used in GLIM. It is interesting, however, that more complex three-parameter models can be far less easy to fit, due to high correlation between parameter estimators. This is a topic which we discuss in greater detail in Chapter 4. In these cases careful selection of starting values for the iterations remains important.

We have seen that satisfactory confidence-interval construction can be dependent on whether parameter estimators can be assumed to be multivariate normal in distribution. Work by Viveros and Sprott (1986) has produced parameter transformations designed to improve the assumption of normality for the profile likelihood (Exercise 2.20).

An alternative method for fitting models to quantal assay data has been presented by Cobb and Church (1983). This is based on parameter estimation of the Spearman–Kärber type, discussed in Chapter 7. Subject to existence, the method has good small-sample properties.

The discussion of goodness-of-fit of models given here in terms of the standard chi-square test is only the first step as far as model assessment and criticism are concerned. More powerful procedures, arrived at through the construction of suitable score tests, are discussed in Chapter 4, and the form and use of generalized residuals are considered in Chapter 3.

As we have seen from the examples of Chapter 1, we frequently need to fit more complicated models than the simple logit. For instance, natural mortality often needs to be taken into account. This is one of several topics considered in Chapter 3 which presents a number of extensions and alternative approaches.

2.12 Exercises and complements

The exercises vary in complexity. It is particularly desirable to attempt the five exercises marked with a †. Exercises marked with a * are generally more difficult or speculative. A number of the following exercises involve the use of a computer.

2.1* In Figure 2.1 we see the shape of the contours of the log-likelihood surface when the (α, β) parameterization is adopted. Check whether the asymptotic estimate of $|\text{Corr}(\hat{\alpha}, \hat{\beta})|$ is usually close to unity, for that and other data sets see for example the data of Exercise 1.3. As an alternative parameterization, consider the model:

$$P_i = 1/(1 + e^{-\{\gamma + \beta(d_i - \bar{d})\}})$$

where

$$\bar{d} = \frac{1}{k} \sum_{i=1}^{k} d_i$$

We see that the model is now parameterized as (γ, β), where $\gamma = \alpha + \beta\bar{d}$.

Consider what the $\mathrm{Corr}(\hat{\gamma}, \hat{\beta})$ will be, and if possible investigate the shape of the corresponding log-likelihood surface. Cf. Exercise 2.15.

2.2[†]   Select a computer optimization routine, and use it to fit the logit model to the data sets presented in Exercise 1.3. Discuss the ease of use of the routine and compare the fitted models with those obtained earlier when the fitting was performed by eye. Investigate the facilities available in a statistics computer package.

2.3   For maximum-likelihood fitting of the logit model, experiment with different termination rules and different starting values.

2.4*   Identify the sufficient statistics for the logit model and quantal assay data. Show that $\partial^2 L/\partial\alpha^2, \partial^2 L/\partial\beta^2$ and $\partial^2 L/\partial\alpha\partial\beta$ are not functions of the sufficient statistics, and use this result to explain why the same iterative procedure results for the Newton–Raphson method and the method-of-scoring. Investigate how these features extend to logistic regression in general. Cf. the case for probit analysis. Consider whether the two iterative procedures are identical for the **complementary log–log** and probit models (Appendix B). Cf. also Exercises 2.6 and 2.41. What distributional form underlies the complementary log–log model?

2.5   Investigate and run the MINITAB program given below for probit analysis. Note that a cubic–logistic approximation to the normal cumulative distribution function is used as an exact model for quantal assay data in Chapter 4. Cf. also Exercise 2.14.

```
DIVI 'Y' 'N' C70
NOTE      C70 IS SMALL P ; METHOD FROM PARADINE & RIVETT
RECODE 0 C70 '*' C71
RECODE 1 C71 '*' C71
NOTE      AVOID ZERO AND 1.0
LET C73=SIGN(0.5-C71)
LET C72=-LOGE(1-C73+2*C73*C71)
LET C72=((4*C72+100)*C72+205)*C72*C72/(((2*C72+56)*C72+192)*C72+131)
LET C72=5-C73*SQRT(C72)
NOTE      PROBIT ; APPROXN., FROM DERENZO, 1977, MAT. COMP. 31:214
REGR      C72 1 'X' C73 C73
NOTE      C73 IS EXPECTED PROBIT. CALCS ALSO FOR MISSING VALUES
LET C74=C73-5
ABSO C74 C75
LET C75=EXPO(-((83*C75+351)*C75+562)*C75/(165*C75+703))
LET C75=0.5*(1+SIGN(C74)*(1-C75))
NOTE      P; APPROX FROM DERENZO, LOC. CIT.
LET C76=2.506628*EXPO(0.5*(C74**2))
NOTE      1/Z
LET C77=C73-C76*(C75-C70)
LET C78='N'/(C75*(1-C75)*C76*C76)
NOTE      WEIGHT
REGR      C77 C78 1 'X' C73 C73 C71
LET C74=C73-5
ABSO C74 C75
LET C75=EXPO(-((83*C75+351)*C75+562)*C75/(165*C75+703))
LET C74='N'*0.5*(1+SIGN(C74)*(1-C75))
LET K70=SUM(('Y'-C74)**2/C74)
LET K72=COUNT('Y')-1
NOTE      CALC CHI-SQ. APPROXN., FROM ABRAMOWITZ AND STEGUN
LET K73=((K70/K72)**(1/3)-1+2/(9*K72))*(SQRT(4.3*K72))
NOTE      NORMAL APPROXN.
LET K74=1-1./(1+EXPO(-1.5976*K73*(1+0.04417*K73*K73)))
NOTE      APPROXN., FOR PROB., FROM PAGE, 1977, APPL. STATS., 26:75
NAME C74 'EXPECT' C71 'COEFFS'
PRINT K70 K72-K74
PRINT 'X' 'N' 'Y' 'EXPECT' 'COEFFS'
END
```

$2.6^†$

(a) Compare and contrast the method-of-scoring and the Newton–
    Raphson method for obtaining maximum-likelihood estimates
    of the location and scale parameters in the two-parameter
    Cauchy density:

$$f(x) = \frac{\beta}{\pi\{\beta^2 + (x-\alpha)^2\}}, \quad -\infty < x < \infty, \beta > 0$$

(b) Investigate the detail of maximum-likelihood fitting of the model
    for quantal-assay data based on the Cauchy distribution. This
    model has been encountered already in Section 1.4, and is
    sometimes called the **Cauchit** model. Produce GLIM code if
    possible. Cf. Exercise 2.4. Note that Silvapulle (1981) has pointed
    out the possibility of the likelihood surface possessing
    multiple maxima when the Cauchy distribution is used. Cf.
    Exercise 2.27.

2.7 Fit the logit model to a set of quantal assay data, by hand

(a) without using tables of working logits and antilogits
and
(b) using such tables.

2.8 We can write the signal detection data of Table 1.11 in general notation as follows:

| | Responses | | | |
Stimulus	No	Not sure	Yes	Totals
N	r_{11}	r_{12}	r_{13}	$r_{1.}$
SN	r_{21}	r_{22}	r_{23}	$r_{2.}$

We may write the log-likelihood, in an obvious notation, as

$$L = \sum_{i=1}^{2} \sum_{j=1}^{3} r_{ij} \log(P_{ij})$$

A signal detection theory model assumes,

$$P_{1j} = F(z_j) - F(z_{j-1}), \qquad 1 \leqslant j \leqslant 3$$

and

$$P_{2j} = F(\beta z_j + \alpha) - F(\beta z_{j-1} + \alpha), \qquad 1 \leqslant j \leqslant 3$$

where $F(\ )$ is a cumulative distribution function, $z_0 = -\infty$, $z_3 = \infty$, and z_1, z_2, α and β are parameters to be estimated.

Show that the maximum likelihood estimates of the parameters result from solving the equations,

$$F(z_1) = r_{11}/r_1$$

$$F(\beta z_1 + \alpha) = r_{21}/r_2$$

$$1 - F(z_2) = r_{13}/r_1$$

$$1 - F(\beta z_2 + \alpha) = r_{23}/r_2.$$

Hence fit this model to the data of Table 1.11 when $F(\ ) = \Phi(\ )$. There is more discussion of this model in Exercise 3.36.

2.9 The importance of weighting inversely with respect to the variance to give the method of minimum logit chi-square depends on how extreme the underlying probabilities are. Investigate this. What are the design implications?

2.10* Experiment with alternative ways of dealing with 0% and 100% responses in minimum logit chi-square estimation. For any dose d, with probability of response p, and R individuals responding out of n, then $\log_e(R/(n-R))$ is called the **empirical logistic transform** of (R, n). Examine how to choose a so that $E\left[\log_e\left(\dfrac{R+a}{n-R+a}\right)\right]$ is approximately $\log_e\left(\dfrac{p}{1-p}\right)$. For further discussion, see Cox and Snell (1989, p. 31), Anscombe (1956) and Gart and Zweifel (1967). Cf. also Exercise 2.20.

2.11 Experiment with estimating ED_{50} and ED_{95} values by eye, and compare your estimates with those resulting from fitting the logit model using maximum likelihood.

2.12 Verify the approximate expression for $\hat{\sigma}^2$ in Example 2.4.

2.13† Aranda-Ordaz (1981) introduced the following model for quantal assay data:

$$P(x) = 1 - (1 + \lambda e^{\alpha + \beta x})^{-1/\lambda} \qquad \text{for } \lambda e^{\alpha + \beta x} > 1$$

$$P(x) = 0, \qquad \text{otherwise}$$

Consider how to fit this model using the method of maximum likelihood. Use the delta-method to construct confidence-intervals for the ED_{50}. Cf. Exercise 2.18. There is more discussion of this model in Chapter 4.

2.14 Morgan (1985) introduced the following model for quantal assay data:

$$P(x) = [1 + e^{-\{\beta(x-\mu) + \gamma(x-\mu)^3\}}]^{-1}$$

Consider how to fit this model using the method of maximum likelihood. How would you estimate the ED_{50}? This model is discussed further in Chapter 4. Cf. also Exercise 2.5.

2.15* Relate the LR confidence interval for θ to that given by Ross (1986), who used the parameterization of Exercise 2.1. Ross wrote: '...limits for ... the LD_{50} are obtained from the reciprocal slopes of tangents from the origin to the contour.' In this case the contour is the log-likelihood contour corresponding to the section shown in Figure 2.2, but in two dimensions, and taken with the (β, γ) parameterization.

2.16* Investigate the use of the bootstrap (Efron, 1982) in obtaining confidence-intervals for ED_{100p} values.

2.17* Design and run a simulation study to compare the maximum-likelihood and minimum logit chi-square methods of parameter estimation for fitting the logit model to quantal assay data. Note here also the results of Little (1968), and Smith *et al.* (1984).

2.18* Aranda-Ordaz (1985b) has produced the GLIM macro, given below, for fitting the model of Exercise 2.13. (Reproduced with permission from the GLIM Newsletter.) Check, and if possible run, his program.

```
$SUBFILE ASYMTRAN !
$M M1 !
$CA A=%EXP(%LP) !
  :  Y=%L*A
  :  C=%IF(%GT(Y,-1),(1+Y)**%J,%U) !
  :  D=1.0-C !
  :  %FV=N*D !
$$E
$M M2 !
$CA C=(N-%FV) !
  :  C=%IF(%EQ(C.O).%U.C) !
  :  C1=C/N !
  :  D=1.0-C1**%L !
  :  %DR=%L/(C*D) !
$$E
$M M3 !
$CA %VA=%FV*C1 !
$$E
$M M4 !
$CA %DI=2*(%YV*%LOG(%YV/%FV)+CR*%LOG(CR/C)) !
$$E
$M INAT !
%CA %U=0.001 !
  :  CR=N-%YV !
  :  %J=-1/%L !
  :  R1=%YV !                                    (Cont.)
```

```
  :   R1=%IF(%EQ(%YV,N),N-.5,R1)  !
  :   R1=%IF(%EQ(%YV,0),.5,R1)  !
  :   NMR=N-R1  !
  :   QU=N/NMR  !
  :     QUL=QU**%L  !
  :   %LP=%LOG((QUL-1.0)/%L)  !
$PR 'INITIAL VALUES VECTOR %LP     '  !
  :   %LP  !
$$DEL R1 NMR QU QUL  !
$$E
$OWN M1 M2 M3 M4  !
$SCALE 1  !
$WA  !
$M LLIK  !
$CA V1=%IF(%EQ(%YV,0),1.0,%YV/N)  !
  :   V2=%IF(%EQ(%YV,N),1.0,(N-%YV/N))  !
  :   VR=%YV*%LOG(V1)  !
  :   VN=(N-%YV)*%LOG(V2)  !
  :   %A=%CU(VR)  !
  :   %B=%CU(VN)  !
  :   %M=%A+%B  !
$PR 'MAXIMUM LOGLIKELIHOOD VALUE ACHIEVABLE:   ' $  !
$PR %M  !
$$DEL V1 V2 VR VN %A %B %M  !
$$E
$RETURN
```

2.19* A standard algorithm for solving non-linear regression problems is the Gauss–Newton algorithm (Draper and Smith, 1981, pp. 462–465; Jennrich and Ralston, 1978). This is an iterative procedure in which, at stage $t + 1$ ($t \geqslant 0$), the vector of estimates $\zeta^{(t)}$ is updated to

$$\zeta^{(t+1)} = \zeta^{(t)} + \delta^{(t)},$$

where

$$\delta^{(t)} = \left\{ \left(\frac{\partial \mu}{\partial \zeta} \right)' \Sigma^{-1} \frac{\partial \mu}{\partial \zeta} \right\}^{-1} \left[\frac{\partial \mu}{\partial \zeta} \right]' \Sigma^{-1} (\mathbf{y} - \mu)$$

where for a random variable $\mathbf{R}$, $\mu = E[\mathbf{R}]$ $\Sigma = \text{Var}[\mathbf{R}]$ and $\partial\mu/\partial\zeta$ is a matrix with (i, j)th element $\partial\mu_i/\partial\zeta_j$, all quantities involved being evaluated at $\zeta^{(t)}$. Show that this iterative procedure can be expressed as iterated weighted least-squares, when used to minimize:

$$S(\zeta) = \sum_{i=1}^{k} \{r_i - n_i P_i(\zeta)\}^2 / [n_i P_i(\zeta)\{1 - P_i(\zeta)\}]$$

2.20   Viveros and Sprott (1986) suggest the transformation,

$$\phi = \tan^{-1}\left[(\theta V_{22} + V_{12})/(V_{11}V_{22} - V_{12}^2)^{1/2}\right]$$

where θ is the ED_{50} and the $\{V_{ij}\}$ form the dispersion matrix for (α, β) in the usual parameterization of the logistic model. The aim of the transformation is to improve the normal approximation to the profile likelihood.

If $I_\phi = \hat{\beta}^2/[V_{22}\cos^2(\hat{\phi})]$, then, assuming that we can treat $(\phi - \hat{\phi})I_\phi^{1/2}$ as an approximate standard normal random variable, write down a confidence-interval for θ. Cf. Exercise 2.10.

2.21*   Using the BMDP package routine LR, or otherwise, verify the logistic fit of equation (2.12) to the data of Table 2.2. Check in particular the expression for degrees of freedom, and provide a probability plot of residuals.

2.22*   The standard linear model is written as:

$$y_i = \mathbf{x}_i'\boldsymbol{\beta} + \varepsilon_i, \qquad 1 \leqslant i \leqslant k$$

and $\{\varepsilon_i\}$ are independent with a $N(0, \sigma^2)$ distribution. Show that the maximum-likelihood estimate $\hat{\boldsymbol{\beta}}$ of $\boldsymbol{\beta}$ is obtained from solving the normal equations:

$$\mathbf{X}'\mathbf{r} = 0$$

where $\mathbf{X}$ is the design matrix with $\mathbf{x}_i'$ as its ith row and $\mathbf{r}$ is the vector of residuals: $\mathbf{r} = \mathbf{y} - \mathbf{X}\hat{\boldsymbol{\beta}}$, and $\mathbf{y}' = (y_1, \ldots, y_k)$.

2.23[†]   (Milicer, 1968) The data opposite result from a survey of age of menarche of rural girls in Poland in 1966.

Describe the salient features of these data using a logit model. Compare the results with those from fitting the logit model to the data of Table 1.4. We return to this comparison in Chapter 3. For relevant discussion, see Milicer (1968) and Burrell *et al.* (1961).

2.24*   (Little, 1968) Write down the equations to be solved for maximum-likelihood and minimum chi-square estimation of the location and scale parameters α and β in the quantal assay model for which the probability of response to dose d is given by:

$$P(d) = F(\alpha + \beta d)$$

Age in months (midpoint of class interval)	No. menstruating	No. questioned	Percentage
131.5	0	3	0
137.5	0	15	0
143·5	0	12	0
149.5	1	13	8.3
155.5	3	16	18.8
161.5	3	21	14.3
167.5	14	31	45.2
173.5	17	28	60.7
179.5	26	38	68.4
185.5	26	32	81.2
191.5	33	41	80.5
197.5	29	33	87.9
203.5	18	19	94.7
209.5	23	23	100.0
215.5	1	1	100.0
221.5	2	2	100.0

for any c.d.f. $F(\ )$. Consider how you might investigate the relationship between $\hat{\beta}$ values from the two different estimation procedures. Check your conclusions by trying out both methods in practice, for a variety of data sets and different models assumed for $F(\ )$.

Comment on the figures given overleaf relating to bias in the estimation of β (table taken from Smith *et al.*, 1985).

2.25* (Moore and Zeigler, 1967) Consider the non-linear regression model:

$$y_i = g(x_i; \boldsymbol{\beta}) + \varepsilon_i, \qquad 1 \leqslant i \leqslant k$$

where $\boldsymbol{\beta}$ denotes a vector of parameters, and the uncorrelated error terms, ε_i, have $\mathrm{E}[\varepsilon_i] = 0$.

One approach to estimating $\boldsymbol{\beta}$ is to choose $\boldsymbol{\beta}$ to minimize:

$$Q(\boldsymbol{\beta}) = \sum_{i=1}^{k} w_i \{y_i - g(x_i; \boldsymbol{\beta})\}^2$$

for suitable weights $\{w_i\}$.

Comparison of observed bias and second-order approximation[a] to the bias of maximum likelihood (MLE) and minimum logit chi-square (MLCS) estimators of the slope coefficient in a simple logit regression model

	Design	MLE	MLCS
Berkson[b]	Exact	0.095	0.048
	Approx.	0.077	0.060
Smith et al.[c]	Observed	0.025	−0.011
	Approx.	0.027	0.001
Smith et al.[d]	Observed	0.027	−0.044
	Approx.	0.029	−0.026

[a] Amemiya (1980).
[b] True response probabilities $P = (0.3, 0.5, 0.7)$, $\beta_1 = 0$, $\beta_2 = \log(7/3)$, 10 subjects per dose (Berkson, 1955).
[c] True response probabilities $P = (0.2, 0.35, 0.5, 0.65, 0.8)$, $\beta_1 = 0$, $\beta_2 = 1$, 24 subjects per dose (Smith et al., 1984).
[d] True response probabilities $P = (0.2, 0.3, 0.4, 0.45, 0.55, 0.6, 0.7, 0.8)$, $\beta_1 = 0$, $\beta_2 = 1$, 15 subjects per dose (Smith et al., 1984).

Show that the resulting equations to be solved for β are the same, with suitable equivalences, as the equations $\partial l/\partial \zeta = 0$ in section 2.2.

2.26* (Moore and Zeigler, 1967) Explain how you would provide estimates of error when you fit the non-linear model of Exercise 2.25 by minimizing $Q(\beta)$.

2.27[†] Silvapulle (1981) presents necessary and sufficient conditions for the existence of maximum likelihood estimates for probit and logit models. His work was initiated from consideration of the following data.

The following table gives the numbers of male and female patients classified by a General Health Questionnaire score (GHQ), and the outcome of a standardized psychiatric interview (the result here was 'case' or 'non-case')

Explain why it is possible to fit a logit model by maximum likelihood to the data for the females, but not to the data for the males. Cf. Exercise 2.30. We return to Silvapulle's conditions in Chapter 7.

							GHQ score								
		0	1	2	3	4	5	6	7	8	9	10	11	12	Total
Males	Cases	0	0	1	0	1	3	0	2	0	0	1	0	0	8
	Non-cases	18	8	1	0	0	0	0	0	0	0	0	0	0	27
Females	Cases	2	2	4	3	2	3	1	1	3	1	0	0	0	22
	Non-cases	42	14	5	1	1	0	0	0	0	0	0	0	0	63

2.28   For the data of Table 2.1, plot the data and the fitted logit curve. Discuss the lack of fit of the model and consider how, if at all, the fit might be improved.

2.29*   (Williams, 1986b) Discuss the following GLIM approach to constructing likelihood-ratio confidence intervals. After reading in the data as in Example 2.3, set the scalar $\%M$ equal to one value for θ, say θ_0. The following sequences will then give $L(\theta_0)$, the difference between the deviance with $\theta = \theta_0$, and the deviance with $\theta = \hat{\theta}$, the maximum-likelihood estimate. (See Appendix B for the definition of deviance.)

```
%YVAR R $
$ERROR   B N $
$FIT   X $
$CALCULATE   %D=%DV:XM=X-%M $
$FIT XM-%GM $
$CALCULATE  %L=%DV-%D $
```

2.30   Hamilton (1979) presents the following fictitious examples of quantal assay data, with 20 individuals exposed at each dose:

Dose	1	2	3	4	5	6	7	8	9	10
Example A	0	0	0	0	0	1	20	20	20	20
Example B	0	0	0	1	0	1	20	20	20	20

In comparison with examples already considered, these data may appear odd (though cf. Exercise 2.27). However, data of this kind appear frequently in routine drug testing, when large dose ranges are used. Hamilton found a maximum-likelihood logistic fit to Example B, but could not obtain a fit to Example A. Discuss this finding, particularly in relationship to the solution to Exercise 2.27. Discuss the effect of the responses to the high and low doses on the precision with which parameters are estimated. We continue discussion of this kind of data in Chapters 7 and 8.

2.31   Consider how you would fit the following model to quantal assay data:

$$P(x) = \zeta F_1(x) + (1 - \zeta)F_2(x)$$

where $F_1(\ )$ and $F_2(\ )$ are suitable c.d.f.s. We discuss this model at length in section 3.2.

2.32*   The following data, on the presence of absence of congenital sex organ malformation, categorized by alcohol consumption of mothers, appeared in Graubard and Korn (1987):

Malformation	Alcohol consumption (average no. of drinks/day)				
	0	<1	1–2	3–5	≥6
Absent	17 066	14 464	788	126	37
Present	48	38	5	1	1
Total	17 114	14 502	793	127	38

Discuss how you would analyse these data.

2.33*   Van Ryzin and Rai (1987) present the following model for the probability of response to dose d:

$$P(d) = 1 - \exp\left[-\left\{\zeta_1 + \zeta_2\left(\frac{d}{1 + d\zeta_4}\right)^{\zeta_3}\right\}\right]$$

with four parameters $\{\zeta_i, 1 \leqslant i \leqslant 4\}$. Provide the details of the method-of-scoring for fitting this model to quantal assay data.

2.34   Copas (1988) considers bias in the maximum-likelihood estimator of ζ where the probability of response to dose d is given by:

$$P(d) = (1 + e^{-\zeta d})^{-1}$$

An approximation to the bias is given by

$$b = \frac{\frac{1}{2} \sum_{i=1}^{k} d_i^3 P(d_i)\{1 - P(d_i)\}\{2P(d_i) - 1\}}{\left[\sum_{i=1}^{k} d_i^2 P(d_i)\{1 - P(d_i)\} \right]^2}$$

Evaluate b for a number of examples and discuss your findings. How might you use this result to correct for bias?

2.35* In the data below, 117 voters were classified according to their sex, their voting intentions, and their attitude to a housing development scheme (with responses: approve or disapprove).

	Male		*Female*	
	Approve	*Disapprove*	*Approve*	*Disapprove*
Conservative	12	3	18	17
Labour	30	13	4	10

Provide analyses based on log-linear and logistic models. A convenient tool is provided by the GLIM package (e.g. Adena and Wilson, 1982, Chapter 7; Healy, 1988, p. 98).

2.36 We saw in section 2.2 that the equations to be solved for maximum-likelihood fitting of the logit model simplify because in that case,

$$f(x) = P(x)\{1 - P(x)\}$$

Verify that the logit is the only distribution with this property. Evaluate the ratio $f(x)/[P(x)\{1 - P(x)\}]$ for normal and complementary log–log link functions.

2.37 An early paper on non-explicit parameter estimation resulting from the method of maximum-likelihood is that of Fisher (1922). In it he considers the following dilution experiment. It is desired to estimate the density of organisms per unit volume, λ, say, in a sample. Direct counting of these is impossible, but it is possible to detect the presence or absence of the organism in any sub-sample. Accordingly, a series of dilutions of the original sample is taken. At the ith dilution n_i sub-samples are tested, in each of which v_i denotes the volume of the original sample present. Assuming a random distribution of organisms in sub-samples, write down the distribution of X_i, the number of sub-samples at the ith dilution which are found to contain the organism. Write down the likelihood for the entire experiment, and consider how to obtain the maximum likelihood estimate of λ using GLIM. For further discussion, see Ridout (1990).

2.38   Consider whether the hypothesis of parallelism is appropriate for the data below, analysed in Finney (1952, p. 69) and Green (1984). See also Lewis (1984) for relevant discussion. We shall return to these data in Example 3.4 and in Chapter 4.

Poisson[a]	$Log_{10}(dose)$	No. in experiment	No. dead
R	0.41	50	6
R	0.58	48	16
R	0.71	46	24
R	0.89	49	42
R	1.01	50	44
D	0.71	49	16
D	1.00	48	18
D	1.31	48	34
D	1.48	49	47
D	1.61	50	47
D	1.70	48	48
M	0.40	47	7
M	0.71	46	22
M	1.00	46	27
M	1.18	48	38
M	1.31	46	43
M	1.40	50	48

[a]R-rotenone (data set 4 of Exercise 1.3), D-deguelin (data set 10 of Exercise 1.3), M-mixture (data set 11 of Exercise 1.3).

2.39   (Tarone, 1985) Show that the Cochran–Armitage test results from a score test of $\beta = 0$ in the standard quantal assay model with probability of response, $P(d) = F(\alpha + \beta d)$, irrespective of the form of F, subject to its being monotone and twice differentiable.

2.40*   Consider how you would devise a test to investigate at which dose levels response was not significantly different from control response. Cf. the work of Williams (1971, 1972, 1986a) and Shirley (1987).

2.41   Show that if the model of section 2.2 is parameterized as

$$P_i = F\{\beta(d_i - \theta)\}, \text{ rather than } P_i = F(\alpha + \beta d_i)$$

then the expected Fisher information matrix is given by:

$$\mathbf{A} = \sum_{i=1}^{k} n_i w(d_i) \begin{bmatrix} \beta^2, & -\beta(d_i - \theta) \\ -\beta(d_i - \theta), & (d_i - \theta)^2 \end{bmatrix}$$

where θ denotes the ED_{50}, and

$$w(d) = \frac{[F'\{\beta(d - \theta)\}]^2}{F\{\beta(d - \theta)\}[1 - F\{\beta(d - \theta)\}]}$$

with $F'(z) = dF(z)/dz$ for any z.

The ED_{100p} may be written as $ED_{100p} = g(\theta, \beta)$. How would you estimate the ED_{100p} and its variance? (equation (8.12).)

CHAPTER 3

Extensions and alternatives

3.1 Introduction

We have seen in Chapter 2 that the basic characteristic of standard quantal assay data is that of binomial response variables, with probability of response modelled as a monotonic increasing function of dose. Several of the examples of Chapter 1 are similar to this pattern, but they need a variety of elaborations to the basic model. Some of these require extensive discussion and so are dealt with at length in later chapters. It is the aim of this chapter to cover the simpler extensions and alternative approaches. For the most part we shall deal with them singly, but of course for any particular problem they may appear in combination. Inevitably, the sections of this chapter will not follow each other in the same logical sequence as in the other chapters of the book.

3.2 Natural or control mortality: EM algorithm and mixture models

In many studies, responses may be due to a 'natural' response, separate from response which may be attributed to the substance being tested. The example of Table 3.1 indicates responses from untreated animals.

Several instances of natural response have already been encountered in the examples of Chapter 1: in the examples of Tables 1.7 and 1.8 instances of natural mortality are observed in the control group; in the example of Table 1.5, beetles were fed during the course of the experiment in the hope that this would eliminate death from natural causes; the deaths of Table 1.10 refer entirely to control populations; a control experiment was used in Table 1.12.

Hoekstra (1987) emphasizes the distinction between **natural** mortality, and **control** mortality, which may well not equate to

Table 3.1 *Data presented by Rai and Van Ryzin (1981), and reported by Tomatis et al. (1972). The toxic agent is DDT, in dose units of p.p.m. in the diet for lifetime feeding. The toxic response is liver hepatoma, and the responding animals are CF-1 female mice*

Dose (ng)	No. of mice responding	No. of mice tested
0	4	111
2	4	105
10	11	124
50	13	104
250	60	90

natural mortality. For ease of exposition we shall use the terms interchangeably. Preisler (1989a) comments on the possibility of insect death due to handling, and suggests that natural mortality in insecticide bioassay is usually in the range, 0–20%. The customary way to incorporate natural or control mortality is to make use of Abbott's formula (Abbott, 1925), and write

$$P(d) = \lambda + (1 - \lambda)\tilde{P}(d) \tag{3.1}$$

where $P(d)$ is the probability of overt response to dose d, λ is the probability of natural response and $\tilde{P}(d)$ is the probability of response to dose d for individuals whose outcome is not due to natural causes, normally with $\tilde{P}(0) = 0$. For example, $\tilde{P}(d)$ could be of logistic form, with dose measured on a logarithmic scale to ensure that $\tilde{P}(0) = 0$. We can see that Abbott's formula results in a simple mixture model, and we shall encounter further mixture models in sections 5.5.2 and in 6.6.2. Preisler (1989a) provides an approach which uses the idea of generalized additive models (Hastie and Tibshirani, 1990) and also separately provides GLIM code for making comparisons in the presence of natural mortality.

Crump *et al.* (1976) suggest an alternative way to model natural mortality. They suppose that natural mortality is equivalent to being exposed to a **base-line** dose of level d_0, say, and then when a dose of level d is applied, it is supposed that the overt response is that which would result from a combined dose of level $(d + d_0)$. In this model, therefore, the probability of natural response is: $\lambda = F(d_0)$, for a suitable c.d.f, $F( )$, and the probability of response to dose d is $F(d_0 + d)$. For illustration, if we use the logit model, with a logarithmic transformation of the doses, we have

$$P(d) = \frac{1}{1 + e^{-\{\alpha + \beta \log(d + d_0)\}}}$$

$$= \frac{1}{1 + e^{-\alpha}(d + d_0)^{-\beta}}$$

Now in this case,

$$\lambda = \frac{1}{1 + e^{-\alpha}(d_0)^{-\beta}}$$

and so

$$P(d) = \frac{1}{1 + \left(\lambda^{-1} - 1\right)\left(\dfrac{d}{d_0} + 1\right)^{-\beta}}$$

$$= \frac{1}{1 + (\lambda^{-1} - 1)\{de^{\alpha/\beta}(\lambda^{-1} - 1)^{1/\beta} + 1\}^{-\beta}} \tag{3.2}$$

which is far less simple than the expression of equation (3.1). But see also Exercise 3.41. It is equation (3.1) which has received the most attention, mainly on account of its simplicity, rather than greater biological plausibility. An exception, however, arises in connection with the incorporation of control information from past experiments, a topic discussed in section 6.7.4. The expression of equation (3.1) can be justified directly when used as a model for the m-alternative forced-choice experiment, sometimes used in psychological tests. In such an experiment a subject selects one of m alternative answers to a question. The one correct response can either be obtained from knowledge of the correct response, or by guessing. Hence in this case we can take $\lambda = m^{-1}$, the probability of guessing the correct response.

If a control group of individuals is sufficiently large, it will provide a precise estimate of λ in equation (3.1), in which case the observed proportions may be suitably 'adjusted' for natural mortality and then the standard analysis can proceed as in Chapter 2. This approach is unlikely to be of much use in practice, now that computer maximization of a likelihood based on equation (3.1) is a straightforward procedure. For discussion, see Barlow and Feigl (1985), and Exercises 3.1 and 3.2.

Example 3.1   (Hoekstra, 1987)

Correction by Abbott's formula is sometimes not considered necessary if control mortality is slight, say less than 5 or 10% (American Public Health Association *et al.*, 1981). For the data below, Hoekstra (1987) evaluated the ED_{50} three ways, viz. (i) completely ignoring the control group, (ii) using 'adjusted' data, as described above, and (iii) by direct, three-parameter maximum-likelihood; a FORTRAN program for doing this is provided by Russell *et al.* (1977). Her results follow the table of data.

Data from Hoekstra (1987) describing the mortality of aphids exposed to nicotine.

% concentration	Number tested	Number dead
Control	45	3
0.0025	50	5
0.005	46	4
0.01	50	3
0.02	46	11
0.03	46	20
0.04	49	31
0.06	50	40
0.08	50	43
0.10	50	48
0.15	50	48
0.20	50	50

Method	ED_{50}	Estimated asymptotic 95% confidence interval
(i)	0.027	(0.023, 0.031)
(ii)	0.034	(0.030, 0.038)
(iii)	0.036	(0.032, 0.042)

We can see the importance of a correct analysis, using the control group information, and a full maximum-likelihood approach. □

The presence of natural mortality can prevent proportions responding dropping to zero, as dose levels are reduced. Natural **immunity** can have the effect of preventing proportions reaching unity, as dose levels are increased. An instance of this has been encountered in Example 1.10. Maximum-likelihood fitting when natural mortality and/or natural immunity are present can be accomplished using GLIM, and the necessary code and theory are provided by Roger (1985) who uses a composite link function approach. For further discussion, see Exercise 3.3 and Appendix E. A more complex application of the idea underlying equation (3.1) will now be described in Example 3.2.

Example 3.2

The data of Table 3.2 form a subset of data from Larsen *et al.* (1977). The objective was to examine how exposure to air pollution (NO_2), and the duration of this exposure, might affect response to an infection. The experiment was performed on mice, for whom the response was death. Increased mortality in mice can be expected to translate into increased human morbidity. The concentrations of NO_2 used corresponded to measurements made in downtown

Table 3.2 *Mortality of mice exposed to three doses of NO_2, for variable amounts of time before being challenged with a viable micro-organism (Streptococcus pyogenes)*

NO_2 (ppm)	Time (hours)	Number tested	Number dead
1.5	96	120	44
	168	80	37
	336	80	43
	504	60	35
	Controls	340	74
3.5	0.5	100	29
	1	200	53
	2	40	13
	3	200	75
	5	40	23
	7	280	152
	14	80	55
	24	140	98
	48	160	121
	Controls	280	56
7.0	0.5	120	52
	1	120	62
	1.5	120	61
	2	120	86
	Controls	120	27

Chicago in 1974. Control animals only breathed clean air. Mortality of treated and control animals was observed for 15 days after a 15-minute exposure to an aerosol of streptococcus, **after** the previous exposure to NO_2.

The subset of data in Table 3.2 was presented and analysed by Hasselblad et al. (1980), who fitted the following probit model, corresponding to an extension of equation (3.1):

$$Pr\left(\begin{array}{c} \text{death from streptococcus infection after exposure for time} \\ t_i \text{ to dose } d_j \text{ of } NO_2 \end{array} \right)$$

$$= \lambda_j + (1 - \lambda_j)\Phi(\mu_{ij}) \tag{3.3}$$

where $\mu_{ij} = \alpha_j + \beta_j \log(t_i)$ for $1 \leq j \leq 3$, and appropriate ranges for i, and λ_j is the probability of response to the streptococcus for the control group corresponding to dose d_j of NO_2. Here, control mortality is not natural mortality, which is ignored, but represents mortality without exposure to NO_2. A priori, one might expect to be able to replace the control mortalities λ_j by a constant λ, and, as shown below, this is in fact possible here.

Thus in the full model of equation (3.3) we have the nine parameters:

$$(\lambda_j, \alpha_j, \beta_j, 1 \leq j \leq 3)$$

A more parsimonious model is given by:

$$\mu_{ij} = \alpha + \beta \log(t_i) + \gamma \log(d_j) + \delta \log(t_i) \log(d_j) \tag{3.4}$$

The results of fitting models (3.3) and (3.4) are shown in Table 3.3, taken from Hasselblad et al. (1980). The selected model is model C, specified by:

$$Pr(\text{death after exposure for time } t_i \text{ to dose } d_j)$$
$$= 0.21 + 0.79\Phi\{ -2.94 + 0.309 \log_e(t_i) + 1.374 \log_e(d_j)$$
$$+ 0.138 \log_e(t_i) \log_e(d_j)\}$$

and the fitted curves are shown in Figure 3.1.

The model C in Table 3.3 appears to provide an adequate description of the data, though it is natural to question whether the fit might be improved to result in a better pattern to the residuals, and this is a topic to which we shall return in Chapter 4. ☐

It is easy to discuss results such as these without regard to the numerical means by which the models are fitted. As discussed in

Likelihood ratio tests for the mice mortality data

Test	Asymptotic chi-square	d.f.	P-value
Common λ (A vs. B)	0.7	2	0.7
Separate lines vs. linear model with interaction (B vs. C)	0.4	2	0.8
Interaction term (C vs. D)	11.9	1	< 0.001

Table 3.3 *Results taken from Hasselblad* et al. *(1980), describing the fits of models (3.3) and (3.4) to the data of Table 3.2*

Fits of various models to the mice mortality data

	Model	Log likelihood	Number of parameters	Goodness of fit[a]	d.f.	P-value
(A)	Separate λ values, intercepts, slopes	-1717.2	9	14.5	11	0.2
(B)	Common λ, separate intercepts, slopes	-1717.6	7	15.2	13	0.3
(C)	Common λ, intercept, $\log_e (NO_2)$, $\log_e (time)$, $\log_e(NO_2) \times \log_e(time)$	-1717.8	5	15.6	15	0.4
(D)	Common λ, intercept, $\log_e (NO_2)$, $\log_e (time)$	-1723.7	4	27.5	16	0.04

[a] Tested using likelihood ratios against the general multinomial model.

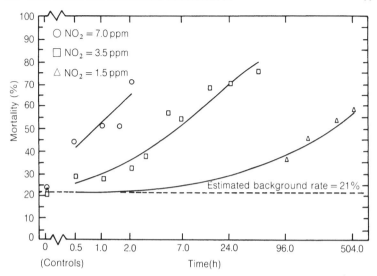

Figure 3.1 *Mortality in mice exposed to NO_2 with fitted model,* P(NO_2, *Time*) $= 0.210 + 0.790\Phi\{-2.94 + 0.309 \times \log_e$ (*Time*) $+ 1.374\log_e$ (NO_2) $+0.138\log_e$ (*Time*) $\times \log_e$(NO_2)}. *Reproduced from Hasselblad et al.* (1980), *with permission from The Biometric Society.*

section 2.1, for many statisticians today the answer is provided by a numerical analysis optimization procedure, such as the simplex method, for maximizing the likelihood:

$$L \propto \prod_j \prod_i \{\lambda_j + (1 - \lambda_j)\Phi(\mu_{ij})\}^{r_{ij}} \times [(1 - \lambda_j)\{1 - \Phi(\mu_{ij})\}^{n_{ij} - r_{ij}}]$$

$$\times \prod_{j=1}^{3} \lambda_j^{r_j}(1 - \lambda_j)^{n_j - r_j} \tag{3.5}$$

where r_{ij} mice die out of a total of n_{ij} exposed for a time t_i to dose d_j and r_j control group deaths occur out of n_j, for $1 \leqslant j \leqslant 3$. In equation (3.5) we have omitted binomial coefficients for simplicity. Hasselblad *et al.* (1980) use an alternative approach, known as the EM algorithm, which we shall now describe.

If we could label dead mice as those which died after breathing clean air, (r_{ijc}), and those which died after breathing polluted air, (r_{ijd}), then we could write the total number deal as

$$r_{ij} = r_{ijc} + r_{ijd} \tag{3.6}$$

corresponding to time t_i and dose d_j. The likelihood under model (3.3) is now no longer given by equation (3.5), but has the much simpler form given below.

$$L \propto \prod_j \prod_i [(1 - \lambda_j)\{1 - \Phi(\mu_{ij})\}]^{n_{ij} - r_{ij}}$$

$$\times \{(1 - \lambda_j)\Phi(\mu_{ij})\}^{r_{ijd}}\lambda_j^{r_{ijc}} \prod_{j=1}^{3} \lambda_j^{r_j}(1 - \lambda_j)^{n_j - r_j}.$$

For the experimental groups the basic distribution is now trinomial, rather than binomial as there are now supposedly three identifiable outcomes, viz. alive, dead due to infection alone, and dead due to infection after breathing polluted air. We shall look at trinomial responses in a more general framework in section 3.5. In the above example we see that the likelihood simplifies to give:

$$L \propto \prod_j \prod_i \{1 - \Phi(\mu_{ij})\}^{n_{ij} - r_{ij}}\Phi(\mu_{ij})^{r_{ijd}}$$

$$\times \prod_j (1 - \lambda_j)^{n_j - r_j + n_{.j} - r_{.j} + r_{.jd}} \prod_j \lambda_j^{r_j + r_{.jc}} \tag{3.7}$$

where we have used the usual dot-notation for summation, viz. $n_{.j} = \sum_i n_{ij}$, etc.

Now the likelihood of equation (3.6) is readily maximized: explicitly we have

$$\tilde{\lambda}_j = \frac{(r_j + r_{.jc})}{(n_j + n_{.j})} \tag{3.8}$$

and

$$\hat{\mu}_{ij} = \Phi^{-1}\left(\frac{r_{ijd}}{n_{ij} - r_{ij} + r_{ijd}}\right)$$

In restricted models such as equation (3.4), then parameter estimates for the components of equation (3.4) follow from iterative maximum-likelihood, as described in Chapter 2.

However, we do not have the division of equation (3.6). Starting from an arbitrary division, then following the above analysis, conditional upon the $\{r_{ij}\}$, we could form the estimated expected (E) values,

$$r_{ijc} = \left(\frac{\lambda_j r_{ij}}{\lambda_j + (1 - \lambda_j)\Phi(\mu_{ij})}\right) \tag{3.9}$$

and the maximization (M) procedure can be repeated again, followed by the E stage of equation (3.9), and so on until convergence is obtained. This approach can be found in a number of papers – see for example, Orchard and Woodbury (1972), who termed it the missing information principle. During the M part of the iteration it may suffice, when iteration is necessary to obtain the maximum-likelihood estimates, to take a single-step of the iterative procedure adopted (Appendix A). This is the approach adopted by Hasselblad *et al.* (1980). The paper by Dempster *et al.* (1977) discusses the EM algorithm and provides proofs of convergence. Although convergence of the EM algorithm is guaranteed it is often slow, and ways of accelerating convergence are sometimes sought (e.g. Titterington and Morgan, 1977). Cox (1984b) reaches the same conclusion as Hasselblad *et al.* (1980), but using a logit model. He shows how the model may be fitted directly by means of an iteratively reweighted Gauss–Newton algorithm, executed by the BMDP program, BMDPAR. See also Barlow and Feigl (1985).

Farewell (1982b) makes the astute observation that mixture models may describe a homogeneous population by means of a mixture. One clear example of a heterogeneous population resulting from a mixture is mentioned by Milicer (1968) in the context of girls reaching the age of menarche, as in the example of section 1.2.2. It was pointed out that samples of girls taken widely throughout a country can be expected to result in heterogeneity, for example due to ethnic differences, or differences between urban and rural populations. One example was cited in which data were pooled from a number of localities widely distributed over Hungary, and resulting in over 7000 girls in all. The next example is another in which clear biological justification was available for the assumed mixture model.

Example 3.3

Lwin and Martin (1989) describe experiments on the use of the anthelminitic drug *thiabendazole* for clearing the worms from sheep infected with worms, *ostertagia* spp. The minimum time required for one generation of *ostertagia* spp. is approximately 3 weeks, and so regular dosing of sheep at a rate of 17 doses per year should, in the absence of resistant strains, result in worm-free sheep. A 5-year experiment was carried out to compare the effects of this régime with one in which no anthelminitic was applied, and also with another experiment in which just five doses of anthelminitic were given at regular intervals each year. After three years of the experiment, the *ostertagia* spp. had developed resistance to the anthelminitic. In year 4 of the experiment, worms were obtained from sheep of each group, cultured to produce sufficient numbers of eggs and then tested with anthelminitic, given at a variety of dose levels. Experimental details are provided in Martin *et al.* (1984); similar experiments are described in Hewlett and Plackett (1979, p. 60). For the zero dose and 17 doses per year groups a single distribution was sufficient to describe the data. However for the five doses per year group it might be supposed that the egg population is composed of susceptible types (as in the zero dose group), resistant types (as in the 17 doses per year group), and hybrid types.

Table 3.4 presents the results, from Lwin and Martin (1989) of fitting a single distribution and two- and three-component mixtures of normal distributions. As can be seen from Table 3.4, in this application there was control mortality as well as an anticipated mixture of tolerance distributions. Thus the model they fitted had, for treated groups, probability of death when exposed to dose d given by:

$$P(d) = \lambda + (1 - \lambda) \sum_{j=1}^{r} \zeta_j F_j \{\log_e(d)\} \tag{3.10}$$

where here r is the number of tolerance distributions fitted, $1 \leqslant r \leqslant 3$, $\{\zeta_j, 1 \leqslant j \leqslant 3\}$ are the mixing probabilities, and the F_j are normal, with mean μ_j and standard deviation σ_j. The expected values of Table 3.4 resulted from maximum-likelihood fitting using the method-of-scoring, but Lwin and Martin also employed the EM algorithm for comparison.

In Table 3.5 are shown the maximum likelihood parameter estimates obtained by the two methods. The slight variation sometimes observed is due to numerical differences, as may result, for example, from different stopping rules for the iterative procedures. An added complexity here, in comparison with Example 3.1, is that if the EM algorithm is used to fit the mixture of distributions, then this whole procedure has to be nested within the estimation of λ. In fact, as we can see from Table 3.5, Lwin and Martin estimated λ by means of a crude proportion, taking the control data and (to increase precision, and at the expense of only a small amount of bias) the lowest dose group, which is therefore also treated as a control group. As a result, the error estimates of Table 3.5 will be conservative

Table 3.4 (Lwin and Martin, 1989) Observed and expected post-treatment egg counts resulting from maximum-likelihood estimation and direct optimization of the likelihood function. The group given the lowest dose was also treated as a control group – see text

Concentration (μg/ml)	Observed eggs killed (total)	Expected One component	Two components	Three components
Control	79(1117)			
0.023	74(909)	–	–	–
0.0469	125(1067)	152.99	121.66	114.39
0.0625	184(1110)	207.61	199.55	197.00
0.0781	288(1091)	253.56	278.83	285.22
0.0938	361(1088)	301.87	351.75	361.39
0.125	474(1113)	402.12	461.45	459.09
0.156	462(1061)	461.14	492.65	476.45
0.1875	512(1065)	531.68	530.00	514.01
0.25	624(1077)	648.21	598.41	612.34
0.3125	680(1077)	730.86	665.40	691.73
0.375	748(1077)	793.50	732.93	745.83
0.50	845(1090)	890.31	859.63	843.80
0.625	947(1103)	956.72	954.18	937.07
0.75	946(1052)	947.84	962.34	955.06
1.00	1044(1079)	1015.09	1041.82	1046.34
1.25	1060(1067)	1026.47	1051.49	1057.06
1.5	250(251)	244.57	249.38	250.28

Table 3.5 Estimates of unknown parameters

Parameters	Estimates (standard errors)					
	Direct optimization technique			EM + GLIM technique		
	C1	C2	C3	C1	C2	C3
ζ_1	1	0.44 (0.0428)	0.41 (0.0946)	1	0.45	0.38
ζ_2	—	0.56	0.23 (0.2052)	—	0.55	0.30
ζ_3	—	—	0.36 (0.0180)	—	—	0.32
μ_1	−0.68 (0.0059)	−1.08 (0.0307)	−1.11 (0.0581)	-0.68	−1.07	−1.12
μ_2	—	−0.36 (0.0295)	−0.61 (0.1469)	—	−0.35	−0.60
μ_3	—	—	−0.22 (0.1170)	—	—	−0.21
σ_1	0.45 (0.0070)	0.19 (0.0307)	0.16 (0.0449)	0.45	0.20	0.16
σ_2	—	0.23 (0.0180)	0.12 (0.1093)	—	0.23	0.19
σ_3	—	—	0.17 (0.0474)	—	—	0.17

C1 – single component analysis
C2 – two-component analysis
C3 – three-component analysis.
Note: λ is estimated as $\hat{\lambda} = (79 + 74)/(117 + 909) = 0.0755$ and is taken as known in determining other parameters.

(Appendix C). GLIM was used within the EM approach, although a one-step procedure was also outlined (Appendix A). For discussion of which mixture model provides the best description of the data, see Exercise 3.7. □

The message of the last three examples is clear: natural mortality is frequently present, and it then needs to be included in the models used for the data. This should be borne constantly in mind in the work of subsequent chapters, which effectively deals with the $\tilde{P}(d)$ part of equation (3.1). As emphasized by Barlow and Feigl (1985) and Hoekstra (1987), it is essential that when natural mortality is anticipated, experiments should contain control groups of appreciable size.

3.3 Wadley's problem; use of controls

In an experiment considered by Wadley (1949), fruit were infested with immature stages of fruit-flies. The larvae are killed by exposure to cold, and treatment consisted of varying lengths of exposure to low temperature. For another experiment on the effect of cooling, see Cannon (1983). Fictitious data presented by Wadley are shown in Table 3.6. In this example the precise extent of initial infestation of the fruit was not known, and so the binomial distribution may not be used. Wadley investigated using the Poisson distribution, and discussion of these data is considered in Exercise 3.8. A more elaborate analysis is required for the data of Example 3.4.

Table 3.6 *Fictitious data from Wadley (1949)*

Number of days exposure to low temperature	Number of fruit-flies seen to emerge
1	715
2	396
3	158
4	55
5	32
6	4

Example 3.4

A modern approach to the analysis of such data is provided by Baker *et al.* (1980), who use GLIM, the goodness-of-link approach of Pregibon (1980), which we discuss in Chapter 4 (in particular Exercise 4.26), and a composite link function to accommodate a control group. Cf. the approach of Roger (1985). The data they analyse are shown in Table 3.7. Certain unusual patterns in the data were acknowledged, and are due in part to the use of optical densities of cell suspensions, rather than a direct cell count. For the purpose of illustration we shall ignore this aspect of the data.

We assume that the number of control organisms in any replication has a Poisson distribution, with mean ζ. If we also assume that the organism does not divide during the course of the experiment, that any deaths observed are due to the toxicant, and independence of response, then it is simple to show (Exercise 3.9) that for a replicate given concentration x_i, the count observed is the realization of a Poisson random variable, of mean $\zeta(1 - P_i)$, where

$P_i = $ Pr(an organism exposed to concentration x_i for 7 days dies)

Baker *et al.* (1980) set $P_i = \Phi\{\alpha + \beta \log(x_i)\}$.

Suppose the total number of organisms in the experiment is n, n_0 of which are controls. Each of the remaining $(n - n_0)$ organisms is therefore exposed to the toxicant, with the ith organism receiving concentration x_i, and resulting in the observed count y_i, $n_0 + 1 \leqslant i \leqslant n$.

The likelihood under the Poisson model is given by

$$L(\alpha, \beta, \zeta) \propto \left(\prod_{i=1}^{n_0} e^{-\zeta} \zeta^{y_i} \right) \left[\prod_{i=n_0+1}^{n} \{\zeta(1 - P_i)\}^{y_i} e^{-\zeta(1 - P_i)} \right]$$

Table 3.7 *Data from Baker* et al. *(1980), on the environmental impact of chemicals. Counts of the unicellular organism, Selenastrum capricoruntum following 7 days' exposure to the toxicant*

Concentration toxicant ($\mu g/ml$)	Replicate counts				
	1	2	3	4	5
Control	219	228	202	237	228
1	167	158	158	175	167
5	105	123	105	105	105
10	88	88	61	61	88
50	61	44	35	35	44

The maximum likelihood results obtained by GLIM from Baker *et al.* (1980) are shown below:

$$\hat{\zeta} = 224.6\,(6.629)$$
$$\hat{\alpha} = -0.5489\,(0.0773)$$
$$\hat{\beta} = 0.3727\,(0.0230)$$

resulting in a residual deviance of 34.21 on 22 degrees of freedom, which is just significant at the 5% level. □

Wadley's (1949) paper was followed by that of Anscombe (1949), who provided a model based on the negative binomial distribution, which may provide a better description of data which may be **over-dispersed** relative to the Poisson distribution (i.e. with variances greater than mean values). An illustration is provided by the data of Table 1.12 which require an analysis based on over-dispersion, as we can see from a check of the ratios of sample means to variances. The topic of over-dispersion is covered in detail in Chapter 6, and there is more discussion of Wadley's problem in Exercises 6.21 and 6.22. The above analysis effectively ignores variation between replicates. One way of making use of this information will be encountered in Chapter 6.

3.4 Influence and diagnostics

In the analyses considered so far, the data have been taken at their face value, without regard to the influence of the responses to individual doses on the analysis and conclusions drawn. In the context of linear regression, the study of the influence of individual points on the analysis is now well established and the procedures available to us are described by Cook and Weisberg (1982), for example, with further possibilities discussed in Cook (1986). In the work of this section we shall see how existing procedures may be used and in some cases modified, to cater for logistic regression in general, with illustrations provided by sets of quantal assay data.

In linear regression a central role in determining the influence of individual points is provided by the 'hat' matrix,

$$\mathbf{H} = \mathbf{X}(\mathbf{X}'\mathbf{X})^{-1}\mathbf{X}' \tag{3.11}$$

where $\mathbf{X}$ is the design matrix of equation (2.20). The hat matrix was so named by John Tukey, as it maps the observed values $\mathbf{Y}$ into the fitted values, $\hat{\mathbf{Y}}$:

$$\hat{\mathbf{Y}} = \mathbf{HY}$$

The diagonal elements of $\mathbf{H}$ provide measures of leverage, indicating points with high potential influence, due to their location in the design configuration. The paper by Hoaglin and Welsch (1978) provides a rule of thumb, based on the diagonal elements of $\mathbf{H}$, for detecting influential points. In addition we can construct measures which assess the effect of individual points on parameter estimates, on measures of goodness-of-fit, and so forth. Once a linear regression model has been fitted to data then all of these diagnostic measures may be computed, examined and plotted without recourse to any further model-fitting. We can measure the effect of individual points on various aspects of the fit either by deletion of points (usually singly), or by means of infinitesimal perturbation. Here we concentrate on assessment by deletion, though we mention here the paper by James and James (1983), which adopts the latter approach, with particular reference to the ED_{50}. We shall discuss this work in Chapter 7.

A scalar measure summarizing the effect of deleting the lth observation on all of the parameters specified in the model is the so-called Cook's distance, given by

$$c_l = s_l^2 h_{ll}/(1 - h_{ll})^2 \qquad (3.12)$$

where h_{ll} is the lth diagonal entry in the hat matrix, $\mathbf{H}$, and s_l is the lth residual:

$$s_l = y_l - \mathbf{x}_l'\hat{\boldsymbol{\beta}}$$

$\mathbf{x}_l'$ being the lth row of the design matrix $\mathbf{X}$.

In the quantal assay context we might, for illustration, take the Irwin data of Table 1.3 and consider whether individual doses have large influence on model-fitting.

Model-fitting by maximum-likelihood, of say the logit model, is not explicit, as we have seen in Chapter 2, and so assessment of individual case influence by deletion is not as straightforward as in the linear regression case. We shall return to this point shortly, but for the moment we note that of course model-fitting by minimum logit chi-square is equivalent to weighted linear regression, and so measures of leverage and Cook's distance, as well as many other diagnostics, are readily available through an appropriate linear regression package, such as the one that is available within the

Table 3.8 *Measures of leverage (h_{ii}) and Cook distance (c_i) for the Irwin data of Table 1.1. Results taken in part from Williams (1987a)*

$10.158 + \log_2$ (serum dose)	r_i	n_i	$h_{ii}^{(1)}$	$h_{ii}^{(2)}$	$c_i^{(1)}$	$c_i^{(2)}$	$c_i^{(3)}$
1	0	40	0.16	0.26	0.05	0.24	0.25
2	2	40	0.28	0.37	0.25	0.23	0.26
3	14	40	0.51	0.35	1.51	1.80	1.28
4	19	40	0.37	0.39	0.08	0.09	0.08
5	30	40	0.68	0.63	0.09	0.37	0.36

Note: $h_{ii}^{(1)}$ and $c_i^{(1)}$ denote the values resulting from using MINITAB and minimum-logit-chi-square; $h_{ii}^{(2)}$ and $c_i^{(2)}$ denote the values resulting from maximum-likelihood, and leaving-one-out; $c_i^{(3)}$ denote the values obtained from the one-step approximation to the maximum-likelihood procedure.

MINITAB package. The results, for leverage and Cook's distance obtained for the Irwin data, are shown in Table 3.8, together with the equivalent values obtained by maximum-likelihood procedures to be described shortly. It seems clear that the third dose is exerting a large amount of influence on the fit. This is not surprising when we consider the plot of the data given in Figure 1.1, and the fitted curves shown later in Figure 4.10. We continue the discussion of these data in Chapter 4.

For many statisticians the preferred method of model-fitting is by maximum-likelihood, and in any case, the above approach to assessing influence using the method of minimum logit chi-square is not suitable for dealing with binary data, such as are illustrated in Table 2.2 (cf. Exercise 2.2). As we have seen in Chapter 2, a problem with maximum-likelihood as a method of model-fitting is that it frequently requires a numerical iterative procedure. In order to assess influence by means of single case deletion this problem is then magnified k-fold times, if k is the number of doses used in the experiment, or in general the number of cases. However, in many cases the power of computers and numerical optimization procedures should suffice to remove the objection to exact assessment of influence by single case deletion. Pregibon (1981) wrote:

'After fitting a logistic regression model, and prior to drawing inferences from it, the natural succeeding step is that of critically assessing the fit. In practice, however, this assessment is rarely considered, and seldom carried out.'

In order to encourage the critical assessment of fit in logistic regression, Pregibon (1981) produced an approximation to the maximum-likelihood assessment of influence, using single case deletion, and as a result of this approximation the non-linear model-fitting is carried out only once. Further discussion of this approach is to be found in Williams (1987a), and the details are now outlined below.

In the notation established in section 2.8 for general logistic regression, the Newton–Raphson iteration for the parameters $\boldsymbol{\beta}$ is given by:

$$\boldsymbol{\beta}^{t+1} = \boldsymbol{\beta}^t + (\mathbf{X'VX})^{-1}\mathbf{X's}$$

This corresponds to maximizing the log-likelihood:

$$l(\boldsymbol{\beta}) = \sum_{i=1}^{k} w_i l_i(\boldsymbol{\beta}) \tag{3.13}$$

with $w_i = 1$, for all i, and $l_i(\boldsymbol{\beta})$ being the contribution to the log-likelihood from the responses of the ith case. If we are interested in the influence of the lth case then we may set $w_l = w$, for some $0 \leqslant w \leqslant 1$, while retaining all the other values of $w_i = 1$, $i \neq l$. If we write the diagonal matrix $\mathbf{W} = \mathrm{diag}(w_i)$, then the Newton–Raphson iteration for the parameters, which we may write as $\boldsymbol{\beta}_l(w)$, to maximize $l(\boldsymbol{\beta})$ given in equation (3.13) is as follows:

$$\boldsymbol{\beta}_l^{t+1}(w) = \boldsymbol{\beta}_l^t(w) + (\mathbf{X'V}^{1/2}\mathbf{WV}^{1/2}\mathbf{X})^{-1}\mathbf{X'Ws} \tag{3.14}$$

(Exercise 3.12).

If we set $w = 0$, and complete the iteration of equation (3.14) for each value of l in turn then we perform the exact assessment of the influence of each case on the maximum-likelihood fit, and this is clearly time-consuming. Pregibon's (1981) approximation results from making a one-step approximation only to the iterations in equation (3.14) (Appendix A) starting from the maximum-likelihood estimate $\hat{\boldsymbol{\beta}}$, obtained from the complete data set, and, in that case, terminated iterations. Writing $\mathbf{z} = \mathbf{X}\hat{\boldsymbol{\beta}} + \mathbf{V}^{-1}\mathbf{s}$, as in section 2.8, then the resulting approximation to $\hat{\boldsymbol{\beta}}_l(w)$ is:

$$\hat{\boldsymbol{\beta}}_l(w) \approx (\mathbf{X'V}^{1/2}\mathbf{WV}^{1/2}\mathbf{X})^{-1}\mathbf{X'V}^{1/2}\mathbf{WV}^{1/2}\mathbf{z} \tag{3.15}$$

(Exercise 3.13). We see from section 2.8 that equation (3.15) is what would result from linear regression, when the response variable is $\mathbf{V}^{1/2}\mathbf{z}$, and the design matrix is $\mathbf{V}^{1/2}\mathbf{X}$, and consequently measures of influence used in linear regression may now be used in this approximation to non-linear, logistic regression.

In particular from equation (3.15), the effect of omitting the lth case on the parameter vector $\boldsymbol{\beta}$ is:

$$\hat{\boldsymbol{\beta}} - \hat{\boldsymbol{\beta}}_l(0) = (\mathbf{X'VX})^{-1}\mathbf{X'Vz} - (\mathbf{X'V}^{1/2}\mathbf{WV}^{1/2}\mathbf{X})^{-1}\mathbf{X'V}^{1/2}\mathbf{WV}^{1/2}\mathbf{z}$$

(3.16)

where $\mathbf{W}$ is now the identity matrix, apart from a zero entry at the (l,l)th diagonal term. After some matrix algebra (Exercise 3.15) this expression reduces to:

$$\hat{\boldsymbol{\beta}} - \hat{\boldsymbol{\beta}}_l(0) = (\mathbf{X'VX})^{-1}\mathbf{x}_l s_l/(1 - h_{ll})$$

(3.17)

where $\{h_{ll}\}$ are now used to denote the diagonal terms of the analogue of the hat matrix:

$$\mathbf{H} = \mathbf{V}^{1/2}\mathbf{X}(\mathbf{X'VX})^{-1}\mathbf{X'V}^{1/2}$$

(3.18)

For properties of $\mathbf{H}$, see Exercise 3.16. We obtain the analogue of Cook's distance when we form a scalar summary of the vector displacement of equation (3.17), viz.

$$c_l = \{\hat{\boldsymbol{\beta}} - \hat{\boldsymbol{\beta}}_l(0)\}'\mathbf{S}^{-1}\{\hat{\boldsymbol{\beta}} - \hat{\boldsymbol{\beta}}_l(0)\}$$

(3.19)

where $\mathbf{S} = \hat{\mathrm{V}}\mathrm{ar}(\hat{\boldsymbol{\beta}}) = \mathbf{X'VX}$.

After further algebra, equation (3.19) reduces to:

$$c_l = \tilde{s}_l^2 \, h_{ll}^2$$

analogous to the linear regression case (Exercise 3.17) and where the $\tilde{s}_l$ are chi-square residuals defined below.

The measure of equation (3.19) has an interesting interpretation in terms of likelihood ratio confidence regions, discussed in section 2.7. In order to obtain confidence regions for $\boldsymbol{\beta}$, we set

$$\psi(\boldsymbol{\beta}) = 2\{l(\hat{\boldsymbol{\beta}}) - l(\boldsymbol{\beta})\} = c$$

where $\hat{\boldsymbol{\beta}}$ is the maximum-likelihood estimate of β, and c is a value determined by the size of the region. We have seen this done in Figures 2.2–2.4. Alternatively, we may evaluate $\psi\{\hat{\boldsymbol{\beta}}_l(0)\} = \tilde{c}_l$, say, to obtain a scalar measure of the influence of the lth case on $\hat{\boldsymbol{\beta}}$. We may compare $\tilde{c}_l$ with the percentage points of χ^2_{m+1} where $(m+1)$ is the dimension of $\boldsymbol{\beta}$, and this indicates, roughly, to which contour of the confidence region $\hat{\boldsymbol{\beta}}$ is displaced due to the lth observation being omitted.

A second-order Taylor series expansion of the log-likelihood, which is equivalent to assuming asymptotic normality (Appendix A) then gives

$$\tilde{c}_l = \{\hat{\boldsymbol{\beta}} - \hat{\boldsymbol{\beta}}_i(0)\}' S^{-1} \{\hat{\boldsymbol{\beta}} - \hat{\boldsymbol{\beta}}_i(0)\}$$

as in (3.19), where $S = X'VX$. Other interpretations of the c_i measure are also possible (Pregibon, 1981).

Morgan *et al.* (1989) describe a computer-package called QUAD which incorporates much of the work of Chapters 2 and 4 for dealing with single sets of quantal assay data, with a single x variate. Designed to run on micro-computers, QUAD could be slow if it evaluated influence measures by single-case-deletion and maximum-likelihood. A particularly attractive feature of Pregibon's one-step approximations is that they make use of 'building blocks' which are all available following the maximum-likelihood fit to the entire data set. For computation, in addition to the Cook distance measures, the diagnostic measures used in QUAD are:

(i) *Leverage values*

$$h_{ii} = \frac{n_i \hat{P}_i \hat{Q}_i \left(\sum_j z_j^2 n_j \hat{P}_j \hat{Q}_j - 2x_j \sum_j x_j n_j \hat{P}_j \hat{Q}_j + x_j^2 \sum_j n_j \hat{P}_j \hat{Q}_j \right)}{\left(\sum_j n_j \hat{P}_j \hat{Q}_j \right) \left(\sum_j x_j^2 n_j \hat{P}_j \hat{Q}_j \right) - \left(\sum_j x_j n_j \hat{P}_j \hat{Q}_j \right)}$$

for fitted probabilities $\hat{P}_i$ and $\hat{Q}_i$, with $\hat{P}_i + \hat{Q}_i = 1$.

(ii) *Chi-square residuals*

$$\tilde{s}_i = \frac{r_i - n_i \hat{P}_i}{\sqrt{n_i \hat{P}_i (1 - \hat{P}_i)}}$$

(iii) *Deviance residuals*

$$e_i = \text{sign}(\tilde{s}_i) \left[2 \left\{ r_i \log\left(\frac{r_i}{n_i \hat{P}_i} \right) + (n_i - r_i) \log\left(\frac{n_i - r_i}{n_i - n_i \hat{P}_i} \right) \right\} \right]$$

(iv) *Individual case influences for* X^2

$$\Delta_i X^2 = \tilde{s}_i^2 / (1 - h_{ii})$$

(v) *Individual case influences for the deviance*

$$\Delta_i D = e_i^2 + \tilde{s}_i^2 h_{ii} / (1 - h_{ii})$$

How to obtain these diagnostics in GLIM is explained by Collett and Roger (1988), and there are also relevant macros in the GLIM macro library (Appendix E). Extensions to the multinomial case (considered in Chapter 5) are provided by O'Hara Hines *et al.* (1992). We shall now provide several examples of these measures at work. The first point to make is that both Pregibon (1981) and Williams (1987a) comment that the one-step approximation appears to perform well in identifying influential points, but tends to underestimate large influence values; we already have examples of this in Table 3.8.

Of the two types of residuals defined above, it is deviance residuals that see more use (Landwehr *et al.* 1984). Williams (1987a) presented a compromise between the two forms, discussed in Exercise 3.24. An index plot for the deviance residuals obtained from fitting models to the Bliss data of Exercise 1.4 is shown in Figure 3.2 and is useful in displaying the overall best performance of the three-parameter Aranda-Ordaz model, defined in Exercise 2.13 and discussed fully in section 4.2.

Example 3.5

We now return to the logistic regression example presented in section 2.8. In Figure 3.3 are shown the plots (index plots) of $m_{ii} = (1 - h_{ii})$,

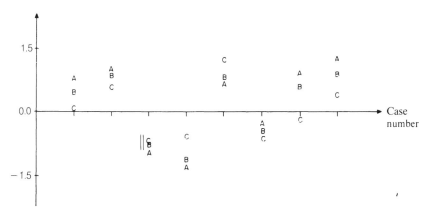

Figure 3.2 *Index plots for the deviance residuals from fitting the Bliss data of Exercise 1.4. C: fitting the Aranda–Ordaz asymmetric model to dose values; A: fitting the logit model to dose values; B: fitting the logit model to the squares of the dose values.*

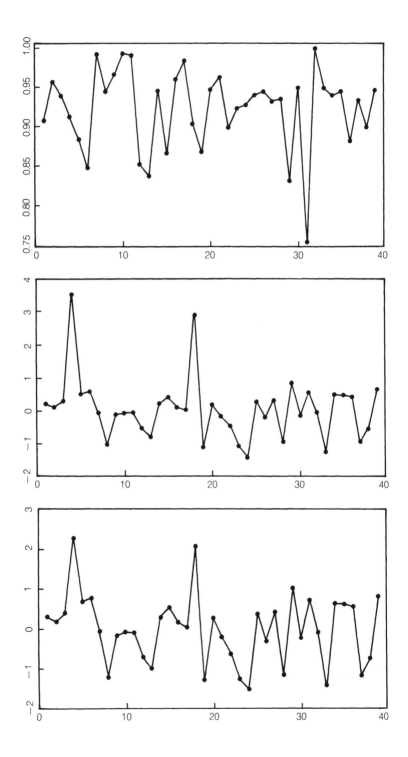

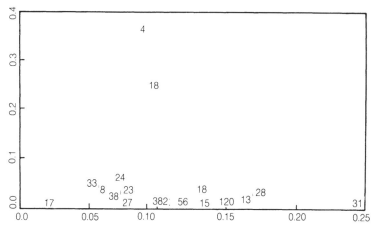

Figure 3.4 *Taken from Pregibon (1981). Scatter plot of* $\tilde{s}_i^2/[\sum_i \tilde{s}_i^2]$ *versus* h_{ii} *for the data of Table 2.2. Reproduced with permission from the Annals of Statistics.*

the chi-square residuals, $\tilde{s}_i$, and the deviance residuals, e_i, in each case plotted against the index number, i. The 31st observation has the highest leverage, but observations 4 and 18 are clearly not well fitted by the model (recall that overall the goodness-of-fit statistics were satisfactory). These points are further emphasized by the plot of Figure 3.4. The Cook distances are shown in Figure 3.5, and emphasize the influence of the 4th and 18th observations. The influence of the 31st point is negligible, despite its extreme location in the design space, as identified by the leverage values (Exercise 3.18). In Figure 3.5 we can see again that the one-step approximation is working well in identifying influential points, but the values of the influence are underestimated. This example is discussed at far greater length in the papers by Pregibon (1981) and Wang (1985). Its unusual features are emphasized by Copas (1988). See also Exercises 3.18–3.21.

Example 3.6

The data of Table 3.9 appeared originally in Martin (1942) and are discussed by Finney (1947, p. 169). They are part of a larger data

Figure 3.3 (opposite) *Taken from Pregibon (1981). Index plots of* $m_{ii} = 1 - h_{ii}$ *versus* i *(top);* $\tilde{s}_i$ *versus* i *(middle) and* e_i *versus* i *(bottom), for the data of Table 2.2. Reproduced with permission from the Annals of Statistics.*

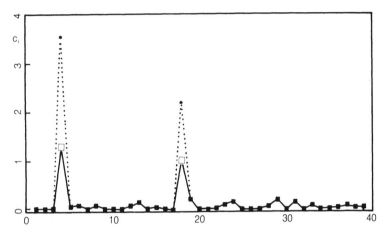

Figure 3.5 *Taken from Pregibon (1981). Index plot of Cook distances for the data of Table 2.2. The symbols (·) and (□) refer respectively to the fully iterated and one-step values. Reproduced with permission from the Annals of Statistics.*

set presented in Exercise 2.38. In this case the substance applied was deguelin, given to groups of *Macrosiphoniella sanborni*, the chrysanthemum aphis. In this example the case with the highest leverage (the first case corresponding to the lowest dose) also has the largest X^2 residual (though not deviance residual), and furthermore has the largest influence. While the case of highest leverage is frequently the case with the highest or lowest dose for standard quantal assay data, this is not always true, as we can see from the illustration in Exercise 3.23. There is further discussion of this example in Chapter 4 and Exercise 3.22. Cf. also the analysis in Green (1984). ☐

Two reasons advanced by Pregibon (1981) for studying influence in logistic regression are:

1. Although possessing good asymptotic properties, maximum-likelihood may be sensitive to outlying observations. (Robust procedures are considered in Chapter 7.)

2. As we have seen amply illustrated in the examples of Chapter 1, data which may be analysed by logistic regression, or more general procedures designed for quantal assay data, may be obtained from observational as well as experimental studies. In the former case the design may be less well thought out than in

Table 3.9 Influence diagnostics for the data from Martin (1942)

Case no. (i)	Log (conc.)	Group size	No. responding	X^3 res. $\tilde{s}_i$	Dev. res. e_i	h_{ii}	ΔX^3	$\Delta_i D$	Cook distance, c_i
1	0.71	49	16	1.863	1.772	0.633	9.463	9.132	16.332
2	1.00	48	18	−1.764	−1.773	0.384	5.057	5.089	3.158
3	1.31	48	34	−1.606	−1.529	0.263	3.501	3.261	1.251
4	1.48	49	47	1.454	1.639	0.263	2.868	3.438	1.021
5	1.61	50	47	0.036	0.036	0.245	0.002	0.002	0.001
6	1.70	48	48	1.448	2.025	0.211	2.657	4.664	0.712

the latter case (Exercise 3.25) and hence potentially be more susceptible to points of high leverage for example.

As described above, straightforward techniques are now available to enable us to carry out a variety of critical investigations of the fit of a model to data. But what do we do when influential points are identified? The aim is certainly not to exclude these points, and so produce a kind of wart-free model, but rather to report on the existence of influential points. In an ideal world the experimenter who provided the data might find, from consulting records of the experiment, that there is a clear explanation for seemingly unusual data points, which may then provide an argument for their exclusion. The discussion here is similar to that of section 3.2 and of Chapter 6, on the identification of elements of mixtures.

An important question in connection with influence is whether points which are influential for β are also influential for aspects of the data analysis which are of particular interest. This point is taken up by Williams (1987a), who uses one-step approximations to assess which points may be influential in the context of score tests of whether, for instance, the fit of the logit model may be significantly improved within a particular extended parametric family. Thus for example it comes as no surprise to find that case number 3, already identified as influential for the logit model fit to the Irwin data set of Table 1.3, is responsible for the logit being rejected as a suitable model for those data. We discuss score tests and this particular data set in more detail in Appendix D and Chapter 4. Williams (1987a) also considers influence of individual doses on the ED_{50} and Figure 3.6 presents the variety of different profile log-likelihoods for the ED_{50} in the Irwin data set, when all the data are included, and when single cases are omitted, assessment being again based on one-step approximations. Once more we see the substantial effect of omitting the third case. The diagnostic techniques described above have been extended by Landwehr et al. (1984), who provide methods like the partial residual plot and the local mean deviance plot, which may be employed after the diagnostic methods have identified problems. Smoothing methods which suggest how an inadequate model may be modified are described in Fowlkes (1987). There are links here with the kernel smoothing of Copas (1983), which is discussed in section 7.6.2, and which, in Kay and Little (1986), revealed the need to add a quadratic term to a logistic regression model. For further

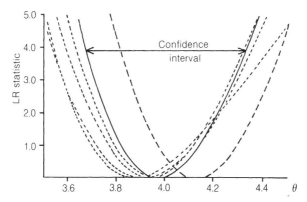

Figure 3.6 *Taken from Williams (1987a), this figure presents profile log-likelihoods of the* ED_{50} *for the Irwin data of Table 1.3. The profile for the complete data is denoted by the continuous curve; the approximate profile when single cases are deleted are shown by dotted curves, except for the case 3 omission, which is denoted by the dashed curve. Reproduced with permission from the Royal Statistical Society. The profiles are given in terms of the likelihood-ratio (LR) static and hence differ slightly from those of Figure 2.2.*

discussion of influence for the ED_{50}, see James and James (1983) and Chapter 7. We note finally that the work of this section has parallels in survival analysis (Cain and Lange, 1984; Pettitt and Bin Daud, 1989). Another investigation of influence, through an added variable plot representation of a score test, is given in section 3.7 and Exercises 3.37 and 3.38.

3.5 Trichotomous responses

An example of data resulting in three responses, rather than the usual two for quantal assay data, has been given in Table 1.7, and further discussion took place in section 3.2.

The standard model for data of this kind was suggested by Aitchison and Silvey (1957), and can clearly be extended to deal with more than three possible responses, but we shall not do that here. The original motivation was of data on insects progressing through different stages of development. For the ith dose, there are probabilities (P_{i1}, P_{i2}, P_{i3}) summing to unity, such that the observed data follow a trinomial distribution (in an obvious notation):

$$\Pr(r_{i1}, r_{i2}, r_{i3} \mid n_i) = \frac{n_i!}{r_{i1}! r_{i2}! r_{i3}!} P_{i1}^{r_{i1}} P_{i2}^{r_{i2}} P_{i3}^{r_{i3}}$$

We may then complete the model with the logistic specification of the appropriate probabilities, viz.

$$P_{i1} = (1 + e^{-\alpha_1 - \beta d_i})^{-1}$$
$$1 - P_{i3} = P_{i1} + P_{i2} = (1 + e^{-\alpha_2 - \beta d_i})^{-1} \qquad (3.20)$$

Aitchison and Silvey used probits, but for simplicity we shall use logits. (This model has the **proportional odds** property. Had we used complementary log–log link function, the resulting model would have the **proportional hazards** property (McCullagh and Nelder, 1989, p. 153).) We need to choose the same slope β, to ensure that the two logistic curves above never cross, so that p_{i2} cannot be negative. We also require that $\alpha_1 < \alpha_2$. We thus here have a logit model for death alone, and also one for death or deformity. On the logit scale, we have,

$$\text{logit}(P_{i1}) = \alpha_1 + \beta d_i$$
$$\text{logit}\{(P_{i1} + P_{i2})\} = \alpha_2 + \beta d_i$$

i.e. parallel lines with the second one above the first by an amount $(\alpha_2 - \alpha_1)$.

An alternative to model-fitting by maximum likelihood, used by Aitchison and Silvey, is minimum logit chi-square, used by Gurland *et al.* (1960). Jarrett *et al.* (1981) employed both of these methods to fit data of the kind shown in Table 1.7. (See also McPhee *et al.* 1984.) For Facey's Paddock Jarrett *et al.* (1981) obtained χ^2 goodness-of-fit statistics: 5.91 (min. logit chi-square) and 6.42 (maximum-likelihood), both clearly not significant when referred to chi-square tables on 5 degrees of freedom. For Tinaroo the corresponding figures were, respectively, 15.40 and 21.81, indicating a poor fit in that case. Figure 3.7 summarizes a study on 12 viruses in all, through the ED_{50} for death and deformity and the ratio of the two ED_{50}s. For discussion, see Exercise 3.27. The model of equation (3.20) has been used by Whitehead and Curnow (1991) in their evaluation of the fixed dose procedure.

Data of the kind displayed in Table 1.7 come under the general heading of contingency-table data when response categories are

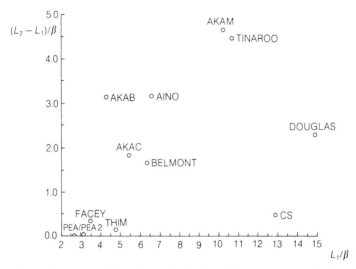

Figure 3.7 *Graph of log* ED_{50} *for death + deformity against the log of the ratio of the two* ED_{50}s *for 12 arboviruses. See also McPhee et al. (1984).*

ordered, and a general approach for analysing such data has been provided by McCullagh (1980). It is assumed that responses within each row result from a latent variable, with cumulative distribution function $F(x)$, falling within intervals specified by cut-off values, z_1 and z_2. In this general model, the probabilities of response for the trichotomous response data are given by:

$$P_{i1} = F(a_i + z_1)$$
$$P_{i2} = F(a_i + z_2) - F(a_i + z_1), \qquad \text{for the } i\text{th dose}, \qquad (3.21)$$

and we would try to model the location parameters, $\{a_i\}$ in terms of the doses.

By comparison with equations (3.20), we see that the model of Aitchison and Silvey is of the general form of equation (3.21), with the correspondences:

Eqn (3.20)	Eqn (3.21)
α	z
βd_i	a_i

We have already encountered a version of McCullagh's general approach in the analysis of the data in Table 1.11, presented in Exercise 2.8 (Exercise 3.36). Model-fitting in GLIM is discussed by Ekholm and Palmgren (1989). Testing for a suitable form for $F(\ )$ is considered in Genter and Farewell (1985), using procedures to be described in Chapter 4. Variation in the $\{\alpha_i\}$ is modelled by Farewell (1982a). The case of the latent variable corresponding to time is considered in Chapter 5.

3.6   Bayesian analysis

A variety of Bayesian approaches have been proposed for the analysis of quantal assay data. We shall defer discussion of procedures for sequential analysis until Chapter 8. The approach of Ramsey (1972), developed further by Disch (1981), is to assume an ordered Dirichlet prior for the set of probabilities, $P(d_i)$, of response to dose d_i. Disch (1981) shows how to form posterior distributions for an effective dose, ED_{100p} and presents approximations to simplify the analysis (Exercise 3.35). There is an interesting link-up with isotonic regression, which will be discussed in section 7.2. This work may be regarded as non-parametric as it does not assume a parametric form for $P(d_i)$. A Bayesian procedure which does is described in Racine *et al.* (1986). This paper contains a number of novel features which are illustrated in the description which follows.

We shall consider the probit model, for which the likelihood is given by:

$$L(\alpha, \beta | \mathbf{r}) = \prod_{i=1}^{k} \binom{n_i}{r_i} \Phi(\alpha + \beta d_i)^{r_i} \{1 - \Phi(\alpha + \beta d_i)\}^{n_i - r_i}, \quad (3.22)$$

and we shall focus our interest on the $ED_{50} = -\alpha/\beta$.

We may have prior information on the values of α and β, possibly from previous experience with drugs similar to the one being tested. From using Bayes theorem we can then obtain the following posterior distribution for α and β:

$$f(\alpha, \beta | \mathbf{r}) \propto f(\alpha, \beta) L(\alpha, \beta | \mathbf{r})$$

where $f(\alpha, \beta)$ denotes the prior distribution for α and β. From an example involving five doses and 35 animals (see Exercise 3.28 for details) the posterior distribution of Figure 3.8 was obtained, resulting from taking $f(\alpha, \beta)$ as constant. This produces what is called

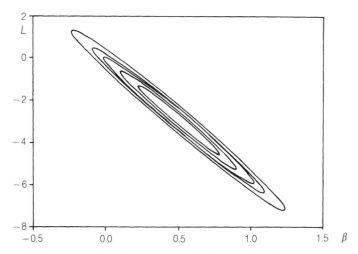

Figure 3.8 *Contours for the posterior distribution* $f(\alpha, \beta | \mathbf{r})$, *assuming a non-informative prior distribution, and the probit-based likelihood of (3.22), for the data of Exercise 3.28. Taken from Racine et al. (1986). Reproduced with permission from the Royal Statistical Society.*

an (improper) non-informative prior, as it gives equal weight to the entire (α, β) space.

For the $ED_{50} = \theta = -\alpha / \beta$, we require a posterior density function for θ: $f(\theta | \mathbf{r})$. From standard transformation-of-variable theory (Exercise 3.29), we obtain:

$$f(\theta | \mathbf{r}) = \int_0^\infty \beta f(-\theta \beta, \beta | \mathbf{r}) d\beta \qquad (3.23)$$

and this then allows us to allocate probabilities to toxicity classes, as specified in Chapter 1, as follows.

$$\Pr(\theta_- < \theta < \theta_+ | \mathbf{r}) = \int_{\theta_-}^{\theta_+} f(\theta | \mathbf{r}) d\theta \qquad (3.24)$$

Numerical integration is required to evaluate the integrals in expressions (3.23) and (3.24). For the data set in Exercise 3.30 the posterior densities for $\log(\theta)$ displayed in Figure 3.9 were obtained, which led to the probabilities given below for a set of toxicity classes (details of the informative prior analysis will be given shortly).

Toxicity class defined in terms of the ED$_{50}$ (measured in mg/ml)	<5	5–50	50–500	500–2000	2000–5000	>5000
Non-informative prior	0.00029	0.00023	0.00281	0.98738	0.00531	0.003098
Informative prior	0.00045	0.00039	0.00457	0.98419	0.00584	0.00456

In order to make use of prior knowledge we need to specify an appropriate informative prior. Two ways of proceeding are suggested by Racine *et al.* (1986).

1. The shape of the contours of Figure 3.8 suggests that we can approximate the likelihood by a bivariate normal distribution, with mean vector μ, and dispersion matrix Σ, which may be estimated from the data (Exercise 3.31). It is then natural to take a bivariate normal form for the prior $f(\alpha, \beta)$, with mean μ_0 and dispersion matrix Σ_0, say, as the posterior distribution will then

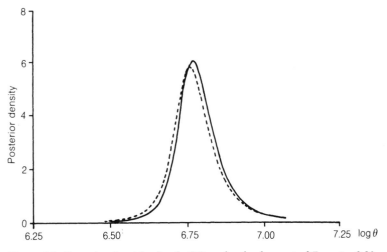

Figure 3.9 *Posterior densities for the* ED$_{50}$, *for the data set of Exercise 3.30, evaluated for the non-informative prior (———) and for the minimal informative prior (----), described below. Taken from Racine* et al. *(1986). Reproduced with permission from the Royal Statistical Society.*

also be bivariate normal, with mean vector:

$$\boldsymbol{\mu}^* = \boldsymbol{\Sigma}^*(\boldsymbol{\Sigma}_0^{-1}\boldsymbol{\mu}_0 + \boldsymbol{\Sigma}^{-1}\boldsymbol{\mu})$$

and dispersion matrix, $\boldsymbol{\Sigma}^* = (\boldsymbol{\Sigma}_0^{-1} + \boldsymbol{\Sigma}^{-1})^{-1}$

(Exercise 3.32). As long as the procedure of building prior beliefs into $\boldsymbol{\mu}_0$ and $\boldsymbol{\Sigma}_0$ can be successfully accomplished, then the above approach provides all that is required.

2. One way in which prior information can be obtained is through the views of an 'expert' experimenter. This expert may *guess* the probabilities π_1 and π_2 of response to some doses (or log doses), d_1 and d_2, and then this information may be used to estimate the parameters, (l_1, l_2, m_1, m_2) of assumed independent beta prior distributions for the pair of probabilities (π_1, π_2):

$$f(\pi_1, \pi_2) \propto \prod_{i=1}^{2} \pi_i^{l_i - 1}(1 - \pi_i)^{m_i - l_i - 1}, \text{ with, say, } \pi_1 < \pi_2$$

One approach is to use the guesses $\hat{\pi}_1$ and $\hat{\pi}_2$ to estimate the beta distribution modes, This gives:

$$l_i = 1 + \hat{\pi}_i(m_i - 2) \qquad (\text{for } l_i > 1 \text{ and } m_i > 2)$$

(Exercise 3.33). Clearly this does not completely specify (l_1, l_2, m_1, m_2) and we can finally arbitrarily decide on values for m_1 and m_2, which allows us to tune the system. In the 'minimalist' prior used above, $m_1 = m_2 = 3$ was taken, and $\hat{\pi}_i = r_i/n_i$. The interesting suggestion has been made that for 'in-house' work, larger values of $\{m_i, i = 1, 2\}$, corresponding to greater prior confidence, may be selected than for 'external' reporting.

Having proceeded as above, we now have a form for the $f(\pi_1, \pi_2)$ prior distribution, while what we need in order to continue the analysis is the prior: $f(\alpha, \beta)$.

Now in the probit model, we have

$$\pi_i = \Phi(\alpha + \beta d_i), \quad i = 1, 2 \tag{3.25}$$

and equations (3.25) simply define a transformation, from (π_1, π_2) to (α, β). Further standard analysis (Exercise 3.34) then results in:

$$f(\alpha, \beta) \propto \{(d_2 - d_1)\phi(\alpha + \beta d_1)\phi(\alpha + \beta d_2)\}$$
$$\times \prod_{i=1}^{2} \{\Phi(\alpha + \beta d_i)\}^{l_i - 1}\{1 - \Phi(\alpha + \beta d_i)\}^{m_i - l_i - 1}$$

The first term in this expression is proportional to a bivariate normal form, and the second term can be approximated by a bivariate normal distribution as before. Consequently the resulting prior distribution is taken to be bivariate normal. For further discussion, see Exercise 3.35.

Further Bayesian work can be found in Tsutakawa (1972, 1980) and in Govindarajulu (1988, Chapter 10), and the references there.

3.7 Synergy and antagonism

The experiment described in Example 1.11 involved testing two insecticides, A and B, separately and as a mixture, in a range of proportions. Of interest was whether performance was enhanced (synergy), or reduced (antagonism) when the substances were presented as a mixture. A preliminary comparison of the performance of the pure substances, A and B, based on the data of Table 1.14, does not reject the hypothesis of parallelism (section 2.9), and we shall use ρ to denote the potency of B relative to A. The work that follows is designed for the case of parallelism of the pure substances – cf. the data of Exercise 2.38. Terminology in the area of the analysis of mixtures of drugs is variable, and here we shall follow Giltinan et al. (1988). If a mixture of a units of A and b units of B results in a response which is equivalent to administering $(a + \rho b)$ units of A, then the two substances are said to exhibit **additive** behaviour, sometimes also described as **simple similar action**. Departures from additive behaviour represent either **synergy**, when the response to the mixture is greater than that to $(a + \rho b)$ units of A, or **antagonism**, when the response to the mixture is less than that to $(a + \rho b)$ units of A.

With reference now to the full data set of Table 1.14, suppose r_{ij} insects die, out of n_{ij} exposed to a_i units of A and b_j units of B. If we adopt a probit model, then r_{ij} has the binomial, Bin (n_{ij}, P_{ij}) model, and if additive behaviour is appropriate,

$$P_{ij} = \Phi\{\alpha + \beta \log (a_i + \rho b_j)\}$$

One of several ways of representing departure from additive behaviour is to set

$$P_{ij} = \Phi[\alpha + \beta \log \{a_i + \rho b_j + \gamma (a_i b_j)^{1/2}\}] \qquad (3.26)$$

Here, $\gamma > 0$ corresponds to synergy, and $\gamma < 0$ represents antagonism, and we now estimate γ by maximum-likelihood, and test whether $\gamma = 0$. For the data of Table 1.14, maximum-likelihood estimates are $\hat{\alpha} = -2.45\,(0.24)$ $\hat{\beta} = 1.06\,(0.09)$, $\hat{\rho} = 0.91\,(0.12)$ and $\hat{\gamma} = -0.98\,(0.15)$. The results suggest a significant amount of antagonism, which is verified also by a score test $(p < 0.001)$ (Appendix D). A significant score or likelihood-ratio test masks the relative contributions of the different mixtures to the significant result. The approach adopted by Giltinan *et al.* (1988) is to use graphical procedures to examine the relative contributions from different mixtures. Thus for the score test, for example, the test of $\gamma = 0$ is shown to be equivalent to testing for adding a variable to a weighted linear regression, and so use may then be made of the appropriate added variable plot (Atkinson, 1982). For further discussion, see Exercises 3.37 and 3.38. For the data of Table 1.14, the evidence of antagonism appeared to be uniform over the range of mixtures used. However, in the corresponding analysis of a non-quantal assay data set from Darby and Ellis (1976) the added variable plot was found to be a useful diagnostic, identifying one treatment as particularly influential in suggesting non-additivity.

The area of joint action of mixtures of drugs is a large one. An introduction to this work is provided by Hewlett and Plackett (1979, Chapter 7), Ashford (1981) and Abdelbasit and Plackett (1982).

3.8 Multivariate bioassays

In the data presented in Exercise 1.24, coal miners were classified not only by whether they wheezed or not, but also by whether they were breathless or not. The response is therefore **multivariate**, in contrast to **polytomous** as in section 3.5. Ashford and Sowden (1970) showed how the threshold model of section 1.5 may be generalized to account for a multivariate response. For a **bivariate** response they assumed an underlying bivariate normal threshold model with cumulative distribution function,

$$\Phi(\omega_1, \omega_2; \rho)$$

where ρ is the correlation coefficient. Both marginal distributions were supposed to be $N(0, 1)$. We can then specify

Pr(breathless and wheeze at age x) $= \Phi(\alpha_1 + \beta_1 x, \alpha_2 + \beta_2 x; \rho)$

Pr(breathless and no wheeze at age x) $= \Phi(\alpha_1 + \beta_1 x)$

$$- \Phi(\alpha_1 + \beta_1 x, \alpha_2 + \beta_2 x; \rho)$$

Pr(wheeze and no breathlessness at age x) $= \Phi(\alpha_2 + \beta_2 x)$

$$- \Phi(\alpha_1 + \beta_1 x, \alpha_2 + \beta_2 x; \rho)$$

Pr(neither symptom present at age x) $= 1 - \Phi(\alpha_1 + \beta_1 x)$

$$- \Phi(\alpha_2 + \beta_2 x)$$

$$+ \Phi(\alpha_1 + \beta_1 x, \alpha_2 + \beta_2 x; \rho).$$

All four possible outcomes are now described and numerical maximum likelihood estimation of the five parameters, $(\alpha_1, \beta_1, \alpha_2, \beta_2, \rho)$, can proceed in the usual way, though it is necessary to be able to evaluate bivariate normal integrals. For the coal miner data the maximum likelihood parameter estimates, together with asymptotic estimates of standard error, are:

$$\hat{\alpha}_1 = -3.611(0.061), \quad \hat{\beta}_1 = 0.0055(0.001),$$
$$\hat{\alpha}_2 = -2.454(0.045), \quad \hat{\beta}_2 = 0.037(0.00098),$$
$$\hat{\rho} = 0.771(0.0088)$$

The X^2 goodness-of-fit statistic is 20.8, clearly not significant when referred to χ^2_{22} tables. An analysis in higher dimensions would require the evaluation of higher dimensional normal integrals. Williams (1986c) suggested an approximate approach based on one- and two-dimensional marginal distributions (Anderson and Pemberton 1985). A computationally simpler approach is given by Grizzle (1971), in terms of multivariate logit analysis, obtaining modified minimum chi-square estimates by the method of weighted least squares, as described in Grizzle et al. (1969). There is a link-up between the multivariate threshold distribution of Ashford and Sowden (1970) and multifactorial models for describing the incidence of certain diseases in sets of relatives (Curnow and Smith 1975). An algorithm for calculating multivariate normal probabilities is given by Schervisch (1984, 1985).

Alternative analyses of the coal miner data are provided by Mantel and Brown (1973) and McCullagh and Nelder (1989, pp. 230–235). In both cases problems with how the miners were selected, and the consequent difficulty of extrapolating from this particular sample to a population, are emphasized (Exercise 1.24). Mantel and Brown

(1973) use a generalized logistic model (Cox and Snell, 1989, p. 155), while McCullagh and Nelder (1989, p. 235) argue in favour of using multivariate logit models when the multiple responses are to be treated symmetrically.

Multivariate bioassay data may result from repeated measures over **time** of the same groups of individuals. Such data are best analysed by particular methods, such as survival analysis, and we defer discussion of such approaches until Chapter 5. Key references to general (non-quantal) multivariate bioassay are the papers by Rao (1954), Armitage *et al.* (1976) and Volund (1980, 1982). We note, finally, that Williams (1986c) remarked that in his experience multivariate bioassay data are relatively rarely encountered in practice.

3.9 Errors in dose measurement

Dose levels provided in experiments are usually taken as fixed. However, this is not always the case. For example, Ridout and Fenlon (1991) described experiments in biological control, in which parasites are treated with substances such as viruses, bacteria or spores, as an alternative to the application of chemicals. With biological control, individual particles may be applied in a liquid suspension, and different treated individuals, treated at the same **nominal** dose, will then in fact probably receive different numbers of infecting particles. The statistical analysis is one of errors-in-variables for binary regression, and has been considered recently by Carroll *et al.* (1984) and Burr (1988). An early paper was by Patwary and Haley (1967), in which it was assumed that the dose received had a Poisson distribution.

We shall use the same notation as in section 2.2. We suppose that the doses $\{d_i, 1 \leqslant i \leqslant k\}$ are not fixed, but that the dose received by the jth individual at the nominal dose level d_i is denoted by d_{ij}, which is assumed to be normally distributed, with mean d_i, and variance σ^2. Ridout and Fenlon (1991) argue that there is evidence in some cases for adopting a log-normal distribution for the amount of dose actually administered at a particular nominal dose level. The normal distribution then results from taking a logarithmic transformation of the dose levels.

The probability of response to dose d_{ij} may be written as:

$$P(d_{ij}) = F(\alpha' + \beta' d_{ij})$$

for a suitable c.d.f. $F(\ )$. However d_{ij} is unknown, and so we derive the probability of response at the nominal dose level d_i very simply as:

$$P(d_i) = \frac{1}{\sigma\sqrt{2\pi}} \int_{-\infty}^{\infty} F(\alpha' + \beta'x)e^{-\frac{1}{2}\left(\frac{x-d_i}{\sigma}\right)^2} dx$$

At this point a computational advantage arises from taking $F(\ )$ to be the c.d.f. of a standard normal random variable, since (Exercise 3.39),

$$\frac{1}{\sigma\sqrt{2\pi}} \int_{-\infty}^{\infty} \Phi(\alpha' + \beta'x)e^{-\frac{1}{2}\left(\frac{x-d_i}{\sigma}\right)^2} dx = \Phi\left\{\frac{(\alpha' + \beta'd_i)}{(1 + (\beta')^2\sigma^2)^{1/2}}\right\} \quad (3.27)$$

Had the dose-response data been analysed without due account being taken of the variation in the administered doses, then standard probit analysis would result in the location-scale pair of parameters, (α, β). We now see from equation (3.27) that these parameters need to be corrected by the expressions:

$$\alpha' = \alpha c; \quad \beta' = \beta c$$

where $c^{-2} = (1 - \beta^2\sigma^2)$, as long as $\beta < \sigma^{-1}$. This does suppose that σ^2 is known. As one might expect, the point estimate of the ED_{50} is clearly unchanged by this correction.

Discussion in Ridout and Fenlon (1991) suggested that the above corrections are only important for large sample sizes, and that the breakdown of the model, which occurs when $\sigma > \beta^{-1}$, was unlikely to occur for the type of biological control assay of interest to them.

Example 3.7   (Ridout and Fenlon, 1991)

Neonate larvae of the diamond-backed moth, *Plutella. xylostella* were fed droplets containing the *P.x. granulosis* virus. Droplets were of a given size, and were taken from suspensions containing fixed quantities of the virus:

Dose (log_{10} (virus/ml))	Number of insects	Number killed
5.48	28	4
6.00	27	11
6.48	27	12
7.00	26	24

A separate study investigated the amount of feed ingested by *Plutella xylostella* larvae, using a fluorimetric technique, and resulted in an estimate of $\sigma = 0.2$ on the log-scale. If we pay no attention to dose variation, we obtain the estimates: $\hat{\alpha} = -8.81$ (1.626) and $\hat{\beta} = 1.403$ (0.2594). After correction, we obtain $\hat{\alpha}' = -9.18$ (2.130), $\hat{\beta}' = 1.462$ (0.3398). □

A different but related problem concerns the transformation of an administered dose into the effective dose which reaches the relevant target organ. One approach is provided by Van Ryzin and Rai (1987), who employ a deterministic compartment model to describe the transformation of the administered dose to an effective dose. The resulting model has already been presented in Exercise 2.33. They also supply several references to related work. The result of equation (3.27) is discussed by Copas (1983) in the context of bias evaluation following smoothing in non-parametric binary regression (section 7.6).

Errors may also result from incorrect assessment of the binary response of each tested individual, and a way of accounting for this is presented by Ekholm and Palmgren (1982) and applied to the wheeze data on coal-miners, given in Exercise 1.24. For discussion of the approach and conclusions, see Exercise 3.40.

3.10 Discussion

As we have seen, all of the developments considered in this chapter have been needed for the analysis of particular data sets. Depending on the context, some of the work of this chapter will undoubtedly be more useful than that of other sections; however, data with natural mortality present seem to be encountered particularly frequently. In many cases simple analyses may suffice, rather than far more complex procedures, which might for instance entail the numerical evaluation of higher-order multivariate normal integrals.

As mentioned in section 3.1, it is not unusual to find that particular applications require a combination of different extensions. Illustrations of this are provided by Ryan (1990), O'Hara Hines *et al.* (1992), Ridout and Fenlon (1990), Watson and Elashoff (1990) and Chen (1990), and additional examples will be encountered in later chapters. For instance, the logit model of section 3.5 may not be the best for any particular data set, and this has been investigated by

Genter and Farewell (1985), using procedures that will be described in the next chapter.

3.11   Exercises and complements

The exercises vary in complexity. It is particularly desirable to attempt the five exercises marked with a †. Exercises marked with a * are generally more difficult or speculative.

3.1[†]   Consider how a control group of individuals may be used to provide an estimate of natural mortality, λ, and then how this may be used in equation (3.1) to allow the standard maximum-likelihood fitting of $\tilde{P}(d)$. Illustrate your approach with the data of Table 3.1.

3.2   Consider whether the approach of Exercise 3.1 may be adapted to enable the model of equation (3.2) to be fitted to data. Fit the model to the data of Table 3.1 and compare the alternative descriptions of the data.

3.3[†]   Write down an extension of Abbott's formula to account for natural immunity. Consider how this approach may be useful for describing the data of Table 1.13.

*3.4**   Morgan (1986) analysed data on hypersensitivity reactions to a drug. In all there were 215 subjects, on each of which there were eight measured variables. The administration of the drug was spread over four centres, and so the first four variables were binary site indices. The remaining four variables were the recorded time (either the time until a reaction was displayed or a censored time, after which the subject was investigated no further), a binary variable indicating whether a reaction was exhibited or not, a sex variable and a dose variable. An illustration of the data was given in Table 1.9. A mixture model was adopted of the following form: we write the probability of an individual having time on test t, without exhibiting a reaction, as:

$$PS(t) + 1 - P \tag{3.27}$$

where P is the probability of ever exhibiting a reaction, and for individuals in this category, $S(t)$ is the **survivor** function, i.e. if **reactors** have a time T to the reaction from the administration of the drug, then

$$S(t) = \Pr(T > t).$$

Thus equation (3.27) is analogous to equation (3.1), but here we also consider times to response. If an individual is observed to react at time t, then the likelihood contribution is taken as $pf(t)$, where

$$f(t) = \frac{d}{dt}\{1 - S(t)\}.$$

If we denote the covariates measured on an individual by the vector x, then we may allow $S(t)$ and p to be functions of x. Farewell (1982b) used this model and adopted the Weibull form for $S(t)$; extensions are given by Larson and Dinse (1985), who used the EM algorithm, described in section 3.2.

The likelihood may now be written as:

$$L = \prod_{i \in R_0} \{P(x_i)S(t|x_i) + 1 - P(x_i)\} \prod_{j \in R_1} P(x_j)f(t|x_j)$$

where R_0 is the set which indexes the non-reactors, and R_1 is the set which indexes the reactors. The likelihood was maximized using a Nelder–Mead simplex algorithm (section 2.1).

A simple first analysis assumed that only P was a function of x, and that the Weibull distribution was parameterized as:

$$f(t) = (\lambda\delta)(\lambda t)^{\delta - 1} \exp\{-(\lambda t)^\delta\}$$
$$S(t) = \exp\{-(\lambda t)^\delta\}$$

for $t \geqslant 0$, with $\delta > 0$ and $\lambda > 0$

For simplicity a logit form was adopted for $P(x)$, viz.:

$$P(x) = \{1 + \exp - (\phi + \zeta' x)\}^{-1}$$

Variables were all **centred** before analysis by subtraction of their mean values. If we simply include sex and dose as covariates, and ignore possible site differences obtain the results below.

$\hat{\lambda}$	$\hat{\delta}$	$\hat{\zeta}_{sex}$	$\hat{\zeta}_{dose}$	$\hat{\phi}$	max. log likelihood
0.05	1.57	0.33	0.84	−2.48	−165.53
(0.007)	(0.23)	(0.45)	(0.38)	(0.68)	—

Discuss these results. How might you evaluate the goodness-of-fit of the model? The work of this exercise relates to the material of Chapter 5.

3.5* (Continuation of Exercise 3.4.) The analysis presented in Exercise 3.4 ignores site differences. If we include site variables, with ζ_i denoting the coefficient for a dummy variable indexing site i, we obtain:

$\hat{\lambda}$	$\hat{\delta}$	$\hat{\zeta}_1$	$\hat{\zeta}_2$	$\hat{\zeta}_3$	$\hat{\zeta}_{sex}$	$\hat{\zeta}_{dose}$	$\hat{\phi}$	Max. log likelihood
0.04	1.69	4.94	2.46	2.80	0.78	0.33	-5.96	-150.90
(0.007)	(0.25)	(1.44)	(1.34)	(1.14)	(0.59)	(0.35)	(1.40)	

A likelihood-ratio test, following a separate fit of the model without ζ_2, suggested that it needs to be retained in the model. However, the dose coefficient is **now no longer significant**. How can such a result arise?

3.6* The data below appear in a paper by Pierce et al. (1979) and summarize the daily mortality in groups of fish subjected to three levels of zinc concentration. Half the fish at each concentration level received one week's acclimatization. There were therefore six treatment groups, and 50 fish were randomized to each group.

Day	Log zinc concentration	Acclimatization time					
		One week			Two weeks		
		0.205	0.605	0.852	0.205	0.605	0.852
1		0	0	0	0	0	0
2		3	3	2	0	'1	3
3		12	17	22	13	21	24
4		11	16	15	8	8	10
5		3	5	7	0	5	4
6		0	1	1	0	0	1
7		0	0	2	0	0	0
8		0	1	0	0	0	0
9,10		0	0	0	0	0	0

Suggest appropriate initial plots of the data. In a follow-up paper,

Farewell (1982b) proposed the following mixture model. If x denotes the two-dimensional vector indicating acclimatization time and zinc concentration, we write the probability that a fish with covariate x is a 'short term' survivor as:

$$P(x) = (1 + e^{-\beta' x})^{-1}$$

and for such fish the time to death, measured from the start of the experiment, is determined by the Weibull distribution, with density function,

$$f(t|x) = \delta\lambda(\lambda t)^{\delta-1} \exp\{-(\lambda t)^\delta\}, \quad t \geqslant 0$$

where

$$\lambda = \exp(-\gamma' x)$$

This model therefore contains five parameters: $\zeta = (\delta, \gamma, \beta)$. 'Long term' survivors are assumed not to die during the course of the experiment.

Write down the likelihood for a general set of data, and indicate, in outline only, how you would proceed to obtain the maximum-likelihood estimate of ζ and associated measures of standard error.

Draw conclusions from the following estimates presented by Farewell for the above data: $\hat{\delta} = 3.47$.

Covariate	β	Estimated standard error	$\hat{\gamma}$	Estimated standard error
Acclimatization	-0.94	0.30	-0.13	0.04
Concentration	3.59	0.56	0.17	0.08

Discuss any further analyses which you would carry out. The work of this exercise involves time to response, a subject which is discussed fully in Chapter 5.

3.7 Provide suitable plots of the data of Table 3.4 and hence discuss which model provides the best description of the data.

3.8 Fit a model to the data of Table 3.6, based on the Poisson distribution. Does an ED_{50} have any meaning in the context of these data? Discuss whether over-dispersion needs to be incorporated in the model.

3.9 With reference to Example 3.4, the number of organisms in a sample has a Poisson distribution, with mean ζ. Let $p = \text{Pr}(\text{an organism dies})$. Show that the distribution of live organisms is Poisson, of mean, $\zeta(1 - p)$.

3.10* (Griffiths, 1977a,b). Consider how to model the data below, which provide the distribution of eggs laid by wasps on host eggs:

No. of eggs	No. of hosts
0	16
1	169
2	50
3	6
4	1
5	1
>5	0

3.11* (Morgan, 1982). Consider how to model the data below, which describe how many fertilizations have taken place in sea-urchin eggs, exposed to sperm for varying lengths of time.

	No. of fertilizations			
Time (seconds)	0	1	2	3
5	97	3		
8	83	16	1	
15	55	40	3	2
40	32	58	9	1
90	5	73	18	4

3.12 Verify the result of equation (3.14)

3.13 Verify the result of equation (3.15)

3.14 Explain why, in Table 3.8, $\sum_i h_{ii} = 2$.

3.15 Verify that equation (3.16) reduces to equation (3.17).

3.16 Let $\mathbf{M} = \mathbf{I} - \mathbf{H}$, and $\mathbf{U} = \mathbf{V}^{-1/2}\mathbf{s}$.
Show that $\mathbf{M}$ is symmetric, idempotent, and spans the residual ($\mathbf{U}$) space.

3.17 Verify that equation (3.19) reduces to:

$$c_l = \tilde{s}_l^2 \, h_{ll}^2.$$

3.18 Discuss the influence of the 31st observation in the data set of Table 2.2, in the light of Figure 3.3.

3.19 Investigate the use of standard multiple regression/linear discriminant analysis to describe the data of Table 2.2.

3.20 Extend the picture provided by the plots of Figure 3.5 by providing separate index plots of the standardized change in each of the components of $\hat{\boldsymbol{\beta}}$.

3.21* Discuss whether deletion of objects should always result in a reduced X^2 goodness-of-fit statistic.

3.22[†] For the data set of Table 3.9, we give below the influence diagnostics when the logit model is based on the dose levels, rather than their logarithms.

Case No. (i)	$\tilde{s}_i$	e_i	h_{ii}	$\Delta_i \chi^2$	$\Delta_i D$	c_i
1	0.667	0.658	0.530	0.947	0.935	1.067
2	−0.746	−0.750	0.394	0.917	0.924	0.595
3	−0.314	−0.311	0.318	0.144	0.143	0.067
4	1.280	1.429	0.361	2.566	2.968	1.450
5	−1.356	−1.183	0.263	2.494	2.055	0.891
6	0.624	0.881	0.134	0.450	0.836	0.070

Discuss these results and comment on the effect of changing the model by removing the logarithmic transformation of the dose. There is more discussion of this feature of logit modelling in Chapter 4.

3.23   The logit model is fitted to the data of Exercise 1.4, for which deviance residuals are plotted in Figure 3.2. Discuss the diagnostics shown below.

Case no (i)	h_{ii}	$\Delta_i\chi^2$	$\Delta_i D$	c_i
1	0.277	1.243	1.141	0.476
2	0.332	1.570	1.499	0.779
3	0.303	1.302	1.329	0.567
4	0.238	2.288	2.261	0.715
5	0.288	0.651	0.672	0.263
6	0.251	0.119	0.116	0.040
7	0.194	0.884	1.064	0.213
8	0.117	0.957	1.790	0.127

3.24   Williams (1987a) proposed the following set of residuals

$$g_i = \text{sign}(\tilde{s}_i) \sqrt{\{(1 - h_i)e_i + h_i\tilde{s}_i\}}$$

which were called **likelihood** residuals. He showed how the $\{g_i\}$ were related to outlier detection, and the measurement of case influence on likelihood ratio tests. Evaluate the $\{g_i\}$, $\{e_i\}$ and $\{\tilde{s}_i\}$ for the data of Exercise 1.4. Give an index plot for the three sets of residuals and discuss how they compare. Williams (1987a) showed how a half-normal plot of the $\{g_i\}$ can be enhanced by the addition of envelopes of such plots from a number of simulated data sets, each simulated from the model fitted to the original data.

3.25[†]   Consider ways in which data resulting from observational and experimental studies may differ. Compare, for example, the data of Tables 1.3, 1.4 and Exercise 2.23.

3.26   Use Figure 3.6 to investigate how single case deletion can influence likelihood-ratio confidence intervals for the ED_{50}.

3.27   Consider how you would present the results of Figure 3.7 in a scientific paper. What other plots might be useful? How would you indicate confidence regions?

3.28 The hypothetical data below were presented by Racine *et al.* (1986) and result in the posterior distribution shown in Figure 3.8

Dose (mg/ml)	No. of animals	No. of deaths
50	5	1
200	10	4
300	5	2
400	10	6
2000	5	4

Provide a non-Bayesian investigation of the data and compare your results with the picture presented in Figure 3.8.

3.29 Verify the form of equation (3.23).

3.30 The data below, presented by Racine *et al.* (1986), result from a routine inhalation acute toxicity test.

Dose (mg/ml)	No. of animals	No. of deaths
422	5	0
744	5	1
948	5	3
2069	5	5

Estimate the ED_{50} and compare your conclusions with those resulting from the posterior densities of Figure 3.9. Note here also the discussion in Ross (1986), Bailey and Gower (1986), Cox (1986) and Healy (1986).

3.31 Consider how to approximate the likelihood, $L(\alpha, \beta | \mathbf{r})$ in equation (3.22), by a bivariate normal distribution, $N(\boldsymbol{\mu}, \boldsymbol{\Sigma})$.

3.32 Verify that if the likelihood is of the bivariate normal form given in Exercise 3.31, and the prior density of (α, β) is bivariate normal, $N(\boldsymbol{\mu}_0, \boldsymbol{\Sigma}_0)$, then the posterior distribution willl also be bivariate normal,

$$N\{\boldsymbol{\Sigma}^*(\boldsymbol{\Sigma}_0^{-1}\boldsymbol{\mu}_0 + \boldsymbol{\Sigma}^{-1}\boldsymbol{\mu}), \boldsymbol{\Sigma}^*\}$$

where $\boldsymbol{\Sigma}^* = (\boldsymbol{\Sigma}_0^{-1} + \boldsymbol{\Sigma}^{-1})^{-1}$.

3.33 We saw in section 3.6 that modes of the beta distribution play an important role in Bayesian analysis. Verify that the beta distribution:

$$f(x) = \frac{x^{l-1}(1-x)^{m-l-1}}{B(l, m-l)}, \quad \text{with } l > 1, \, m < 2$$

has mode given by $x = \left(\dfrac{l-1}{m-2}\right)$

3.34 Starting from the bivariate distribution,

$$f(\pi_1, \pi_2) \propto \prod_{i=1}^{2} \pi_i^{l_i - 1}(1 - \pi_i)^{m_i - l_i - l}$$

verify the form given in section 3.6 for the bivariate distribution $f(\alpha, \beta)$, when we transform: $\pi_i = \Phi(\alpha + \beta d_i)$, for $i = 1, 2$, as in equation (3.25). Investigate a bivariate normal approximation to $f(\alpha, \beta)$.

3.35* (Disch, 1981). Denote the probability of response to dose d_i by $Q_i = P(d_i)$, $0 \leqslant i \leqslant m + 1$, where $Q_0 = 0$ and $Q_{m+1} = 1$. Discuss the use of the Dirichlet prior distribution:

$$f(Q_1, \ldots, Q_n) \propto \prod_{i=1}^{m+1} (Q_i - Q_{i-1})^{\alpha_i - 1}$$

with particular reference to the choice of the $\{\alpha_i\}$ and how this choice will affect degrees of prior belief. Write down the posterior distribution. Consider how to use the following results to derive marginal posterior distributions (s is a positive integer):

$$\int_0^a q^{b-1}(1-q)^s (a-q)^{c-1} dq = \sum_{j=0}^{s} (-)^j \binom{s}{j} B(b+j, c) a^{b+c+j-1}$$

$$\int_0^1 q^2 (1-q)^{b-1}(q-a)^{c-1} dq = \sum_{i=0}^{s} (-)^i \binom{s}{i} B(b+i, c)(1-a)^{b+c+i-1}$$

How might one form distributions for an ED_{100p} value?

3.36 Consider the relationship between the model of equation (3.21) and the model used in Exercise 2.8, to describe signal-detection data.

3.37 As in equation (2.20), we can write the multiple linear regression model as:

$$\mathbf{Y} = \mathbf{X}\boldsymbol{\beta} + \boldsymbol{\varepsilon}$$

Might we improve the fit by taking

$$\mathbf{Y} = \mathbf{X}\boldsymbol{\beta} + \phi\mathbf{Z} + \boldsymbol{\varepsilon}$$

for some vector $\mathbf{Z}$? We can test whether $\phi = 0$ by an F test in the usual way. However, we can use an **added variable plot** to check for points which may be exerting particular influence on the conclusion. Show that

$$(\mathbf{I} - \mathbf{H})\mathbf{Y} = \phi(\mathbf{I} - \mathbf{H})\mathbf{Z} + (\mathbf{I} - \mathbf{H})\boldsymbol{\varepsilon}$$

where $\mathbf{H}$ is the hat matrix of section 3.4. The added variable plot results when we plot $(\mathbf{I} - \mathbf{H})\mathbf{Y}$ versus $(\mathbf{I} - \mathbf{H})\mathbf{Z}$. Provide an interpretation of each of these vectors.

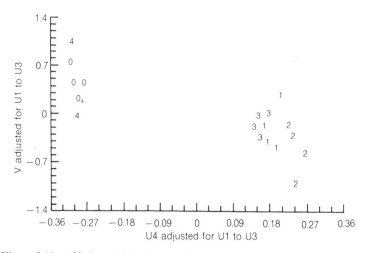

Figure 3.10 *Added variable plot for the data of Table 1.14. The symbols used represent mixture numbers, with higher numbers indicating greater proportions of insecticide B. Reproduced from Giltinan* et al. *(1988), with permission from the Royal Statistical Society.*

3.38   With reference to the model of equation (3.26), under the hypothesis that $\gamma = 0$, denote the maximum likelihood estimates of the parameters as $\hat{\alpha}, \hat{\beta}, \hat{\rho}$. We now define the following variables: $u_1 = 1$, $u_2 = \log(a + \hat{\rho}b)$, $u_3 = \hat{\beta}b/(a + \hat{\rho}b)$, $u_4 = \hat{\beta}(ab)^{1/2}/(a + \hat{\rho}b)$. These are respective derivatives of P_{ij} in equation (3.26), with respect to α, β, ρ and γ, evaluated when $\gamma = 0$. The score test statistic for $\gamma = 0$ is the test statistic for adding the vector of values u_4 to the weighted regression of v on (u_1, u_2, u_3), with weights w, where, in the notation of section 3.7,

$$\hat{\phi}_{ij} = \phi\{\hat{\alpha} + \hat{\beta}\log(a_i + \hat{\rho}b_j)\}$$
$$\hat{P}_{ij} = \Phi\{\hat{\alpha} + \hat{\beta}\log(a_i + \hat{\rho}b_i)\}$$
$$v_{ij} = r_{ij}/(n_{ij}\hat{\phi}_{ij}) \text{ and}$$
$$w_{ij} = \hat{\phi}_{ij}^2 n_{ij}/\{\hat{P}_{ij}(1 - \hat{P}_{ij})\}$$

See Giltinan *et al.* (1988). For the data of Table 1.14 the resulting added-variable plot is shown in Figure 3.10. Provide a suitable interpretation.

3.39[†]   Verify the relationship of equation (3.27), viz.

$$\frac{1}{\sigma\sqrt{2\pi}} \int_{-\infty}^{\infty} \Phi(\alpha' + \beta'x)e^{-(x - d_i/\sigma)^2/2}dx = \Phi\left\{\frac{(\alpha' + \beta'd_i)}{(1 + (\beta')^2\sigma^2)^{1/2}}\right\}$$

[Note: while this may be solved analytically, it is also profitable to approach the problem through using the idea of a tolerance distribution – see section 1.5.]

3.40   (Ekholm and Palmgren, 1982). It might be supposed that assessment of wheeze in coal-miners is subject to error. Introduce a latent binary response variable $\eta(x)$, so that $\eta(x) = 1$ corresponds to wheeze being present for a miner of age x, and $\eta(x) = 0$ otherwise. We may adopt a standard logit model for $\eta(x)$ but then add misdiagnosis through the two additional parameters:

$$P(\text{diagnosis is wheeze} \mid \eta(x) = 0) = \varepsilon_0$$
$$P(\text{diagnosis is no wheeze} \mid \eta(x) = 1) = \varepsilon_1$$

Indicate how you would form the likelihood under this model. Ekholm and Palmgren (1982) contrast direct maximum-likelihood estimation with an EM approach using GLIM. Their parameter estimates, together with estimated standard errors, are given below. What conclusions may be drawn? For similar work, which relates to the material of section 6.7.2, see Ridout and Morgan (1991), and Exercise 6.35. For further discussion and GLIM code, see Burn (1983). Copas (1988) presents a model for **resistant** fitting, in which a binary random variable Y might be misrecorded, with probability γ. Show how this results in the model:

$$\Pr(Y = 1) = (1 - \gamma)p + \gamma(1 - p)$$

for suitable γ and p. Compare this approach with that of Ekholm and Palmgren. Cf. also the material of Exercise 3.3. See also Palmgren and Ekholm (1987) and Palmgren (1987).

| Model | Residual deviance (d.f.) | Parameter | | | |
		ε_0	ε_1	α	β
Logit	8.20 (7)	–	–	−4.258 (0.0847)	0.0652 (0.00177)
Logit with mis-diagnosis	3.73 (5)	0.00053 (0.0168)	0.322 (0.131)	−4.2 (0.151)	0.0765 (0.0129)

3.41 Show that under the assumption of an exponential tolerance distribution, expressions (3.1) and the corresponding approach leading to equation (3.2) (under the assumption of a logit model), for incorporating natural mortality are identical. Discuss this result. An application is provided by Hoel and Yanagawa (1986)–see also Exercise 6.38.

CHAPTER 4

Extended models for quantal assay data

4.1 Introduction

The topic of this chapter is work that has taken place over the 20-year period since 1970 on extending standard models for quantal assay data. We shall review this work, explain the motivation for these models and make recommendations for those scientists and statisticians who may be interested in using extended models.

All of these models can be regarded as extensions of simple models such as the logit, probit and complementary log–log models, which may occur as particular cases of the extended new models. The idea of embedding simple two-parameter models within more complex models with three or four parameters is not a new one (Cox, 1961). This procedure is used also in other areas such as survival analysis (Aranda-Ordaz, 1983, Bennett, 1983, Ciampi *et al.*, 1982, Farewell and Prentice, 1977, and the material of section 5.4). This approach allows us to test the adequacy of simpler models through score or likelihood-ratio tests, as we shall see later. More sensitive goodness-of-fit tests then result for the simpler models than would arise from the usual chi-square goodness-of-fit test of section 2.5, because we are restricting the family of alternative models to the simpler model in question.

It is natural for statisticians to want to improve the fit of models to data if that is feasible. However, it is also important not to lose sight of the aim of modelling in the first place. If the objective is to make simple summaries of quantal assay data, using ED_{50} values for example, or to make simple comparisons, based on the assumption of parallelism as in section 2.9, then simple models, providing reasonable fits to data, may well be all that is necessary. In some cases, however, we may require more from our model, such as the estimation of an ED_{90} or an ED_{10}, for example. It has long been recognized that in certain safety regulatory contexts the estimation of extreme doses such as these may be far more relevant than estimation of an ED_{50}. (For example, Brown, 1967, and the FDA Advisory Committee on Protocols for Safety Evaluation, 1971). Unfortunately, precise estimation of extreme doses can be a difficult, if not impossible, task. In an ideal world, a number of pilot studies may be necessary in order to establish a range of doses to span the required ED_{100p} value. Alternatively a full-blown formal sequential approach may be adopted, as discussed in Chapter 8. In practice statisticians are often required to estimate extreme doses using a single standard set of quantal assay data, in which case the goodness-of-fit of models used for estimation and inference can be crucial. Typically the new extended models which have been proposed improve the fit of models to data for extreme doses, as they employ tolerance distributions with more flexible tails and of more flexible shape than those of the simpler two-parameter models. We shall see examples of this later in the chapter. If estimation of extreme doses involves **extrapolation** outside the range of doses employed in the experiment, then no amount of improvement of fit of the model within the range of doses in the experiment can remove the inevitable uncertainties associated with extrapolation. When extrapolation is unavoidable, then Prentice (1976b) has argued in favour of using a three- or four-parameter model, since the addition of extra parameters frequently increases imprecision in estimators. It is unusual to find statisticians arguing in favour of reducing precision, but we can see that in the case of extrapolation to extreme doses we would not want to convey a false impression of (relatively) too high precision.

As discussed in section 1.7, early work on modelling quantal assay data involved a variety of alternative two-parameter models, most of which are now no longer used. In a similar way, we are now presented with a rich family of alternatives for extending simple models. In this chapter we shall describe this recent work and suggest a strategy for deciding which approaches to adopt in practice. We shall start in the next section by considering a three-parameter asymmetric model.

4.2 The Aranda-Ordaz asymmetric model

The improvement of fit that may be obtained by using extended models can best be appreciated by illustrative examples. The first of several that we shall consider in this chapter is given in Example 4.1, and is typical of many that are often used for illustration of fitting extended models in that it results from an experimental study many years old.

Example 4.1

When we fit a logit model to the data of Table 3.9 using maximum likelihood we obtain the fit of Figure 4.1. This is a poor fit on a number of grounds: there is a clear pattern to the residuals, the fitted curve provides a poor description of the responses to the three largest doses, and the group sizes are sufficiently large that the chi-square goodness-of-fit statistic is significant at the 1% level: $X^2 = 13.38$ referred to χ_4^2 tables. The residuals suggest a possible systematic departure from the logit model, rather than **over-dispersion**, as cause of the poor fit. However, it is difficult to argue this convincingly on the basis of just six dose levels. The data therefore seem well suited

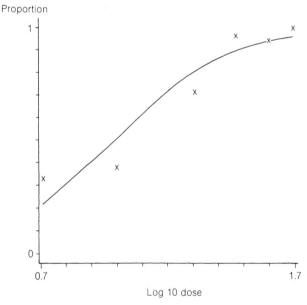

Figure 4.1 *The maximum likelihood fit of the logit model to the data of Table 3.9. Reproduced from Morgan* et al. *(1989) with permission from* Computer Methods and Programs in Biomedicine.

for experimenting with extended models. We return to this example in section 6.4, when we take the alternative approach of improving the fit by adding over-dispersion, i.e. by allowing the variance to exceed that predicted by the binomial model.

We saw from the influence statistics of Table 3.9 that if the first observation were omitted, the fit would be greatly improved. Finney (1947, p. 69) found that for a number of data sets from Martin (1942), the response to the lowest dose appeared atypical, if a symmetrical model is to be fitted. He surmised that this could be due to contamination of the deguelin and proceeded by omitting the lowest dose and the corresponding response. If contamination is present it is likely to affect low doses more than high doses. The general undesirability of doing this was acknowledged, but it was felt to be reasonable in this case as he was most interested in the fit at the highest doses. For further discussion, see Green (1984) and Lewis (1984).

Aranda-Ordaz (1981) has suggested two extended, three-parameter models for quantal assay data. One of these has, as probability of response to dose d,

$$P(d) = 1 - (1 + \lambda e^{\alpha + \beta d})^{-1/\lambda} \quad \text{for } \lambda e^{\alpha + \beta d} > -1$$
$$= 0, \quad \text{otherwise.}$$

This model was originally proposed by Darroch (1976), and again, independently, by Pregibon (1980). It has been used also by Aranda-Ordaz (1983) in the context of survival analysis. A random variable with cumulative distribution function $P(d)$ can be shown to have a log Burr distribution (Exercise 4.30).

The **link-function** (Appendix B) corresponding to $P(d)$ is given by

$$h(\mu) = \log_e \left\{ \frac{(1 - \mu/n)^{-\lambda} - 1}{\lambda} \right\}$$

where n is the index of the binomial distribution.

We see that this model contains the logit as a special case, when $\lambda = 1$, and we also obtain as a special case the complementary log–log model, as $\lambda \to 0$ (Exercise 4.1). Apart from the case $\lambda = 1$, these models have asymmetric tolerance distributions, with probability density function, $f(d) = P'(d)$ and we see from Figure 4.2 that both positively skewed and negatively skewed distributions are included in this family; see also Exercise 4.2.

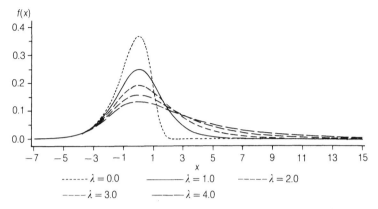

Figure 4.2 *Examples of the density functions for the Aranda–Ordaz tolerance distributions, with* $\alpha = 0$ *and* $\beta = 1$. *Reproduced from Morgan* et al. *(1989) with permission from* Computer Methods and Programs in Biomedicine.

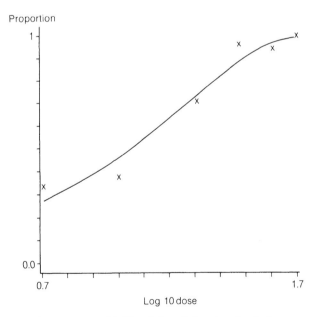

Figure 4.3 *The maximum likelihood fit of the Aranda–Ordaz asymmetric model to the data of Table 3.9. Reproduced from Morgan* et al. *(1989) with permission from* Computer Methods and Programs in Biomedicine.

When maximum-likelihood is used to fit this model to the data of Table 3.9 we obtain the fit of Figure 4.3. Details of the fit are given in Table 4.1.

The fit of Figure 4.3 is clearly superior to that of Figure 4.1, and the goodness-of-fit test statistic is now acceptable. There is now a slightly better pattern to the residuals, and the point made above regarding better estimation of extreme doses resulting from a better-fitting model is clearly seen from comparing Figures 4.1 and 4.3.

The correlation values in Table 4.1 are high in absolute terms, and suggest that an iterative procedure to obtain maximum-likelihood estimates, using the (α, β, λ) parameterization, may have difficulty converging. The approach adopted in Morgan and Pack (1988) was to use the minimum-chi-square procedure to estimate α and β for each of a range of λ values. The best (α, β, λ) triple, producing the lowest overall minimum, was then taken as the starting point for the maximum-likelihood search. Checks that the parameters are within the valid range, i.e $\lambda e^{\alpha + \beta d} > -1$ for d in the dose range were made at regular intervals. The first- and second-order derivatives required for the Newton–Raphson approach are required in Exercise 2.13.

When the logit model is fitted to the data of Table 3.9, the maximum log-likelihood is -119.09, and we can perform a likelihood-ratio test of the hypothesis that $\lambda = 1$, i.e. we test that it is not necessary to extend the logit model within the Aranda-Ordaz asymmetric family. The test statistic for this example obtains the

Table 4.1 *The maximum likelihood fit of the Aranda-Ordaz asymmetric model to the data of Table 3.9. Also given is the estimated asymptotic matrix and standard errors (on the diagonal)*

Parameter	Estimate		Estimated errors (in parentheses) and correlations		
			$\hat{\alpha}$	$\hat{\beta}$	$\hat{\lambda}$
α	-2.640	$\hat{\alpha}$	(0.510)		
β	2.027	$\hat{\beta}$	-0.970	(0.499)	
λ	-0.408	$\hat{\lambda}$	-0.786	0.906	(0.234)

$X^2 = 6.49$ ($p \approx 10\%$, referred to the χ^2_3 tables); deviance $= 7.05$; maximum log-likelihood $= -114.91$.

value of $2(119.09-114.91) = 8.36$, which is significant at the 0.5% level. The likelihood-ratio test requires us to fit both the logit and the Aranda-Ordaz models. An alternative is to perform a score test instead (Buse, 1982) which only requires the fitting of the simpler, logit model. (See Exercise 4.3 and Appendix D for details.) Also referred to χ_1^2 tables, the score test provides a statistic of 10.28, leading us clearly to the same conclusion regarding these data.

So at this stage we conclude our discussion of Example 4.1 with an extended, three-parameter asymmetric model providing a satisfactory fit to the data, a significant improvement over the logit model, and using all of the data. However, as we shall see in section 4.4, there is far more that can be done with these data, but we shall first consider another family of extended models, resulting in **Quantit** Analysis.

4.3 Extended symmetric models

We now consider the extended symmetric model proposed by Copenhaver and Mielke (1977). This is called the quantit model, based on a distribution called the omega distribution.

For the quantit model the link function is given in terms of an

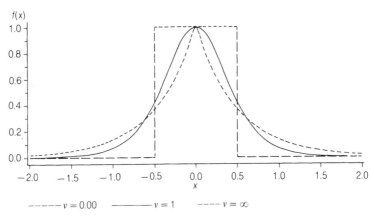

Figure 4.4 *Illustrations of the omega distribution density function for a range of values of the shape parameter v. Reproduced from Morgan* et al. *(1989) with permission from* Computer Methods and Programs in Biomedicine.

integral:

$$h(\mu) = \int_{0.5}^{\mu/n} \frac{dz}{1 - |2z - 1|^{v+1}}, \quad \text{for } 0 \leqslant \mu \leqslant n, \quad v > -1$$

where n is the index of the binomial distribution, and v is the shape parameter. Whatever the value of v, the omega distribution is symmetric (Exercise 4.7).

When $v = 1$ we obtain the logistic distribution, with cumulative distribution function,

$$F(x) = \frac{1}{(1 + e^{-4x})}, \quad -\infty \leqslant x \leqslant \infty$$

when $v = 0$ we obtain the double-exponential form:

$$f(x) = F'(x) = e^{-2|x|}, \quad \infty \leqslant x \leqslant \infty$$

and as $v \to \infty$ we obtain the uniform distribution, with

$$f(x) = 1 \quad \text{for } -0.5 < x < 0.5$$
$$= 0 \quad \text{otherwise.}$$

The derivations of these limiting forms are required in Exercise 4.10. These and other examples are illustrated in Figure 4.4.

Standard manipulations allow us to write

$$h(\mu) = \frac{\delta}{2} \sum_{i=0}^{\infty} \frac{|2p - 1|^{i(v+1)+1}}{i(v+1) + 1}, \quad \text{where } p = \mu/n, \text{ and where}$$

$$\delta = \begin{cases} +1 & \text{if } p > \frac{1}{2} \\ 0 & \text{if } p = \frac{1}{2} \\ -1 & \text{if } p < \frac{1}{2} \end{cases}$$

and this expression may be used for computation when combined with a suitable truncation rule. Unfortunately there is no closed form for $P(d)$, and this is evaluated using a Newton–Raphson procedure nested within the global iteration for (α, β, v). (Exercise 4.11.)

When we consider the examples of the omega distribution presented in Figure 4.4, we can see that there is a relatively large change in shape between $v = -\frac{1}{2}$ and $v = 4$, but a relatively small change in shape between $v = 4$ and $v = \infty$. This can give rise to convergence problems, for instance with the likelihood being very flat with respect to v for large v. Reparameterization in terms of $\omega = \log(1 + v)$ has not provided any great improvement, but other possibilities, such as $\omega = v^{-1}$ may prove useful.

Box and Tiao (1973, pp. 156–159) consider and illustrate the three-parameter exponential power distribution with probability density function given as:

$$f(x) = k\phi^{-1} \exp\left(-\frac{1}{2}\left| \frac{x - \zeta}{\phi} \right|^{2/(1+\beta)} \right), \quad -\infty < x < \infty$$

where

$$k^{-1} = \Gamma\left(1 + \frac{1 + \beta}{2} \right) 2^{1 + (1 + \beta)/2}$$

and the three parameters have ranges:

$$\phi > 0, \ -\infty < \zeta < \infty, \ -1 < \beta \leqslant 1.$$

It is interesting to note that this distribution also contains the uniform and double exponential as special cases, as well as the normal (Exercise 4.8). Although it is possible to write down the probability density function, the cumulative distribution function has to be evaluated numerically, and this distribution does not appear to have been used for the analysis of quantal assay data. In contrast, quantit analysis is now widely used (e.g. Hubert, 1992, p. 119).

We can think of v as the parameter determining the kurtosis of the omega distribution. The same role is played by the parameter β of the exponential power distribution. However, in that case β ranges more evenly over the range of distributions ($\beta \to -1$ gives uniform; $\beta = 0$ gives normal; $\beta = 1$ gives double exponential: Exercise 4.8).

Because of the complexities of quantit analysis, both with regard to the evaluation of $h(\mu)$ and $P(d)$, and possible convergence problems, Morgan (1985) introduced a much simpler cubic-logistic model as an approximation to the omega distribution (Exercise 2.5). For the cubic-logistic model,

$$P(x) = (1 + e^{-\{\beta(x - \theta) + \gamma(x - \theta)^3\}})^{-1}$$

so that the parameter θ is the ED_{50}. A similar approach was employed by van Montfort and Otten (1976), who used simple regression to fit their model, to obtain good starting values for fitting

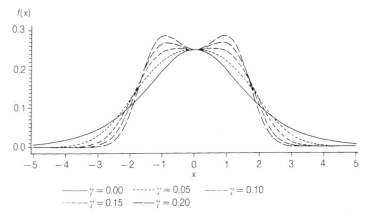

Figure 4.5 *Illustrations of the cubic-logistic density functions, with $\theta = 0$ and $\beta = 1$. Reproduced from Morgan* et al. *(1989) with permission from* Computer Methods and Programs in Biomedicine.

a model based on the λ-distribution, which is considered later. The cubic-logistic model was also used in a simulation study by Wu (1985). When $\gamma = 0$ we obtain the logit form. Examples of the corresponding probability density functions are given in Figure 4.5.

The cubic-logistic model above requires $\gamma \geqslant 0$ in order that $P(x)$ is a monotonic increasing function of x. Morgan (1985) found that the case $\gamma < 0$ for the cubic-logistic model was useful for data sets which required $v < 1$ in the quantit model. As we can see from Figure 4.6(a), although $P(x)$ may not be monotonic increasing in such a case, it may be acceptable as a description of responses within the range of doses available. However, it would be of little use for extrapolation, and extreme dose levels, such as the ED_{99}, either do not exist or are not unique.

Copenhaver and Mielke (1977) applied quantit analysis to 22 published data sets, 10 of which are listed in Exercise 1.3. For only six did they find $v < 1$, resulting in heavy tails and one of these cases corresponded to $v = 0.99$. In most cases, light-tailed forms of the omega distribution, with $v > 2$, were found to be useful. Heavy-tailed forms with $-1 < v < 0$ were needed for only three data sets. We therefore anticipate that the cubic-logistic model will be most useful when $\gamma \geqslant 0$.

Copenhaver and Mielke (1977) fitted the quantit model using maximum-likelihood, and optimization nested with respect to v: for

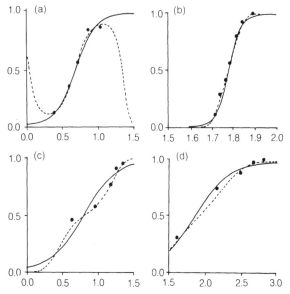

Figure 4.6 *Logit and cubic-logistic models fitted to data sets 4 (resulting in (a)), 9(b), 11(c) and 14(d) of Copenhaver and Mielke (1977), reproduced in Exercise 1.3. For corresponding* ED_{50} *estimates, see Table 4.4. Only for case (a) do we find* $\hat{v} < 1$ *for quantit analysis. Reproduced from Morgan (1985), with permission from the Royal Statistical Society.*

each of a range of fixed values for v, the likelihood is maximized with respect to all of the other parameters, and then v is estimated by the value which provides the largest maximum likelihood. It was shown by Morgan (1983) how direct three-parameter optimization could take place, using the method-of-scoring. For computational details, see Exercise 4.12. Initial estimates of (α, β, v) can be obtained from setting $v = 1$ and using minimum logit chi-square estimates of α and β, or directly by using the method of minimum quantit chi-square, which proceeds as follows.

Quantits, $h_v(q)$, are defined as follows:

$$h_v(q) = \int_{0.5}^{q} (1 - |2z - 1|^{v+1})^{-1} \, dz.$$

If a proportion, $p_i = r_i/n_i$ of individuals respond to a dose x_i, then equating observed and theoretical quantits gives

$$\alpha + \beta x_i = h_v(p_i)$$

By analogy with minimum logit chi-square estimation (section 2.6), for any fixed v we can obtain explicit minimum quantit chi-square estimates of α and β by simply minimizing with respect to α and β the weighted sum-of-squares,

$$S_v(\alpha, \beta) = \sum_{i=1}^{k} w_i \{\alpha + \beta x_i - h_v(p_i)\}^2$$

where $\omega_i^{-1} = V\{h_v(p_i)\}$.

We can approximate $V\{h_v(p_i)\} \approx V(p_i)\{h_v'(p_i)\}^2$ (Appendix A) and so estimate $V\{h_v(h_i)\}$ by

$$\frac{p_i(1 - p_i)}{n_i}(1 - |2p_i - 1|^{v+1})^{-2}.$$

Here we benefit from the integral form for $h_v(x)$, which is ideal for the formation of $h_v'(x)$ required here. It is the specification of quantit analysis via the inverse of the cumulative distribution function $F^{-1}(x)$, or equivalently, $h_v(x)$, which helps to make this method of parameter estimation, which requires $F^{-1}(x)$, so much simpler than maximum-likelihood. The overall minimum is obtained by nested minimization with respect to v. However, if the objective of minimum

Table 4.2 *The results of quantit analysis of data sets 4, 10, 11 and 14 from Copenhaver and Mielke (1977), using maximum-likelihood; estimated asymptotic standard errors are given in parentheses*

Data set	Maximum likelihood parameter estimates			Estimated asymptotic correlation matrix of estimators		
	$\hat{\alpha}$	$\hat{\beta}$	$\hat{v}$			
					$\hat{\beta}$	$\hat{v}$
4	−1.42	2.07	0.37	$\hat{\alpha}$	−0.999	0.977
	(0.96)	(1.41)	(1.91)	$\hat{\beta}$		−0.979
					$\hat{\beta}$	$\hat{v}$
10	−0.81	0.81	9.88	$\hat{\alpha}$	−0.969	0.532
	(0.12)	(0.10)	(10.02)	$\hat{\beta}$		−0.662
					$\hat{\beta}$	$\hat{v}$
11	−0.68	0.83	9.86	$\hat{\alpha}$	−0.938	0.495
	(0.08)	(0.09)	(17.33)	$\hat{\beta}$		−0.685
					$\hat{\beta}$	$\hat{v}$
14	−1.32	0.69	12.99	$\hat{\alpha}$	−0.983	0.481
	(0.12)	(0.05)	(8.40)	$\hat{\beta}$		−0.579

quantit chi-square analysis is just to obtain starting values for a maximum-likelihood analysis, then the nested optimization need only be done in a cursory manner. Examples of the results of maximum likelihood fits of the quantit model to data sets 4, 10, 11 and 14 of Copenhaver and Mielke (1977) are given in Table 4.2.

Estimates of the ED_{100p} values, θ_p, are given by

$$\hat{\theta}_p = \left\{ h_v \left(\frac{p}{100} \right) - \hat{\alpha} \right\} \bigg/ \hat{\beta}$$

Thus for example, $ED_{50} = -\hat{\alpha}/\hat{\beta}$, irrespective of v, as we would expect. We estimate $V(\hat{\theta}_p) \approx \mathbf{a'Sa}$ (Appendix A) where $\mathbf{a'}$ is the estimate of

$$\left(\frac{\partial \theta_p}{\partial \alpha}, \frac{\partial \theta_p}{\partial \beta}, \frac{\partial \theta_p}{\partial v} \right)$$

and $\mathbf{S}$ is the estimate of the variance–covariance matrix of $(\hat{\alpha}, \hat{\beta}, \hat{v})$. Here we find, after some algebra, that

$$\frac{\partial \theta_p}{\partial \alpha} = -\frac{1}{\beta}$$

$$\frac{\partial \theta_p}{\partial \beta} = \left\{ \frac{\alpha - h_v(p)}{\beta^2} \right\}$$

and

$$\frac{\partial \theta_p}{\partial v} = \int_{0.5}^{p} \frac{|2z - 1|^{v+1}}{(1 - |2z - 1|^{v+1})^2} \log(|2z - 1|)\, dz$$

See Exercise 4.13 for the evaluation of this integral.

Returning now to the cubic-logistic model, the details for the method-of-scoring are given in the solution to Exercise 2.14. For the case $\gamma > 0$, if we require an ED_{100p} dose level, then we can write

$$-\log_e(p^{-1} - 1) = \beta(\theta_p - \theta) + \gamma(\theta_p - \theta)^3$$

and if we set $\eta_p = \log_e(p/(1 - p))$, $0 < p < 1$, it is readily shown that as $\gamma > 0$,

$$\theta_p = \theta + y - \beta/(3\gamma y)$$

where

$$y^3 = \left(\frac{\eta_p}{2\gamma} \right) \left\{ 1 + \left(1 + \frac{4\beta^3}{27\gamma\eta_p^2} \right)^{1/2} \right\}$$

In order to estimate the precision of $\hat{\theta}_p$, we again use

$$V(\hat{\theta}_p) \approx \mathbf{a}'\mathbf{S}\mathbf{a}$$

where $\mathbf{S}$ is now the estimate of the variance/covariance matrix of $(\hat{\gamma}, \hat{\beta}\,\hat{\theta})$, and $\mathbf{a}'$ is the estimate of

$$\left(\frac{\partial \theta_p}{\partial \gamma}, \frac{\partial \theta_p}{\partial \beta}, \frac{\partial \theta_p}{\partial \theta} \right)$$

where

$$\frac{\partial \theta_p}{\partial \gamma} = -\frac{\beta}{3\gamma\gamma^2} + \frac{\partial y}{\partial \gamma}\left(1 + \frac{\beta}{3\gamma y^2} \right)$$

$$\frac{\partial \theta_p}{\partial \beta} = -\frac{1}{3\gamma y} + \frac{\partial y}{\partial \beta}\left(1 + \frac{\beta}{3\gamma y^2} \right)$$

$$\frac{\partial \theta_p}{\partial \theta} = 1$$

and

$$\frac{\partial y}{\partial \gamma} = -\frac{y}{3\gamma} - \frac{\beta^3}{81\gamma^3 y^2 \eta_p}\left(\frac{2\gamma y^3}{\eta_p} - 1 \right)^{-1}$$

$$\frac{\partial y}{\partial \beta} = \frac{\beta^2}{27\gamma^2 \eta_p y^2}\left\{ \frac{2\gamma y^3}{\eta_p} - 1 \right\}^{-1}$$

In Table 4.3 we compare the fits of the logit, quantit and cubic-logistic models to eight of the data sets of Exercise 1.3. In Table 4.4 we present a comparison of the ED_{50} and ED_{99} values under the three models.

From likelihood ratio tests based on the results of Table 4.3, we can see that the cubic-logistic model provides a significantly better fit than the logit model for just the same four data sets as for the quantit model. The similarity between the fits of the quantit and cubic-logistic models is seen further from the ED_{50} and ED_{99} values of Table 4.4. Point estimates of ED_{50} values do not vary greatly between models. However, with the improvement of fit to tails, resulting from using extended models, quite large disparities arise when we consider the ED_{99} values. Note that we use these data sets solely as a means of comparing the performance of the three models. From a practical viewpoint, any discussion of the ED_{99} in the context of published data sets is unrealistic unless the experiments were designed with the aim of estimating an extreme dose level. Thus the

Table 4.3 *Maximum log likelihood values from fitting three models to eight sets of quantal assay data. The number of the data sets is that used by Copenhaver and Mielke (1977), and the data are presented in Exercise 1.3. The figures given below do not include the redundant constant terms resulting from binomial coefficients. Also listed are score test statistics for testing the logit model versus the cubic-logistic model, and which are asymptotically χ_1^2 if $\gamma = 0$*

| | Model | | | |
Data set	Logit	Quantit	Cubic Logistic	Score test
4	119.86	119.79	119.60[b]	0.50
7	71.78	71.74	71.74	0.07
9	89.92	89.19	89.15	1.22
[a]10	119.09	116.14	115.53	6.21
[a]11	129.84	127.96	127.22	4.59
[a]14	158.36	155.51	156.04	3.89
[a]18	86.93	84.56	84.17	3.06
22	29.83	29.47	29.59	0.34

[a]indicates a significant improvement (at the 5% level) in the fit of both quantit and the cubic logit model over the fit of the logistic model.
[b]The fit here is obtained for $\hat{\gamma} = -20.04$, $\hat{\beta} = 8.53$, $\hat{\theta} = 0.68$. The resulting non-monotonic curve is illustrated in Figure 4.6(a).

'evil of extrapolation' mentioned by Finney (1971, p. 210) is present for five out of the eight data sets considered above, with the estimated ED_{99} lying outside the range of doses employed in the experiment. From Table 4.4 we see that the estimated standard errors for the ED_{50} values are very similar for all three models. For the case of the ED_{99} values, however, quite substantial differences are evident.

For data set 14, $\hat{\nu} = 12.99$, and we would expect the fitted quantit model to be similar to that of a uniform distribution. We can see from Figure 4.6(d) that in this case the fit of the cubic-logistic model is of this qualitative form, similar to one which would result from using a uniform tolerance distribution. For data set 11, however, $\hat{\nu} = 9.86$, again suggesting that a uniform distribution might be appropriate, but in this case the cubic-logistic model provides a quite different type of fit. In such cases care would undoubtedly need to be exercised in using the cubic-logistic model. But it remains a useful and simple substitute for the quantit model, in demonstrating an improvement over the logit model.

Table 4.4 *A comparison of estimates of* ED_{50} *and* ED_{99} *values for the three different models, for the selected eight data sets. Also given are estimated standard errors, using the expected information matrices. As explained by Morgan (1983), the errors given by Copenhaver and Mielke (1977) for the quantit model* ED_{99} *case are less than those given here as they did not account for the variability of* v. *The* ED_{99} *estimates obtained here under the quantit model sometimes differ slightly from those given by Copenhaver and Mielke, as they were obtained by the simultaneous estimation approach of Morgan (1983)*

Data set	ED_{50} values Model		
	Logit	Quantit	Cubic-logistic
4	0.68 (0.02)	0.68 (0.02)	0.68 (0.02)[a]
7	2.00 (0.18)	2.01 (0.17)	2.01 (0.17)
9	1.77 (0.01)	1.77 (0.01)	1.77 (0.01)
10	1.00 (0.04)	1.00 (0.04)	1.00 (0.04)
11	0.80 (0.04)	0.81 (0.03)	0.82 (0.03)
14	1.84 (0.04)	1.90 (0.04)	1.89 (0.04)
18	3.98 (0.16)	3.92 (0.16)	3.85 (0.14)
22	0.70 (0.02)	0.69 (0.03)	0.69 (0.03)

Data set	ED_{99} values Model		
	Logit	Quantit	Cubic-logistic
4	1.33 (0.08)	1.41 (0.31)	[a]
7	5.95 (0.64)	5.52 (1.56)	5.24 (1.47)
9	1.91 (0.02)	1.88 (0.02)	1.87 (0.01)
10	2.03 (0.11)	1.69 (0.08)	1.67 (0.06)
11	1.89 (0.12)	1.49 (0.15)	1.46 (0.06)
14	2.95 (0.08)	2.68 (0.04)	2.73 (0.06)
18	7.75 (0.61)	6.16 (0.45)	6.05 (0.39)
22	1.05 (0.10)	0.89 (0.08)	0.92 (0.09)

[a]Here the $\hat{\gamma} < 0 -$ see text.

We saw in section 4.2 that the data of Table 3.9 were well-fitted by the Aranda-Ordaz asymmetric model. The score test for the cubic-logistic model applied to these data is 6.21 and the cubic-logistic fit is shown in Figure 4.7.

The goodness-of-fit statistic is $\chi_3^2 = 8.03$, implying that this fit is also unsatisfactory.

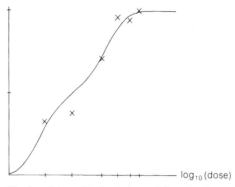

Figure 4.7 *The fit of the cubic-logistic model to the data of Table 3.9.*

To summarize the material of this section, the quantit model provides a flexible family of symmetric extended models for quantal assay data. It can prove difficult to fit when parameterized in terms of v. A convenient alternative symmetric extended model is the cubic-logistic which is much simpler to fit and, empirical evidence suggests, is a useful proxy for the quantit model with regard to detecting improvement to the logit fit and, to a certain extent, in estimating ED_{100p} values and their errors. A model based on the exponential power distribution is worth considering, but this has not yet been done. As we have seen, care needs to be taken with the cubic-logistic model as it may overfit the data (Exercise 4.14). We note finally that there are situations when the types of fit shown in Figures 4.6(a) and (c) could correspond to expected behaviour. Finney (1971, p. 265) presents data resulting from a mixture of substances, exhibiting behaviour that is an extreme form of the curve shown in Figure 4.6(c). More recently Simpson and Margolin (1986) discuss procedures for quantal assay data involving toxicity of the substance tested at high doses and a resulting non-monotonic dose-response relationship, as results for the high doses in Figure 4.6(a).

4.4   Transforming the dose scale

The dose scale in the data of Table 3.9 has been transformed by a $\log_{10}$ transformation. As we have seen in Chapter 1, such transformation of the dose scale may improve the fit of a logit model. In terms of tolerance distributions, a small fraction of individuals may have much higher tolerances than usual, giving rise to positively

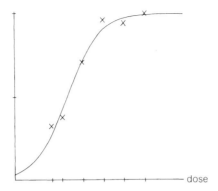

Figure 4.8 *The fit of the logistic model to the data of Table 3.9, but fitting to doses rather than* $\log_{10}$ *(dose) values.*

skewed tolerance distributions. It is in such cases that we might expect the logarithmic transformation to be useful. However, it is a mistake to suppose that a logarithmic transformation is always advantageous. If we return to the data of Table 3.9, and fit a logistic model to the **doses**, rather than the $\log_{10}$ (dose) values, we obtain the fit of Figure 4.8. The goodness-of-fit statistic is now $\chi_4^2 = 4.967$ and the score test statistics are 0.07, for the Aranda-Ordaz model, and 0.00 for the cubic-logistic model. In this example, therefore, if we do not take logarithms of the doses then there is no need to discard points, or to fit extended models. Another example of this kind is given in Example 4.2.

Example 4.2

Bliss (1935) presented the data given in Exercise 1.4, which describes the mortality of adult beetles after 5 hours' exposure to gaseous carbon disulphide. Usually following a logarithmic dose transformation, this example has been frequently used to illustrate the performance of asymmetric models – e.g. Prentice (1976b), Aranda-Ordaz (1981), Looney (1983), Pregibon (1980), Dobson (1990, p. 111) and Genter and Farewell (1985). When we fit the logit model to the logarithms of the doses, we obtain the score-statistics of: Aranda-Ordaz 7.60; cubic-logistic: 1.25. If we fit the logit model to the untransformed doses, we obtain score-statistics of: Aranda-Ordaz: 4.17, and cubic-logistic: 1.11. There therefore remains an

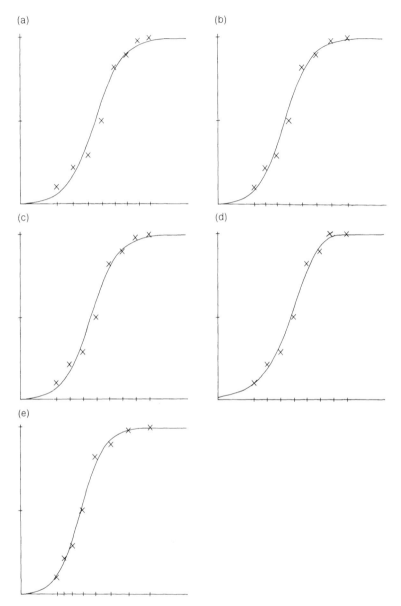

Figure 4.9 *Models fitted to the Bliss (1935) data:*
(a) the logit model fitted to the untransformed doses;
(b) the Aranda-Ordaz asymmetric model fitted to the untransformed doses;
(c) the logit model fitted to $\log_e$ (dose);
(d) the logit model fitted to $dose^2$;
(e) the logit model fitted to $dose^4$

improvement of the Aranda-Ordaz model over the logit. However, if we fit the logit model to the squares of the doses, we obtain the respective score statistics of 1.83 and 0.68. In this example, the logit fit to the logarithms of the doses was not, in fact, rejected by the chi-square test: $\chi_6^2 = 10.02$; fitted to the doses themselves, the logit model gives $\chi_6^2 = 6.72$. Figure 4.9 displays the logit and Aranda-Ordaz fits to the data on the untransformed dose scale, as well as the logit fit to the log (dose), dose^2 and dose^4. The improvement in fit, for instance between the logit fit to the log (dose) and dose^2, is quite apparent. For fitting to the dose^2, we obtain the chi-squared goodness-of-fit statistic of $\chi_6^2 = 4.48$.

The three dose transformations above are all particular instances of the general Box–Cox family of transformations, and Egger (1979) and Guerrero and Johnson (1982) have applied this family to the dose scale. The idea for this approach can be found in an exercise in Cox (1970, p. 110) (cf. Exercise 4.19), and it was also suggested in van Montfort and Otten (1976) and Prentice (1976b). In the Box–Cox transformation, doses $\{d_i\}$ are transformed to:

$$z_i = (d_i^\lambda - 1)/\lambda, \quad \lambda \neq 0$$
$$= \log_e (d_i), \qquad \lambda = 0$$

This transformation is sometimes referred to as the Box–Tidwell transformation, since we are transforming an explanatory, rather than a response, variable (Box and Tidwell, 1962). As an alternative, Kay and Little (1987) fit logit models to $\log(x)$, $\log(1-x)$ and $\log\{x/(1-x)\}$ – Exercise 4.35. For the Bliss data of Example 4.2 Egger (1979) found that the maximum-likelihood estimate of the Box–Cox transformation parameter λ was $\hat{\lambda} = 4.13$, resulting in a goodness-of-fit statistic of $\chi_5^2 = 2.81$ Fitted values are shown in Table 4.5, for the logit model fitted to (dose)^2 values and for the Aranda-Ordaz model fitted to dose values. A number of ED_{100p} values, from different fits, are shown in Table 4.6.

With reference to Table 4.6, the logit fit improves as the dose transformation ranges from log (dose) through dose, to dose^2. We see that for $p = 10\%$ and $p = 50\%$, as the fit improves, the confidence interval width increases. For $p = 90\%$, however, as the fit improves, the confidence interval width decreases. This same feature can be seen in comparing the logit and Aranda-Ordaz fits to the doses. We see therefore that increasing the number of parameters in a fitted model does not necessarily result in a corresponding decrease in

Table 4.5 *Fitted values for the Bliss data for the logit model applied to the square of the dose-scale, and the Aranda-Ordaz asymmetric model applied to the dose-scale*

Dose CS_2 (mg/l)	Observed number responding	Expected number responding, logit model applied to dose^2	Expecter number responding, Aranda-Ordaz model applied to dose
49.06	6	4.89	5.91
52.99	13	10.30	11.23
56.91	18	20.81	20.55
60.84	28	31.83	30.09
64.76	52	49.45	47.94
68.69	53	54.01	54.36
72.61	61	60.23	61.10
76.54	60	59.48	59.91

estimated precision of ED_{100p} values. But see also Exercise 4.27 on this point.

We conclude this section with a third example, for which a symmetric extended model is found to be useful.

Example 4.3

The data of Table 1.3 are taken from Table IV of Irwin (1937). If we fit the logit model to these data, we obtain the poor fit of Figure 4.10(a), with $\chi_3^2 = 15.69$.

In this case, taking logarithms improves the fit, as we can see from Figure 4.10(b), with $\chi_3^2 = 5.00$. Neither of the score tests is significant (with values of 1.74 for the Aranda-Ordaz model, and 3.10 for the cubic-logistic model), but we would clearly be unhappy with the fit of Figure 4.10(b), which is **overfitting** the data. In fact a likelihood-ratio test of the hypothesis that $\gamma = 0$ in the cubic-logistic model is significant at the 5% level, and the cubic-logistic fit is shown in Figure 4.10(c). We now have $\chi_2^2 = 0.32$, but we would obviously need to exercise care in using such a fitted model, in view of the variation of the fitted curve. It is interesting to note that the dose transformation, $z = d^{-1}$, produces the picture of Figure 4.10(d) and a fit which is acceptable both on the grounds of the goodness-of-fit

Table 4.6 *A comparison of* ED_{100p} *values for the Bliss data. Here all the confidence interval values result from the delta-method*

(i) *Fitting logit and Aranda-Ordaz models to dose*

	Logit			Aranda-Ordaz	
$p(\%)$	ED_{100p}	Width of confidence interval	$p(\%)$	ED_{100p}	Width of confidence interval
10	50.61	3.77	10	49.05	6.10
50	59.43	2.07	50	60.19	2.46
90	68.25	3.46	90	67.98	2.83

(ii) *Fitting the logit model to* $\log_{10}$ *dose and dose*^{2}

	Logit fitted to $\log_{10}$ (dose)	
$p(\%)$	ED_{100p}	Width of confidence interval
10	51.00	3.28
50	59.12	2.06
90	68.52	3.83

	Logit fitted to dose^2	
$p(\%)$	ED_{100p}	Width of confidence interval
10	50.07	4.38
50	59.74	2.08
90	68.04	3.18

statistic of 2.36, referred to χ_3^2 tables, and the pattern in the residuals. The two score-test statistics here are, 0.07 for the Aranda-Ordaz model, and 0.05 for the cubic-logistic model. Williams (1987a) emphasizes the influence of the response to the third largest dose in the fit of Figure 4.10(b). While this point remains the most influential in the fit of Figure 4.10(d), in this case (and not the case of the logit fit to $\log_{10}$ (dose)) the value of the influence is not exceptional.

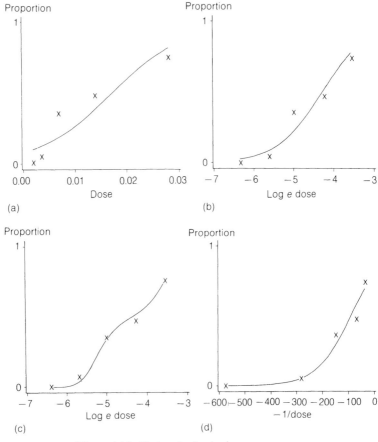

Figure 4.10 *Fitting the Irwin data:*
(a) logit fit to dose; (b) logit fit to $\log_e$ *(dose);*
(c) cubic-logistic fit to $\log_e$ *(dose);*
(d) logit fit to $-1/dose$

4.5   Additional models and procedures; goodness of link

4.5.1   *Other models*

The extended model idea can be traced to the smooth tests of
Neyman (Neyman, 1937; Rayner and Best, 1990) and is analogous
also to Tukey's (1949) test for non-additivity.

An early paper is that by Atkinson (1972). If $P(d)$ denotes the probability of response to dose d, then he set

$$\lambda = \text{logit}\{P(d)\} = \beta(d - \theta)[1 + \exp\{v(d - \theta)\}][1 + \exp\{k(d - \theta)^2\}].$$

This four parameter model reduces to the logit when $v = k = 0$, and λ is a monotonic increasing function of d for all values of v and k. Score tests were outlined for the parameters v and k. The model was used in the related but more complex Bradley–Terry model for paired comparisons. In Chen (1974), the inverse to $P(d)$ is specified through an incomplete beta integral:

$$P^{-1}(z) = \int_{0.5}^{z} y^{\gamma + \delta - 1}(1 - y)^{\gamma - \delta - 1}\,dy$$

Once again there are two shape parameters, γ and δ, and the logit model results when $\gamma = \delta = 0$. If $\gamma = \delta = 0.5$, we obtain a model based on an exponential tolerance distribution. Both of these models are also special cases of families of models proposed by Whittemore (1983) and Prentice (1976b).

The four-parameter family of models introduced by Prentice (1976b) has already been described in section 1.5. Although the models in this family are difficult to fit by maximum-likelihood, they may be used to provide a framework for score tests of fit, and this has been done by Brown (1982). The resulting tests were available in the BMDP computer package for statistical analysis within program PLR. Brown (1982) emphasized how the score-statistics weight the discrepancies between observed and expected values under the logit model. He also used simulation to check the adequacy and power of the chi-square tests based on score-statistics, coming to the conclusion that the asymptotic chi-square approximation was satisfactory even for quite small samples ($k = 50$, but with just a single binary variate observed for each dose-level). Power varied according to sample size and the assumed underlying model. A clear disadvantage of this approach is that if a score test is significant then it is not a simple matter to fit an appropriate extended model. In one case, Brown augmented a logistic fit to give a quadratic-logistic fit, in the light of the result of a score test, and such *ad hoc* approaches are clearly possible (Exercise 4.16). A useful three-parameter subclass of the Prentice model is provided by the **power logit model**, for which $P(x) = (1 + e^{-\beta(x - \mu)^{-m}})$ (Exercise 4.31). Genter and Farewell (1985) present a log-gamma model, and fit it to the data of Example 3.2 (Exercise 4.37).

In the paper by van Montfort and Otten (1976), the logit model is extended within the λ-distribution family proposed originally in Hastings *et al.* (1947). Here, we have

$$\lambda\left(\frac{x-\mu}{\sigma}\right) = F(x)^\lambda - \{1 - F(x)\}^\lambda, \quad \text{for } \lambda \neq 0$$

and

$$\left(\frac{x-\mu}{\sigma}\right) = \log_e[F(x)/\{1 - F(x)\}] \quad \text{if } \lambda = 0$$

where μ and σ are the usual location and scale parameters, and λ is a measure of kurtosis. This extended family is symmetric, and van Montfort and Otten suggest incorporation of the Box–Cox transformation as a way of coping with asymmetry.

As well as investigating the asymmetric model which we have discussed above, Aranda-Ordaz (1981) also proposed a symmetric extended family, based on:

$$x = \left(\frac{2}{\lambda}\right)\left[\frac{F(x)^\lambda - \{1 - F(x)\}^\lambda}{F(x)^\lambda + \{1 - F(x)\}^\lambda}\right]$$

which has clear similarities with the λ-distribution above. In this case the logit model results as $\lambda \to 0$. Both van Montfort and Otten (1976) and Aranda-Ordaz (1981) considered fits of their extended models to the data on reaching the age of menarche in Polish women, presented in Table 1.4 (Exercises 4.6 and 4.20). Score-tests may also be carried out within the context of the Aranda-Ordaz extended symmetric model. Furthermore, GLIM macros are now available to assist in testing within the Aranda-Ordaz models (Aranda-Ordaza, 1985a, b; Exercise 4.6). However, Goedhart (1985) found the Aranda-Ordaz symmetric model difficult to fit, and also objected to the particular shapes resulting for the tolerance distributions, with some being over a finite range and others with infinite range. Finite range models may be criticized both on the grounds that it is unlikely that the process being modelled would have probabilities that display an abrupt change in value, and (as a consequence) that extrapolation could be especially difficult. (See, for example, Cox and Snell, 1989, p. 22.)

Both Stukel (1985) and Looney (1983) fitted extended models to the Bliss data of Exercise 1.4. Looney (1983) proposed **Weibit** analysis, based on the three-parameter Weibull distribution, with

$$P(x) = 1 - \exp\left[-\{(x - \gamma)/\beta\}^{\alpha}\right]$$

where $x > \gamma$, $\alpha > 0$, $\beta > 0$. In this case, for $\alpha \approx 3.6$, $P(x)$ is well-approximated by a normal distribution. If $\alpha < 3.6$, $P(x)$ is a positively skewed distribution, while if $\alpha > 3.6$, $P(x)$ is a negatively skewed distribution. See also Thompson and Funderlic (1981) and Crump (1984). In addition, Looney (1983) fitted his model to the 22 data sets brought together by Copenhaver and Mielke (1977). A Weibull model with $\gamma = 0$ is considered in section 4.6, for the estimation of very low dose levels. The model proposed by Stukel (1985) contains four parameters, and effectively allows separate fitting to the upper and lower tails of the distribution; GLIM macros are provided for the model-fitting procedure (Exercises 4.17). See also Stukel (1988, 1990), where there is also discussion of score tests, and ED_{100p} estimation, illustrated on the menarche data of Table 1.4. In Singh (1987) an Edgeworth series distribution is proposed for $P(x)$, again involving four parameters, and again fitted to the data of Table 1.4 (Exercise 4.32).

It has been pointed out by Cox (1987) that a synthesis of the Tukey λ-family and the Aranda-Ordaz extended symmetric family is achieved through the form:

$$\frac{g_1 F(x)^{\lambda_1} + g_2\{1 - F(x)\}^{\lambda_2} + g_3}{h_1 F(x)^{\lambda_1} + h_1\{1 - F(x)\}^{\lambda_2} + h_3} = \alpha + \beta x$$

where the $g_i(\lambda_1, \lambda_2)$, and $h_i(\lambda_1, \lambda_2)$, $1 \leqslant i \leqslant 3$ are functions to be specified. Thus, for example, taking $g_1 = \lambda_2, g_2 = -\lambda_1, g_3 = -\lambda_2 + \lambda_1$, $h_1 = h_2 = 0$, and $h_3 = \lambda_1 \lambda_2$, generalizes the λ-family and yields a model proposed by Pregibon (1980), presented in equation (4.4); cf. also, Ramberg *et al.* (1979). Setting $g_1 = 2$, $g_2 = -2$, $g_3 = h_3 = 0$; $h_1 = \lambda_1$, $h_2 = \lambda_2$ generalizes the symmetric Aranda-Ordaz model.

One way of introducing a further parameter into a model with response probability $P(x)$ is to replace $P(x)$ by

$$\lambda + (1 - \lambda)P(x), \quad \text{as described in section 3.2}$$

This is known as Abbott's formula, and provides a way of accounting for natural response, as discussed in section 3.2. When Abbott's formula, with the logistic form for $P(x)$, is fitted to the data of Table 3.9, the residual deviance drops to 4.14, on 3 degrees of freedom (Cox, 1987). However, it is not known whether natural mortality was likely to be an important factor for the experiment.

We conclude this section by noting that a uniform c.d.f., as the basis for $P(x)$, results as a special case of the omega, exponential power and Tukey λ models. When adjusted to allow for an independent background mortality rate this uniform model becomes what is known as a hockey-stick model – applications are described and illustrated in Cox (1987).

In the work considered so far in this chapter we have concentrated on extended forms for the probability, $P(d)$, for response to dose d. Alternatively, we could think in terms of extending the link function, and this is now described in the following section.

4.5.2   Goodness of link

To illustrate the approach, let us consider the Aranda-Ordaz asymmetric model. Here,

$$P(x) = 1 - (1 + \lambda e^{\alpha + \beta x})^{-1/\lambda} \quad \text{for } \lambda e^{\alpha + \beta x} > -1$$
$$= 0 \quad \text{otherwise}$$

and we can think of this as an extension of the logit model which results when $\lambda = 1$. As seen already in section 4.2, the corresponding link function is

$$h(\mu) = \log_e \left\{ \frac{(1 - \mu/n)^{-\lambda} - 1}{\lambda} \right\} \tag{4.1}$$

where n is the appropriate binomial index. If $\lambda = 1$ here we obtain the logit link. Clearly, extending the link function is equivalent to extending the corresponding form for the probability of response, $P(x)$. However, for statisticians who think in terms of GLIM as a way of fitting models, it is more natural to operate in terms of the link function. This is the approach of Pregibon (1980), who also made an important additional contribution which we shall now explain.

Fitting the binomial model with the logit link function in GLIM is easy, as it makes use of the built-in procedures of the package. If we know the value for λ then equation (4.1) provides a fully-specified alternative link function to the logistic. It may be specified using the OWN facility of GLIM, and this is the approach of Aranda-Ordaz (1985b). (But see here the discussion of Appendix B.) Of course usually we do not know λ, and so a variety of values for λ need to be tried and the value of λ corresponding to the overall maximum is then adopted. This procedure nests the GLIM optimization within optimization with respect to λ. The same approach was used by Copenhaver and Mielke (1977), as mentioned in section 4.3. It is slightly time-consuming but conveniently fits models such as that specified by equation (4.1) without the need to write special, purpose-built maximization programs. A further disadvantage, however, is that in order to obtain correct measures of error, one cannot escape all of the algebra, Measures of error are obtained from inverting the three-dimensional Hessian corresponding to parameters (α, β, λ), while GLIM simply produces errors for fixed λ. It is therefore necessary to border the two-parameter Hessian with appropriate second-order derivative terms, as explained in Appendix C. This is not done by Aranda-Ordaz and so his estimates of error are incorrect, as was true also of the original quantit analysis.

The attractive feature of Pregibon's (1980) paper is that by means of a simple approximation based on a first-order Taylor series expansion we can approximate to the fit of model (4.1), and gauge whether we need to take $\lambda = 1$, by using just the binomial error structure and logit link function. In the context of the Aranda-Ordaz asymmetric model the argument is as follows.

We denote the link function of equation (4.1) by $h(\mu; \lambda)$ and by a Taylor series expansion we write,

$$h(\mu; \lambda) \approx h(\mu; \lambda_0) + (\lambda - \lambda_0) \left. \frac{\partial h}{\partial \lambda} \right|_{\lambda = \lambda_0} \qquad (4.2)$$

for some specific value λ_0 of λ. In the context of testing the goodness-of-fit of the logit model (or as Pregibon puts it, the goodness-of-link of the logit link function), we would naturally take $\lambda_0 = 1$ in this example.

Now the model given by the form for $P(x)$ translates in terms of the link function as

$$h(\mu; \lambda) = \alpha + \beta x$$

which we may now write, using expression (4.2) as

$$h(\mu; \lambda_0) \approx \alpha + \beta x + (\lambda_0 - \lambda) \left. \frac{\partial h}{\partial \lambda} \right|_{\lambda = \lambda_0}$$

In this example we can readily show that

$$\frac{\partial h}{\partial \lambda} = \left[\frac{\lambda \log (1 - \mu/n)}{\left\{ \left(1 - \frac{\mu}{n} \right)^{\lambda} - 1 \right\}} - 1 \right] \Big/ \lambda$$

so that for the case $\lambda_0 = 1$, we have

$$h(\mu; 1) \approx \alpha + \beta x + (\lambda - 1) \left\{ \frac{n}{\mu} \log \left(1 - \frac{\mu}{n} \right) + 1 \right\} \qquad (4.3)$$

We now proceed as follows. First of all we fit the model with $\lambda = 1$ to obtain maximum-likelihood estimates, $(\hat{\alpha}, \hat{\beta})$ under the logit model. For any value x we can estimate

$$\hat{\mu} = \frac{n}{(1 + e^{-(\alpha + \beta x)})}$$

which enables us to estimate

$$z = \left\{ \frac{n}{\mu} \log \left(1 - \frac{\mu}{n} \right) + 1 \right\}$$

If the Taylor-series expansion is reasonably good, and if we need to take $\lambda \neq 1$ then a measure of the extent of the improvement in taking $\lambda \neq 1$ is obtained from a further GLIM fit with the binomial error structure and logit link function, but now with **two** covariates, (x, z). The deviance difference between the two fits may be used to assess the improvement in fit.

A comparison of test-statistics is provided in Table 4.7. All are referred to χ_1^2 tables. For discussion, see Exercise 4.39.

This procedure may be iterated in a variety of ways (Exercise 4.22) to give, subject to convergence, the correct maximum-likelihood fit. Pregibon (1980) in fact suggests that the one-step of a fully-iterated procedure, which is all we consider here, is likely to be adequate in most situations for gauging whether or not an extended model needs to be fitted, and for representing the extended model fit. Pregibon's contribution in providing this approximate way of fitting extended models is substantial and not at all limited to the binomial model of quantal assay data. For example, the approach may be used to fit Wadley's problem with controls and a Poisson error distribution, described in Chapter 3, and extended in Chapter 6. See here the

Table 4.7 (Goedhart, 1985) *The relative values of likelihood-ratio, score and goodness-of-link test statistics for 9 data sets. The numbering adopted is that used by Copenhaver and Mielke (1977) and in all cases the analyses were based on using the original dose scale*

Data set	Logit vs. Aranda-Ordaz symmetric			Complementary log log vs. Aranda-Ordaz asymmetric		
	Likelihood-ratio	Score	Link	Likelihood ratio	Score	Link
3	0.35	0.48	0.53	2.44	3.63	3.24
6	5.68	5.38	6.31	0.30	0.17	0.23
9	2.58	2.26	2.44	0.10	0.09	0.09
10	6.93	8.17	9.37	0.75	0.73	0.86
11	1.79	0.85	0.85	0.05	0.05	0.04
15	1.25	0.47	0.52	0.21	0.13	0.17
16	0.11	0.05	0.05	0.84	0.28	0.23
19	0.89	0.22	0.23	1.50	0.03	0.03
23	7.79	7.54	8.28	0.01	0.00	0.00

paper by Baker *et al.* (1980), which also provides the GLIM code (Exercise 4.26).

Pregibon (1980) presented two models; one was that later independently proposed by Aranda-Ordaz for asymmetry. The other model which Pregibon suggests for binomial data has the link function,

$$h(\mu) = \left\{ \frac{\left(\frac{\mu}{n}\right)^{\alpha - \delta} - 1}{(\alpha - \delta)} \right\} - \left\{ \frac{\left(1 - \frac{\mu}{n}\right)^{\alpha + \delta} - 1}{(\alpha + \delta)} \right\} \quad (4.4)$$

This can be seen to be a generalization of the model based on the λ-distribution and fitted by van Montfort and Otten (1976), which results when $\delta = 0$ and we obtain the symmetric case. The details for using the Pregibon approach with the link function of equation (4.4) are outlined in Exercise 4.23. Users of GLIM who are interested in fitting extended models to quantal assay data may well find that the Pregibon (1980) paper provides all the necessary tools, within a familiar computational framework, subject to the procedure converging.

4.6 Safe dose evaluation

Model elaboration should improve the fit to data describing low and high response. One area in which a good description of low responses is important is that of low dose extrapolation, for the estimation of 'virtually safe' doses (VSDs). A virtually safe dose of a substance is one which produces a very low incidence of response, such as 10^{-8}. Of interest here is the acceptable level of the addition of substances such as food preservatives, potentially beneficial but also potentially toxic.

Krewski and Van Ryzin (1981) reported estimates of the VSD at a risk of 10^{-5} above background risk for the logit, two-parameter Weibull, multi-hit (Exercise 1.17) and multi-stage (Exercise 1.19) models, for 14 data sets. In Table 4.8 are presented the ratio R_1, of the largest to smallest VSD for the Weibull, logit and multi-hit models, and the ratio, R_2, of the largest VSD from the Weibull, logit and multi-hit models to that from the multistage model. With the exception of vinyl chloride, the Weibull, logit and multi-hit models produce similar VSD values, which may be partly due to the fact that they are mathematically identical for very low doses (Stukel, 1989; Exercise 4.38). For discussion of the anomalous result for vinyl chloride, see Van Ryzin (1982). The multistage model exhibits different behaviour overall, and Stukel (1989) recommends using the one-hit model, the multistage model and one of the other three. A detailed study of the multistage model for low-dose extrapolation is given in Portier and Hoel (1983). The one-hit model is linear in dose at low-dose levels and is usually (but not always – see Van Ryzin, 1982) the most conservative.

The estimated standard errors of the number of hits, h, in the multi-hit model, from Table 4.8, are usually large relative to the value of $\hat{h}$, and in most cases would suggest that a one-hit model would provide an adequate description of the data. This has been verified in a simulation study by Laurence and Morgan (1990b), and we return to this work in Chapter 5. A test of the one-hit model versus the two-hit model is given by Cox (1962), and a FORTRAN algorithm is provided by Thomas (1972). For many data sets, models which fit the observed data equally well may predict quite different VSDs. Because of this, the use of a range of models is usually recommended, as in the report by the Scientific Committee of the Food Safety Council (1980). This is an area where the use of extended models to improve fit to the observed data would be valuable, if there was a significant improvement in the fit. Additionally, as commented earlier, the reduced precision resulting from the use of extended models can be regarded as advantageous in this area. It

Table 4.8 (Krewski and Van Ryzin, 1981) *The ratio, R_1, of the largest to the smallest VSD from the Weibull, logit and multistage models, and R_2, the ratio of the largest VSD above to that from the multistage model. Also presented is the maximum-likelihood estimate, $\hat{h}$, of the number of hits, h, in the multi-hit model, together with an asymptotic estimate of its standard error*

Substance	$\hat{h}$	R_1	R_2
NTA	28.4 (14.9)	1.3	473
Aflatoxin	3.2 (1.4)	4.9	80
ETU	7.2 (1.2)	2.8	4.5
TCDD	2.5 (1.2)	1.9	2
DMN	2.8 (0.8)	3.0	3
Vinyl chloride	0.41 (0.002)	21.4	9×10^{-6}
HCB	1.0 (0.3)	1.8	2
Botulinum toxin	27.6 (5.9)	2.4	2.4
Bischloromethyl ether	1.8 (0.5)	2.5	75
Sodium saccharin	9.2 (8.0)	1.7	2.5
ETU	8.2 (2.1)	3.8	4.7
Dieldrin	2.3 (0.8)	3.7	91
DDT	1.7 (0.5)	4.1	53
Span oil	1.6 (1.8)	3.9	104

would be attractive to be able to rely on a known biological mechanism to justify the use of a particular mechanistic model. Unfortunately, however, this is usually not possible (Portier and Hoel, 1983). While a conservative approach would seem to be sensible, it may produce unrealistically low VSDs. An illustration of this is provided by Park and Snee (1983), who emphasize that any VSD estimation **must** be tempered by common sense, and incorporate as wide a range as possible of known information regarding the substance under test. Blind use of a conservative procedure has been regarded as scientifically indefensible by the Scientific Committee of the Food Safety Council (1980). Non-parametric approaches have been considered by Schmoyer (1984) and Schell and Leysieffer (1989) but we defer discussion of such methods until Chapter 7.

A key feature in discussion of VSD estimation is how the effect of interest combines with any background effect. Two approaches have already been suggested, in equations (3.1) and (3.2), and in fact this material links up with the work of section 3.7, on the effect of mixtures of substances, as discussed in Armitage (1982).

4.7 Discussion

In this chapter we have encountered a wide variety of approaches designed to improve the fit of simple models, such as the logit, to quantal assay data. A way through the maze of alternatives has been suggested by Goedhart (1985), and this is to fit the logit model and then consider score tests against a single simple asymmetric alternative model, such as the Aranda-Ordaz asymmetric model, and a single simple symmetric alternative model, such as the cubic-logistic model. If a score test suggests fitting the Aranda-Ordaz asymmetric model, it may simply be the case that the logit model cannot cope with the asymmetry in the data, and if the complementary log–log model were then fitted, for example in GLIM, a score test could be carried out to see whether the Aranda-Ordaz asymmetric model is necessary. In this way it may be possible to describe the asymmetry as simply as possible, using a two-parameter model, without in fact having to fit the Aranda-Ordaz three-parameter model – see the results of Table 4.7 for illustrations of this. Similarly, if a symmetric extended model with heavy tails is needed, it may be useful also to try a model based on the Cauchy distribution (Exercise 4.33). The approaches suggested are not restricted to the case of a binary response, and Genter and Farewell (1985) use the log-gamma extended model of Exercise 4.37 to examine goodness-of-link in the ordinal regression models of McCullagh (1980), described in section 3.5.

As we have seen in section 4.4, extended models, as a means of possibly improving the fit over the logit model, may not, in fact, be necessary if a suitable Box-Cox transformation of the dose scale can be found. However, fitting a three-parameter model, with parameters α and β measuring location and scale, and with parameter λ as the Box–Cox parameter can be difficult because of the inevitable correlation between $\hat{\lambda}$ and $\hat{\beta}$. This can result in a pronounced ridge to the likelihood surface, and good starting values are frequently necessary for iterative procedures to converge to the maximum-likelihood solution (Goedhart, 1985). An alternative is to fit the logit model to data transformed by a variety of values for the Box–Cox parameter λ, a useful range values for λ being $\lambda = -1, 0, 1, 2$. (Exercise 4.34.)

A possible disadvantage of transforming the dose scale is that comparisons between different sets of data could involve different scales, though in such a case graphs could be re-drawn to correspond to a common dose scale. If different extended models are fitted to different sets of data then comparisons are more complex, for precisely the same reason. An illustration is provided later in section 5.5.2. Of course, if several data sets are fitted simultaneously, the same dose-transformation or distribution shape parameter could be taken for all of the data sets. In practice we need not be limited to a single technique, and a variety of approaches is likely to be useful. In the context of more general logistic modelling, involving several regressor variables, the approach of using extended models is likely to be more useful than that of using Box–Cox transformations, simply because increasing the number of explanatory variables increases the number of potential transformation variables. Examples are provided by Brown (1982) and Kay and Little (1986), though in the latter case reservations are expressed concerning the use of extended models in general logistic regression. We have seen from the work of this chapter that simple models such as the probit and logit are by no means the only ways of describing and analysing quantal assay data, and many alternative procedures may be carried out with a view to improving fit. This may be important if an objective of the analysis is to estimate extreme dose levels, such as the ED_{95}, for example, when goodness-of-fit of the model can be very important. It should be emphasized that in many cases the data will simply not warrant being described by a model with more than two parameters. Morgan et al. (1989) describe a computer package called QUAD, which fits the logit model to raw doses or to specified transformed doses, performs score tests for the cubic-logistic and Aranda-Ordaz asymmetric models and also fits either of these models if necessary. Written in IBM BASIC, the program makes use of good micro-computer graphics to illustrate fitted models. It also provides likelihood-ratio, as well as delta-method confidence-intervals, for ED_{100p} values for the logit model (Chapter 2), for which the influence measures and generalized residuals of Chapter 3 are also displayed. The emphasis in QUAD is on using simple models which are usually not dificult to fit. A number of the new models presented here are not easy to fit by maximum-likelihood, and there is more discussion of this in Morgan (1988).

Influence analysis was introduced and discussed in Chapter 3. Williams (1987a) has noted that the approach of influence analysis may also be used in the case of testing within the context of extended models. Thus if an extended model provides a significant improvement in fit over a simpler model, this result may mainly be due to a small number of observations, which may be examined critically in case they were due to detectable experimental error. (Exercise 4.25.)

Models which incorporate times for individuals to respond, as well as whether they respond or not, produce alternative models for quantal assay data resulting at any time cross-section. These models are the subject of Chapter 5, and we return to this topic there. We have seen already, in Exercise 1.17, how the multi-hit model corresponds to an underlying gamma tolerance distribution, a special case of the extended models of Prentice (1976b). The competing stochastic model proposed by Puri and Senturia (1972) results in a tolerance distribution of the type III Pareto form, given in section 5.3.2.

In a number of papers, new methods proposed are tried out on a range of sets of data, gleaned from published papers, usually in the distant past – examples have been encountered above. This approach is of limited usefulness as the aims of the original experiment may be lost, the possibility of natural mortality may be ignored, and data may be distorted. An instance of this last possibility occurs in Berkson (1953) and Puri and Senturia (1972), where the data of Table 3.9 are analysed but without reference to the lowest dose, possibly because Finney's analysis mentioned earlier did not explicitly use that dose. Further examples are found in Rai and Van Ryzin (1979).

It is not possible to separate questions of goodness-of-fit and precision of estimators from the full practical context of the experimental data. If a model is selected on the basis of a range of score tests on a single set of data then it is clearly going to result in a reduced goodness-of-fit statistic, and typically estimates of precision will not reflect the conditioning brought about by the score tests. The same difficulties are encountered in many areas. For a good discussion within the context of log-linear modelling of contingency-table data, see Bishop *et al.* (1975, p. 317).

The problems that have been mentioned here can only be answered satisfactorily in discussion with experimenters producing or collecting the data. It is only then that measures of influence, for example, may in some cases be used to omit data points. Tests of significance, measures of fit and of precision are ultimately only useful as guides to model selection and interpretation within the full practical context.

A fundamental assumption in the modelling work of this chapter has been that the binomial distribution assumption is acceptable. If extra-binomial variation, or heterogeneity, may be anticipated, then that may account for a poor logit fit based on the binomial distribution. Examples are given in section 6.5.1. Certain statistical packages routinely assume heterogeneity, and adjust standard errors by a scaling heterogeneity factor. This is undoubtedly wrong, and we end with the good advice of Finney (1971, p. 74): 'One should be reluctant to draw important conclusions from a single experiment, and an experiment such as this ought to be considered in relation to others of a series.'

Extended models have also been used in the area of sequential design (e.g. McLeish and Tosh, 1990) and so we shall encounter them again in Chapter 8.

4.8 Exercises and complements

The exercises vary in complexity. It is particularly desirable to attempt the five exercises marked with a †. Exercises marked with a * are generally more difficult or speculative.

4.1 Show that as $\lambda \to 0$ in the Aranda-Ordaz asymmetric model of section 4.2, we obtain as a special case the complementary log–log model.

4.2 Derive an expression for the skewness of the tolerance distribution in the Aranda-Ordaz asymmetric model.

4.3† Construct the score test of the fit of the logit model within the Aranda-Ordaz asymmetric family. See Appendix D for details for the score test.

4.4 Looney (1983) fitted the shifted Weibull distribution to quantal assay data, using the parameterization for the c.d.f.: $F(x) = 1 - \exp[-\{(x - \gamma)/\beta\}^\alpha]$, for $x > \gamma$. Show that if the shape parameter $\alpha = 3.6$, then the resulting distribution is approximately normal, with mean 0.9011 and standard deviation 0.2780 (Dubey, 1967). Discuss the possible limitations of fitting the Weibull distribution to quantal assay data. Comment on the ED_{100p} estimates and the delta-method confidence intervals given below from Looney (1983), from fitting the data of Exercise 1.4 and Example 4.2. For the Weibit model the untransformed dosages were used but in the other three cases a log transformation was employed.

	$\hat{E}D_{01}$	95% CI	$\hat{E}D_{50}$	95% CI	$\hat{E}D_{99}$	95% CI
Logit	43.55	(41.21, 46.03)	59.16	(58.08, 60.12)	80.54	(76.74, 84.53)
Probit	44.87	(43.75, 45.92)	59.02	(57.94, 59.98)	77.45	(76.03, 78.89)
Prentice (1976b) model (skewed logistic)[a]	36.31	(31.41, 41.98)	60.39	(57.15, 63.83)	74.13	(69.18, 79.25)
Weibit	37.63	(33.60, 41.65)	60.11	(59.04, 61.18)	73.04	(71.07, 75.01)

[a]see Exercise 4.31.

4.5   The symmetric extended family of Aranda-Ordaz (1981) does not contain the normal distribution as a special case. Consider how you might select the shape parameter λ to obtain a good approximation to the normal distribution. In one approach, Aranda-Ordaz suggested taking $\lambda \approx 0.4$. (Cf. Exercise 4.18.)

4.6*   The GLIM macro of Exercise 2.18 is taken from Aranda-Ordaz (1985b), and is for fitting his asymmetric extended family. Using this as a model, write a GLIM macro to fit the Aranda-Ordaz symmetric extended family. Fit this model to the data of Table 1.4 for a range of values for λ. Deduce the maximum-likelihood estimate of λ and construct a 95% likelihood-ratio confidence interval for λ. Derive the correct estimates of error for the location and scale pair of parameters α and β, using the method of Appendix C. The logit model arises as a special case when $\lambda = 0$. Use this result, and that of Exercise 4.5, to compare the fit of the probit and logit models to the data of Table 1.4. A similar comparison is made in van Montfort and Otten (1976) (Exercise 4.20). Cf. the conclusion of Chambers and Cox (1967) regarding the large sample sizes needed to ensure reasonable power for discriminating between logit and probit models. Finally, consider appropriate analyses of the further sets of data on age at menarche given in Burrell et al. (1961) and Milicer (1968) (Exercise 2.23).

4.7   Demonstrate that all of the omega distributions of section 4.3 are symmetric.

4.8 Verify, for a random variable Y with the exponential power distribution, that

$$E[Y] = 0,$$

$$V(Y) = 2^{(1+\beta)} \left\{ \frac{\Gamma[\frac{3}{2}(1+\beta)]}{\Gamma[\frac{1}{2}(1+\beta)]} \right\} \phi^2 = \sigma^2, \text{ say}$$

The parameter β measures kurtosis. Verify that $\beta = 0$ gives the normal distribution, $\beta = 1$ gives the double exponential, and that the uniform distribution over the range, $[\theta - \sigma\sqrt{3}, \theta + \sigma\sqrt{3}]$ results as $\beta \to -1$.

4.9* Write a computer program to fit the exponential power distribution to quantal assay data.

4.10[†] Verify that the logit, double-exponential and uniform distributions result as special cases of the omega distribution (when $v = 1$, 0 and as $v \to \infty$, respectively).

4.11*

(a) Derive a Newton–Raphson procedure for computing $P(d)$, the probability of response at dose d, for quantit analysis, based on the omega distribution.
(b) Verify that the link function within quantit analysis may be written in the infinite summation form given in section 4.3.

4.12*

(a) Write a computer program for fitting the cubic-logistic model to quantal assay data – see Exercise 2.14 for the appropriate derivatives.
(b) Write a computer program to fit the quantit model.

4.13* (Morgan, 1983). Show that we can write:

$$\int_{1/2}^{p} \frac{|2z-1|^{v+1}}{(1-|2z-1|^{v+1})^2} \log(|2z-1|)\, dz$$

$$= \frac{\delta z}{2a} \left\{ \frac{z^a \log z}{1-z^a} - (1 + \log z) \sum_{j=1}^{\infty} \frac{z^{aj}}{(aj+1)} + \sum_{j=1}^{\infty} \frac{z^{aj}}{(aj+1)^2} \right\}$$

where

$$z = |2p - 1|$$
$$a = v + 1$$

and

$$\delta = \begin{cases} 1 & \text{if } p > 0.5 \\ 0 & \text{if } p = 0.5 \\ -1 & \text{if } p < 0.5 \end{cases}$$

4.14 Discuss the cubic-logistic fits shown below for data sets 7, 10 and 18 of the collection of Copenhaver and Mielke (1977).

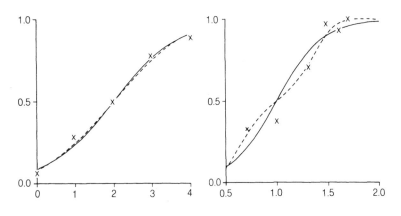

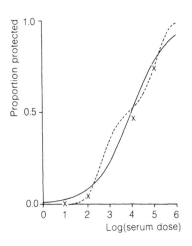

4.15 Discuss how you would fit an extended model to two or more sets of quantal assay data, but maintain comparability between the data sets.

4.16 Brown (1982) extended a logit model by adding a quadratic term to the exponent. This followed a significant score test within the Prentice extended family. Compare and contrast this approach with that of the cubic-logistic model.

4.17* In the logit model, the probability of response at dose d is given by: $P(d) = (1 + e^{-\eta})^{-1}$ where $\eta = \alpha + \beta d$.

Stukel (1985) fitted the model: $P(d) = (1 + e^{-\psi(\eta)})^{-1}$, where $\psi(\eta)$ is a strictly increasing non-linear function of η. Specifically, she assumed the following: For $\eta \geqslant 0$ $(P(d) \geqslant \frac{1}{2})$,

$$\psi(\eta) = \begin{cases} \kappa_1^{-1}(e^{\kappa_1|\eta|} - 1) & \text{for } \kappa_1 > 0 \\ \eta & \text{for } \kappa_1 = 0 \\ -\kappa_1^{-1}\log(1 - \kappa_1|\eta|) & \text{for } \kappa_1 < 0 \end{cases}$$

and for $\eta \leqslant 0$ $(P(d) \leqslant \frac{1}{2})$,

$$\psi(\eta) = \begin{cases} -\kappa_2^{-1}(e^{\kappa_2|\eta|} - 1) & \text{for } \kappa_2 > 0 \\ \eta & \text{for } \kappa_2 = 0 \\ \kappa_2^{-1}\log(1 - \kappa_2|\eta|) & \text{for } \kappa_2 < 0 \end{cases}$$

Here, (κ_1, κ_2), are additional shape parameters to be determined from the data. For this Bliss data of Example 4.2, for instance, estimates of $\kappa_1 = 0.16$ and $\kappa_2 = -0.53$ were obtained from fitting the model to the logarithms of the doses. Draw graphs of $\psi(\eta)$ versus η and $P(d)$ versus η and comment on the role of κ_1 and κ_2. Discuss how this model may be fitted using GLIM.

4.18 The symmetric extended tolerance distributions proposed by Aranda-Ordaz (1981) are defined implicitly by:

$$x = \left(\frac{2}{\lambda}\right)\left[\frac{F(x)^\lambda - \{1 - F(x)\}^\lambda}{F(x)^\lambda + \{1 - F(x)\}^\lambda}\right]$$

Draw graphs to illustrate the distributions for a variety of values for λ.

4.19 (Cox and Snell, 1989, p. 199.) If the probability of response at dose d, $P(d)$ is given by $P(d) = F(d)$, where $F(d)$ is a fully-specified cumulative distribution function, show that it is possible to transform the dose scale to: $w = z(d)$, for a suitable function $z(d)$ so that the probability of response to the transformed dose value w can be given by any cumulative distribution function required. Discuss the implications of this for the work of section 4.4.

4.20   (van Montfort and Otten, 1976.) Derive the necessary algebra for fitting a model based on the λ-distribution to quantal assay data. From fitting the model to the age-at-menarche data of Table 1.4, van Montfort and Otten estimate $\hat{\lambda} = 0.12$, with an estimated standard error of 0.06. Verify that the distribution with $\lambda = 0.14$ is approximately normal (cf. Exercise 4.5) and discuss how these results may be interpreted in the context of choosing between the probit and logit models for these data (cf. Exercise 4.6). Note that the standard Pearson goodness-of-fit statistics are here: $X^2 = 21.87$ for the logit model, and $X^2 = 21.90$ for the probit model (d.f. $= 23$ in each case).

4.21   Use the Pregibon goodness-of-link approximation for fitting the Aranda-Ordaz asymmetric model to the Bliss data set of Example 4.2, using GLIM, and compare the results with the exact fit using the GLIM program of Exercise 2.18.

4.22*   Provide a way of iterating the Pregibon one-step procedure considered in Exercise 4.21. Program the method in GLIM and try it out on the data of Example 4.2, using a variety of different starting values. You may find it useful to consider the GLIM program provided by Baker *et al.* (1980) (Exercise 4.26).

4.23[†]   Use the Pregibon goodness-of-link approximation to fit the extended λ-distribution of equation (4.4) to a variety of data sets.

4.24   (Brown, 1982.) Use the results of Appendix D to construct two score tests within the extended model proposed by Prentice (1976b) and given in section 1.4, of $m_1 = m_2$ (i.e. a test of symmetry) and of $m_1 = m_2 = 1$ (i.e. a test of the adequacy of the logit model). Note that Ciampi *et al.* (1982) used this model in the context of survival analysis.

4.25* (Williams, 1987a). Use the Pregibon (1981) one-step approximation, described in Appendix A and used in Chapter 3, to obtain approximations to the influence of responses to individual doses in determining the significance of tests of fit of simple models within an extended family.

4.26* (Baker *et al.* (1980). Use the Pregibon goodness-of-link approach to fit Wadley's problem with controls and a Poisson error structure, as discussed in Chapter 3. Provide a GLIM program.

4.27 (van Montfort and Otten, 1976). If a logit model is appropriate for a set of data but instead an extended model is fitted which contains the logit model as a special case, one might expect estimates of features of interest, such as ED_{100p} values, to be estimated less precisely due to the introduction of an additional shape parameter, λ, say. Such a loss of precision can be estimated, using ratios of variances of, say, an ED_{100p} value, evaluated when each model is fitted to data. Carry out this calculation for the model based on the λ-distribution, for a variety of types of experiment (i.e. varying doses, the number of doses, and the distribution of individuals over doses).

4.28 (Lwin and Martin, 1989). The assumption of homogeneous groups of individuals responding uniformly to a given dose level of a substance is not always tenable, and ways of relaxing this assumption will be described in Chapter 6. Consider how you would develop a model in which the tolerance distribution is a mixture of two component distributions. (This approach is also discussed in Example 3.3. Cf. also Good (1979), Boos and Brownie (1991) and the work of section 5.5.1.)

4.29 As a further possible model, consider a Box–Cox transformation of the odds ratio

$$\left(\frac{P(d)}{1 - P(d)} \right)$$

rather than the dose-scale, where $P(d)$ is the probability of response to dose d. See Guerrero and Johnson (1982) for further discussion.

4.30

(i) Conditional upon the parameter β, a random variable X has the Weibull distribution, with probability density function

$$f(x|\beta) = \alpha\beta x^{\alpha-1} e^{-\beta x^{\alpha}}, \quad \text{where } \alpha, \beta > 0, \quad x \geqslant 0$$

If β has a gamma distribution with probability density function,

$$g(\beta) = \frac{\delta^{v}\beta^{v-1}e^{-\delta\beta}}{\Gamma(v)}, \quad \text{where } v, \delta > 0, \beta > 0$$

make use of the fact that $\int_0^\infty g(\beta)d\beta = 1$, to show that the unconditional probability density function of X is given by

$$f(x) = \frac{\alpha v \delta^{v} x^{\alpha-1}}{(\delta + x^{\alpha})^{v+1}}, \quad \text{for } x \geqslant 0$$

The random variable X is said to have a Burr distribution. Write down its cumulative distribution function.

(ii) $\{Y_1, \ldots, Y_n\}$ is a random sample from a logistic distribution, with cumulative distribution function,

$$F(y) = \frac{1}{(1 + e^{-y})}, \quad \text{for } -\infty < y < \infty$$

Show that $X = \exp[\text{Min}\{Y_i\}]$ has a Burr distribution.

(iii) Verify that if $T = \log X$ then the distribution of T corresponds to that given by the Aranda-Ordaz asymmetric model. (For an application in survival analysis, see Gould and Lawless, 1988.)

(iv) Explore the potential of the Burr distribution as a model for quantal assay data.

4.31   In the power logit model, the probability of response to dose d is given by:

$$P(d) = (1 + e^{-\beta(d-\theta)})^{-m}$$

Experiment with fitting this model to a number of data sets, in each case making a comparison with the logit model. This model is called a skewed logistic model by Wu (1985) and McLeish and Tosh (1990).

4.32   Singh (1987) presents the following Edgeworth series distribution model for the probability of response to dose d:

$$P(d) = \Phi(z)\{1 - \lambda_3 H_2(z)/6 + \lambda_4 H_3(z)/24 + \lambda_3^2 H_5(z)/72\}$$

where $z = (d - \mu)/\sigma$, and the $\{H_i(z)\}$ are the Hermite Polynomials of order i:

$$H_2(z) = z^2 - 1$$
$$H_3(z) = z^3 - 3z$$
$$H_5(z) = z^5 - 10z^3 + 15z$$

Here the parameters μ, σ, λ_3 and λ_4 measure location, scale, skewness and kurtosis of the tolerance distribution. Discuss this model and its fit to the data of Table 1.4, for which the maximum-likelihood parameter estimates are:

$$\mu = 13.01\,(1.22)$$
$$\sigma = 1.10\,(1.16)$$
$$\lambda_3 = 0.41\,(3.14)$$
$$\lambda_4 = 0.20\,(9.04)$$

4.33 Cauchit analysis results when a model based on the Cauchy cumulative distribution function is fitted to quantal assay data (Exercise 2.6). Experiment with fitting the Cauchit model to a range of data sets, especially to data for which a quantit or cubic-logistic fit provides an improvement over a logit model. In the context of Wadley's problem, the Cauchit link is explored by Morgan and Smith (1992) and Smith (1991).

4.34 Goedhart (1985, p. 54) presented the results of fitting the logit model by maximum-likelihood, together with a Box–Cox transformation, to the data sets of Exercise 1.3. Discuss the results given overleaf for data sets 3 and 6 of Copenhaver and Mielke (1977) – cf. Exercise 1.3.

How do you think inclusion of the parameter λ will affect estimation of the ED_{50}? Would you lose a degree of freedom for estimating λ? Cf. the results of Taylor (1988), who has shown that failure to allow for the shape-parameter variability in the Aranda-Ordaz models results in a relatively small over-estimate of precision, in a particular averaged sense. (Cf. also Exercise 4.27.)

4.35 Kay and Little (1987) fit a range of models to the data of

			Estimated errors (in parentheses) and correlations		
Data set	Parameter	Estimate	$\hat{\alpha}$	$\hat{\beta}$	$\hat{\lambda}$
3	α	-5.89	$\hat{\alpha}$ (3.55)		
	β	2.46	$\hat{\beta}$ -0.996	(2.79)	
	λ	-0.20	$\hat{\lambda}$ 0.884	-0.971	(0.36)
6	α	-3.03	$\hat{\alpha}$ (1.39)		
	β	1.7×10^{-6}	$\hat{\beta}$ -0.956	(1.6×10^{-5})	
	λ	5.03	$\hat{\lambda}$ 0.952	-1.000	(2.32)

Table 1.4. Let $x(\hat{\lambda})$ denote the Box–Cox transformed ages and, for these data, let $z = (x - 9)/9$. Discuss the various fitted models, details of which are given below:

Model	Covariates	Residual deviance	d.f.	X^2
Logit	x	26.70	23	21.87
	$x(\hat{\lambda})$	21.94	22	17.88
	$\log z, \log(1 - z)$	14.66	22	13.11
	$\log(z/(1 - z))$	17.17	23	15.25
Probit	x	22.89	23	21.90
	$x(\hat{\lambda})$	14.50	22	13.03

4.36 Discuss how all the different ways of describing the data of Table 1.4, presented in this chapter, compare, from the points of view of (i) ease of calculation and (ii) conclusions to be drawn. (Refer to 'menarche' in the index.) Note that further analyses of these data appear also later in the book!

4.37 Farewell and Prentice (1977) present the log-gamma distribution, with probability density function,

$$f(x) = \frac{|\lambda|\lambda^{-2\lambda^{-2}}}{\Gamma(\lambda^{-2})} \exp\{\lambda^{-2}(\lambda x - e^{\lambda x})\}, \; \lambda \neq 0$$

$$= (2\pi)^{-1/2} \exp(-x^2/2), \; \lambda = 0$$

Why is this called the log-gamma distribution?

Show that log–log and complementary log–log links result when $\lambda = -1, +1$, respectively. Trivially, $\lambda = 0$ gives the probit link.

When this model is fitted to the data of Example 4.2, for a range of values of λ, the following results are obtained (Genter and Farewell, (1985):

λ	Max. log-likelihood
-2	-36.6
-1	-27.1
0	-18.2
1	-14.8
2	-16.7

Plot a graph of these values. Estimate the maximum-likelihood estimate of λ, and discuss how the fitted model compares with the results of Example 4.2.

4.38 (Stukel, 1989). Investigate the probability of response to low doses in the Weibull, logit and multi-hit models. Show how the slope parameter in the logit model, fitted to log dose values, may be used to approximate the number of hits in the multi-hit model.

4.39 Discuss the relative performance of the likelihood-ratio, score and goodness-of-link tests, based on the results of Table 4.7. Suggest an explanation for why significant improvements of the Aranda-Ordaz asymmetric model over the logit model are not accompanied by significant improvements over the complementary log–log model.

CHAPTER 5

Describing time to response

5.1 Introduction and data

As can be judged from the examples of earlier chapters, in statistical papers it is quite usual for quantal assay data to be presented without any mention being made of the duration of the experiment. However, we note for example that the data of Table 2.1 were collected 1 hour after exposure to the chemical, and the data of Table 1.14 were obtained 96 hours after treatment. When times are cited, multiples of 24 hours are quite usual, as is the 7-day assay. For instance, Sprague (1970) mentions 48- and 96-hour ED_{50} assays. Notation such as $ED\ \omega/\mu$ is sometimes encountered, signifying the dose expected to affect $\omega\%$ of individuals in μ time units (e.g. McLeish and Tosh, 1990; Carter and Hubert, 1984).

There are various different ways in which information can be gathered on how long it takes for a reaction to take place. For instance, Boyce and Williams (1967) describe three different experiments. In the first, five concentrations of N-tritylmorpholine were selected for each of six different exposure times: 1 hour, 2 hours, 4 hours, 8 hours, 17 hours and 32 hours, to give 30 concentration-time combinations. Prior knowledge from pilot studies was used in selecting the concentrations used at each exposure time, to give mortalities in the 10–90% range. Nine hundred snails (*A. glabratus*) were then allocated at random to these 30 combinations. In the second experiment, five **concentrations** were selected, viz. 10, 0.5, 0.25, 0.1 and 0.05 p.p.m. of N-tritylmorpholine, and approximately 1200 snails were allocated at random to a series of exposure times for each of these concentrations. In the third experiment, three groups of 100 snails each were used. One set was a control, another was given a dose of 0.01 p.p.m. and the third set was given a dose of 0.0075 p.p.m. Twice a week, for 10 weeks, the aquaria were cleared out, and the dead snails were counted and removed. This experiment may therefore be seen to be similar to that of section 1.2.3, and Table 1.5.

For experiment 1 above, we could fit a dose-response model for each time separately, and in each case deduce, say, an ED_{50}. For experiment 2, for any concentration we have a range of exposure times and so we may calculate an ET_{50}, say, that being the time, for that concentration, which would correspond to an expected proportion of 50% of snails responding. We might expect there to be a form of balance between dose and time, with a low dose and a long exposure time producing a similar effect to a high dose and a short exposure time. A formalization of this is provided by Ostwald's equation (Ostwald and Dernoschek, 1910): if a given level of response, say $p\%$, is produced by concentration X and time-to-response T, then Ostwald's equation predicts the relationship:

$$XT^\lambda = \kappa_p$$

for constants κ_p and λ or equivalently,

$$\log X + \lambda \log T = \log \kappa_p.$$

Boyce and Williams (1967) did not find their data to be in agreement with Ostwald's equation (Exercise 5.1), but Hairston (1962) found it to be a useful description of experiments on the effect of Bayluscide on snails, and it also tallies with the results of Petkau and Sitter (1989), as will be seen in section 5.5.2.

The difference between experiments 1 and 2 above and experiment 3 is that in experiment 3, for any given dose, the same original group of individuals is examined at successive times. In this case the binomial response random variables at each sampling time are clearly not independent, and similarly ED_{50}s evaluated at different times would no longer be independent. We focus on this experimental paradigm in this chapter. Examples of the kind of data which may result have already been provided in Tables 1.5, 1.6 and 1.8, and analyses suggested in Exercises 3.4–3.6. Two further sets of data are shown in Tables 5.1 and 5.2, and we shall consider analyses of both of these data sets later in the chapter.

A wide range of approaches is available for such data, the spectrum ranging from very crude descriptive measures on the one hand, to very complex stochastic mechanistic models on the other. We shall start by examining examples of both of these procedures, and conclude by presenting methods based on survival analysis, which

Table 5.1 *Data from Kooijman (1981) describing the numbers of Daphnia dying at various times after being exposed to cadmium chloride*

Dose (μg/l)	Group size	Time (days)								
		2	4	7	9	11	14	16	18	21
0.0	50	0	0	0	0	0	0	0	0	1
3.2	50	0	0	0	0	1	1	1	1	2
5.6	49	0	0	0	0	2	3	3	5	7
10.0	50	0	0	0	3	5	8	10	10	12
18.0	53	0	0	13	25	27	36	36	38	42
32.0	50	0	15	36	46	50	50	50	50	50
56.0	59	31	53	59	59	59	59	59	59	59

Table 5.2 *Data from Carter and Hubert (1984), on the toxic effect of a copper substance applied to tanks containing 10 fish each; the data provide the number of dead fish*

Block	Time (days)	Concentration (μg/l)						
		0.10	0.20	0.30	0.50	1.0	2.0	2.5
1	2	0	1	1	2	3	3	4
	3	1	1	2	3	5	6	7
	4	1	1	3	4	8	7	9
2	2	0	1	1	2	3	4	4
	3	1	1	2	3	5	6	7
	4	1	1	2	4	7	8	8

would seem to be more generally useful. Frequently survival times are not known more precisely than the inverval (which could be a day for example) within which response occurred. Examples are provided by the data of Tables 1.5, 5.1 and 5.2. We have here an instance of survival analysis with **interval censoring**. As mentioned in section 3.8, these data may also be considered as multivariate. We note finally that univariate analyses are often carried out on the data at each time point – this neglects the time dependency in the data we are considering (Exercise 5.23).

5.2 Descriptive methods

5.2.1 *The MICE index*

Izuchi and Hasegawa (1982) proposed a MICE index to summarize the behaviour over time of eggs injected with a virus. Here MICE is a mnemonic for **mortality index for chicken embryos**. To illustrate the operation of this index we can take as an example the responses of the 20 eggs given the highest dose of arboviruses in Table 1.8. In this case we have the following responses:

								Day							
	1	2	3	4	5	6	7	8	9	10	11	12	13	14	Totals
Dead	0	1	1	1	2	2	3	6	7	10	11	12	12	14	82
Alive	20	19	19	19	18	18	17	14	13	10	9	8	8	6	198

The MICE index is taken as $82/198 = 0.414$. It is easy to see how this index can vary if there is much early, or late, death; however, it is clearly only a crude way of describing the data (Exercise 5.2). Jarrett (1984) showed how to obtain measures of error for the MICE index and also suggested improvements. We shall return to the MICE index later in the chapter.

5.2.2 *Polynomial growth curves*

We assume that n_i individuals are exposed to the ith dose, and that there are k doses in all, given by: $\{d_i, 1 \leqslant i \leqslant k\}$. Observations are assumed to be taken at times $\{t_j, 1 \leqslant j \leqslant m\}$. Then for each $1 \leqslant i \leqslant k$, we have r_{ij} individuals responding in the time-interval, $(0, t_j)$, for $1 \leqslant j \leqslant m$.

Clearly, as j increases, the cumulative sums, $\{r_{ij}\}$ increase for each i, and so the problem can be made to resemble a **growth-curve** problem. This is the basis of the approach adopted by Carter and Hubert (1984), who utilize the theory of polynomial growth-curve models developed by Potthoff and Roy (1964). Thus Carter and

Hubert (1984) propose the model:

$$\text{arcsine } \sqrt{r_{ij}/n_i} = \sum_{l=1}^{q} (\beta_{l1} + \beta_{l2} \log_{10} d_i) t_j^{-(l-1)} + \varepsilon_{ij}$$

$$\text{for} \quad 1 \leqslant i \leqslant k, 1 \leqslant j \leqslant m \tag{5.1}$$

where q and $\{\beta_{l1}, \beta_{l2}\}$ are parameters to be estimated.

For data such as that of Example 5.2, with the possibility of block effects, they replace the term in brackets on the right-hand-side of equation (5.1) by:

$$(\beta_{l1} + \beta_{l2} \log_{10} d_i + \rho_{ls}), \text{ with } \rho_l = 0, \text{ for all } l;$$

$$s \text{ indicates the relevant block.}$$

It is further assumed that for each i the $\{\varepsilon_{ij}\}$ have a multivariate normal distribution with mean vector $\mathbf{0}$ and dispersion matrix which is not a function of i.

While the model of equation (5.1) may appear to be rather complex, the theory of polynomial growth curves permits explicit estimation of the parameters of the model and tests may be performed in a straightforward way on the value of q. For example, for the data of Table 5.2, the value of $q = 2$ sufficed to provide a good fit to the data, and the form of the fitted model was then that given below:

$$\text{arcsine } \sqrt{r_{ijs}/n_{is}} = 1.193 - \frac{1.286}{t_j} + 0.782 \log_{10} d_i$$

$$- \frac{0.935}{t_j} \log_{10} d_i + \hat{\rho}_{1s} + \frac{\hat{\rho}_{2s}}{t_j}$$

where

$$\hat{\rho}_{11} = -\hat{\rho}_{12} = 0.0302, \text{ and}$$
$$\hat{\rho}_{21} = -\hat{\rho}_{22} = -0.0853$$

and now r_{ijs} individuals respond prior to time t_j out of the n_{is} in block s exposed to dose d_i, for $1 \leqslant i \leqslant k$.

This model has recently been severely criticized by Pack (1986a), Petkau and Sitter (1989) and Williams (1986c), who point out, in particular, that the multivariate normal distribution assumption is untenable (Exercise 5.3). It therefore appears that the advantage of explicit estimators for this model is outweighed by its disadvantages, for the types of data considered here. For more discussion, see Carter and Hubert (1989). We return briefly to the Potthoff and Roy (1964) work in Chapter 8.

5.3 The stochastic model of Puri and Senturia

5.3.1 A mechanistic model

What happens when an insect is sprayed with insecticide is clearly a complex process, involving many different stages. It is clear that in some cases, when single doses of the insecticide are administered, some insects apparently manage to escape being affected, the rest being less fortunate. A crude way to model the behaviour of the insects is therefore to suppose that insects attempt to remove the insecticide from their system, perhaps by some form of metabolism, and that during the time that this takes the insects are at risk. While loss of insecticide is likely to be a continuous process of decrease, an approximate model suggested by Puri and Senturia (1972) is illustrated in the diagram of Figure 5.1. So at time $t = 0$, the insect is given dose d of the substance.

It is supposed that the insect loses insecticide at times which occur in a Poisson process of rate μ, and that the amounts lost at each time i, V_i are independent exponential random variables with probability density function, $\beta e^{-\beta t}$. This is obviously only a rough model – for instance the $\{V_i\}$ might more realistically be assumed to be dependent. If still alive, the insect finally removes all the

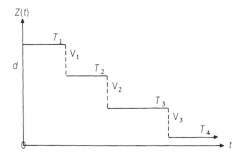

Figure 5.1 *A diagrammatic representation of the model proposed by Puri and Senturia (1972) to describe how insects lose insecticide.*

insecticide when

$$\sum_{i=1}^{u} V_i \geq d \text{ for some } u \geq 1$$

so that in fact the amount V_u of the last 'release' is likely to exceed the remaining insecticide.

If we denote the amount of insecticide remaining in an insect by time t as $Z(t)$, then with $Z(0) = d$, we have,

$$Z(t) = \max\left(0, d - \sum_{i=1}^{N(t)} V_i\right)$$

where $N(t)$ has a Poisson distribution with mean μt.

To complete this model we can specify a **hazard** function by $\Psi(t)Z(t)$, for some parametric function $\Psi(t)$ to be specified. By this we mean that in the small interval of time, $(t, t + \delta t)$, the probability of an insect with amount of insecticide $Z(t)$ dying is given by $\Psi(t)Z(t)\delta t + o(\delta t)$. We are therefore modelling death by means of a continuous-time stochastic process, in fact by a non-homogeneous death process, driven by another stochastic process, $\{Z(t)\}$. Readers who are unfamiliar with the general theory of stochastic processes may find it useful to consult Bailey (1964, pp. 90–91) for analysis of simple death processes, and the paper by Puri (1975) for an analysis of a stochastic birth and death process developing under the influence of another stochastic process. Puri and Senturia (1972) considered the case, $\Psi(t) = \delta$, a constant independent of time. For the flour-beetle data of Example 1.2.3, Diggle and Gratton (1984) found it necessary to take $\Psi(t) = \delta + \gamma t$, in order to obtain the correct qualitative match between model and data. This model is of interest as it proposes a distinct stochastic mechanism for the action of the insecticide – cf. the discussion in section 1.5. This model will therefore be investigated in some detail here.

5.3.2 Fitting endpoint mortality data

Although the original Puri–Senturia model was proposed for data observed over time, they only fitted their model to marginal data, resulting for large t, effectively for $t = \infty$. We can show that in this case the probability of not responding to dose d is

$$1 - P(d) = e^{-\beta d}\left(1 + \frac{\delta}{\mu}d\right)^{\beta(\mu/\delta) - 1} \tag{5.2}$$

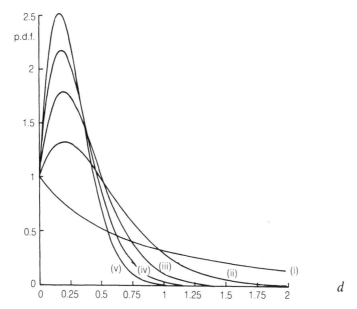

Figure 5.2 *Examples of the type III Pareto distribution density function with scale parameter $\rho = 1$, and β increasing from 0.5(i) to 20.5(iv), in steps of 5. Reproduced from Laurence and Morgan (1989), with permission from the Biometric Society.*

Equation (5.2) therefore provides us with a new model for quantal assay data. In contrast to the extended models of Chapter 4 this is only a two-parameter model, since δ and μ are combined in the single parameter, $\rho = \delta/\mu$. The above form of $P(d)$ is the cumulative distribution function of a type III Pareto distribution (Johnson and Kotz, 1970, p. 234), with scale parameter ρ. Illustrations of the corresponding density function are given in Figure 5.2, for the case $\rho = 1$. We see therefore that this model results in an asymmetric family of distributions for finite β.

In Table 5.3 we present the quantal assay data of Martin (1942), given earlier in Table 3.9, but without the lowest dose and the responses to that dose. As mentioned earlier, Finney (1947) analysed the data with the lowest dose and response missing, and it seems likely that subsequent authors, such as Berkson (1953) and Puri and Senturia (1972), took Finney's lead and so did not use the original entire data set.

Table 5.3 *A comparison of the fits of the logit model and the model of equation (5.2), obtained for endpoint mortality data. For the male flour-beetle data the logit model has also been fitted to log (dose) values*

(a) Martin (1942) data on the toxic effect of Deguelin, without the lowest dose.

Concentration (x_i) (mg/l)	Total number of insects (Aphis ruminus) (n_i)	Number dead (r_i)	Expected no. dead (logit model)	Expected no. dead (model (5.2))
10.1	48	18	18.4	18.2
20.2	48	34	34.4	35.0
30.3	49	47	44.6	44.5
40.4	50	47	48.8	48.7
50.5	48	48	47.7	47.7

(b) Male flour-beetle data (t = 13)

Dose (d_i)	n_i	r_i	Expected no. dead (logit model)	Expected no. dead (model (5.2))	Expected no. dead, (logit model applied to log (dose))
0.2	144	43	49.7	50.3	45.3
0.32	69	50	41.4	41.7	46.0
0.5	54	47	47.4	46.3	48.1
0.8	50	48	49.5	49.3	48.6

(c) Female flour-beetle data (t = 13)

Dose (d_i)	n_i	r_i	Expected no. dead (logit model)	Expected no. dead (model (5.2))
0.2	152	26	29.9	28.0
0.32	81	34	28.1	30.2
0.5	44	27	28.1	28.6
0.8	47	43	43.6	43.0

In Table 5.3 we also give the flour-beetle data for $t = 13$, from Table 1.5; taking $t = 13$ to represent the endpoint mortality corresponding to $t = \infty$ appears to be reasonable in this case. Hewlett (1974) also analysed these endpoint mortality data (Exercise 5.4). In addition, in Table 5.3 we give the maximum-likelihood fitted values under model (5.2), and for comparison we present the logit model fit.

It is possible to test formally whether, say, model (5.2) is better than the logit model for any data set, either by embedding both models within a larger parametric form, and proceeding analytically

Table 5.4 *Parameter estimates for model (5.2) resulting in the fits of Table 5.3, together with goodness-of-fit statistics. For comparison, the goodness-of-fit statistics are also presented for the logit model (see also Exercise 5.4)*

	Model					
	Model (5.2)				Logit	Logit and log dose
	$\hat{\beta}$ (standard error)	$\hat{\rho}$ (standard error)	Correlation $(\hat{\beta}, \hat{\rho})$	X^2 (d.f.)	X^2 (d.f.)	X^2 (d.f.)
Reduced Martin data	0.241 (0.136)	0 (0.013)	0.24	−0.971 (3)	4.18 (3)	4.61
Male flour-beetle data	22.99 (16.6)	0.71 (0.46)	−0.988	8.24 (2)	10.41 (2)	1.72 (2)
Female flour-beetle data	33.62 (37.4)	0.24 (0.24)	−0.995	1.17 (2)	2.84 (2)	—

using score tests, or by using Monte Carlo testing. Neither approach has been adopted here. A computational problem with Monte Carlo testing is simply that for data sets requiring a symmetric model, model (5.2) fits the data by requiring very large values of β. The parameter estimates and goodness-of-fit statistics of both models to the three data sets of Table 5.3 are summarized in Table 5.4. The fits of the two-parameter models to the male flour-beetle data are poor, but the logit fit to the log (dose) values is excellent – see fitted values of Table 5.3.

Also presented in Table 5.4 are the parameter estimates and measures of error and correlation of the estimators, estimated from the expected Hessian at convergence. We see that the parameters are imprecisely estimated, and there is an extremely negative estimated correlation between $\hat{\beta}$ and $\hat{\rho}$ (Exercise 5.6). Confidence regions are also very skewed. Matters are improved by reparameterization, and in Figure 5.3 we present confidence regions for $z = \log(\rho\beta)$ and $y = \log(\rho)$. For the Martin data we obtain corr $(z, y) = -0.047$, while for the male and female flour-beetles, respectively, the corresponding correlations are: -0.453 and -0.693.

A complication with the model of this section is that we cannot write down explicitly the ED_{50}, θ, but rather we need to solve

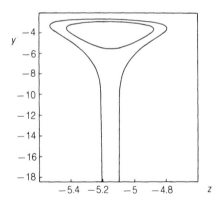

Figure 5.3 *95% and 99% two-dimensional confidence regions for z and y, resulting from fitting the type III Pareto distribution of equation (5.2) to the abbreviated Martin (1942) quantal assay data of Table 5.3(a), where* $z = \log(\beta\rho)$ *and* $y = \log(\rho)$. *The confidence regions result from taking appropriate sections of the log-likelihood surface. Reproduced from Laurence and Morgan (1989), with permission from the Biometric Society.*

numerically the equation:

$$\log_e 2 = \beta\theta + \left(1 - \frac{\beta}{\rho}\right)\log_e(1 + \rho\theta) \qquad (5.3)$$

This is straightforward using Newton–Raphson (Exercise 5.5) but it is an additional inconvenience, in comparison with other models.

It is of interest now to see how we might fit the Puri–Senturia model to a full data set with time to response, and not just to endpoint mortality resulting from a single time cross-section. In the next section we look at the model proposed by Diggle and Gratton (1984) for the time-dependent case.

5.3.3 The Diggle–Gratton extension

Fitting a model with a hazard of the form:

$$(\delta + \gamma t)Z(t) \qquad (5.4)$$

is considered by Diggle and Gratton (1984), specifically motivated by the data of Table 1.5, which they show do not behave as predicted by the $\gamma = 0$ case. They used maximum-likelihood results in quite complex expressions, which is why they chose this model as a vehicle to illustrate their approach to Monte Carlo inference. They experienced difficulty with the procedure, and attributed this to being unable to exclude the possibility that β is very small. Setting $\beta = 0$ may be interpreted as the insecticide all being lost at once, after a time since the administration of the dose which is a realization of a random variable with the exponential density function $\mu e^{-\mu t}$. This is because β^{-1} is the mean amount of insecticide lost at each excretary event. The case $\beta = 0$ allows standard maximum-likelihood to proceed in a relatively straightforward manner, and Diggle and Gratton (1984) found further that fits to the data were also compatible with $\delta = 0$.

If T denotes the lifetime of a beetle given dose d at time $t = 0$, then the **survivor** function, $S(t)$, is defined as

$$S(t) = \Pr(T \geqslant t)$$

In the case $\beta = \delta = 0$ and the hazard of equation (5.4) above, $S(t)$ is

given by
$$S(t) = \exp\{-(\mu + \gamma dt/2)t\} + \sqrt{2\pi}$$

$$\times \frac{\mu}{\sqrt{\gamma d}} \exp\{\mu^2/(2\gamma d)\} \{\Phi(t\sqrt{\gamma d}) + \mu/\sqrt{\gamma d} - \Phi(\mu/\sqrt{\gamma d})\} \qquad (5.5)$$

where $\Phi(.)$ is the standard normal cumulative distribution function.

Diggle and Gratton (1984) fit this model separately for each sex-dose combination. This results in an extravagant number of 16 parameters to describe the data, and one would certainly hope at a later stage to be able to reduce the dimensionality of the overall fitted model in the light of the parameter estimates that result (Exercise 5.7). Parameter estimates are given in Table 5.5, and estimates of the correlation between estimators $\hat{\gamma}$ and $\hat{\mu}$ are provided in Table 5.6.

As $t \to \infty$ in equation (5.5) we obtain as the resulting probability, $P(d)$, of responding under this model to dose d,

$$P(d) = 1 - \sqrt{2\pi} \frac{\mu}{\sqrt{\gamma d}} \exp\{\mu^2/(2\gamma d)\} \left\{1 - \Phi\left(\frac{\mu}{\sqrt{\gamma d}}\right)\right\} \qquad (5.6)$$

In equation (5.6) we can see from the relationship between μ, γ and

Table 5.5 *Fitting the two-parameter model resulting when* $\beta = \delta = 0$. *Point estimates of* γ *and* μ *obtained for the data of Table 1.3 using the methods of conditional mean and zero frequency, and maximum likelihood. Asymptotic estimates of standard error are given in parentheses. (The maximum likelihood estimates of error differ slightly, on occasion, from those given by Diggle and Gratton, 1984)*

	Conditional mean and zero frequency		Maximum likelihood	
Dose	γ	μ	γ	μ
Males				
0.2	0.326(0.068)	0.297(0.037)	0.284(0.058)	0.286(0.036)
0.32	0.610(0.105)	0.119(0.028)	0.589(0.091)	0.116(0.027)
0.5	0.353(0.059)	0.047(0.018)	0.345(0.053)	0.047(0.018)
0.8	0.205(0.033)	0.013(0.009)	0.197(0.029)	0.013(0.009)
Females				
0.2	0.173(0.049)	0.350(0.049)	0.203(0.055)	0.364(0.050)
0.32	0.303(0.068)	0.245(0.040)	0.336(0.069)	0.250(0.041)
0.5	0.117(0.026)	0.101(0.025)	0.107(0.025)	0.099(0.025)
0.8	0.180(0.031)	0.027(0.014)	0.175(0.028)	0.027(0.013)

Table 5.6 *Fitting the two-parameter model resulting when* $\beta = \gamma = 0$. *Asymptotic estimates of* $Corr(\hat{\gamma}, \hat{\mu})$

	Dose d	Estimates using maximum likelihood	Estimates using the method of conditional mean and zero frequency
Males	0.2	0.379	0.387
	0.32	0.065	0.090
	0.5	0.022	0.032
	0.8	0.005	0.010
Females	0.2	0.528	0.537
	0.32	0.231	0.271
	0.5	0.175	0.098
	0.8	0.012	0.019

d that if the data for time-sections $< \infty$ are ignored, then only $\mu/\sqrt{\gamma}$ is estimable from the data, as compared with the combination of μ and δ in the original model of Puri and Senturia (1972). As we would expect, the one-parameter model given by equation (5.6) provides a very poor description of the endpoint mortality data.

5.3.4 A method of conditional mean and zero-frequency

The form for $S(t)$, given by equation (5.5), is complex, but the Laplace transform of $S(t)$ is much simpler:
Let

$$\xi_\zeta(s) = \int_0^\infty e^{-st} S(t) dt$$

where

$$\zeta = (\mu, \gamma)'$$

If $\psi_\zeta(s) = s\xi_\zeta(s)$, then $\psi_\zeta(s)$ may be written as a function of the single parameter

$$\alpha(s) = \frac{s + \mu}{\sqrt{\gamma d}}$$

i.e.

$$\psi_\zeta(s) = \sqrt{2\pi}\,\alpha(s)\exp\{\alpha^2(s)/2\}\Phi\{-\alpha(s)\}, \quad s = 0 \qquad (5.7)$$

The method-of-moments equations

$$\psi_{s_i}(\zeta) = \hat{\psi}(s_i), \quad i = 1, 2$$

obtained by equating the theoretical transform to an appropriate estimator $\hat{\psi}(s)$ –the empirical transform discussed below – for two distinct values of s, s_1 and s_2, say, can be solved for ζ. Such an approach, though in a different transform context, was considered by Press (1972, 1975) and generalized by Feuerverger and McDunnough (1981a, b, 1984). Alternatively, allowing $s_1 \to s_2 = s$, these method-of-moments equations become

$$\psi_s(\zeta) = \hat{\psi}(s)$$

and

$$\frac{\partial \psi_s(\zeta)}{\partial s} = \frac{\partial \hat{\psi}(s)}{\partial s} \tag{5.8}$$

The empirical transform used here has been considered in similar work by, for example, Feigin *et al.* (1983) and Laurence and Morgan (1987), and is based on a Stieltje's integral approximation to $\psi_\zeta(s)$, given by

$$\hat{\psi}(s) = \sum_{i=0}^{m-1} (e^{-sc_i} - e^{-sc_{i+1}})\hat{p}_{i+1}, \quad s > 0$$

where $\hat{p}_{i+1}$ is the observed proportion of individuals surviving beyond sampling point t_{i+1}, $0 \leqslant i \leqslant m - 1$, m is the total number of times at which observations are made, and c_i is a point in the interval (t_i, t_{i+1}), here taken as

$$c_i = \frac{(t_i + t_{i+1})}{2}, \quad 1 \leqslant i \leqslant m - 1$$

with

$$c_0 = 0, \quad c_m = \infty \tag{5.9}$$

The solution to these equations depends on the value assigned to $\mathbf{s} = (s_1, s_2)$. For this example, simulation studies show that the generalized variance of $\hat{\zeta}(\mathbf{s})$ is minimized by setting $s_1 = s_2 = s$ and letting $s \to 0$. This choice for s was also suggested by techniques which explore the diagonal of the $\mathbf{s}$-space, $s_1 = s_2$, and select $\mathbf{s}$, either (a) to maximize the likelihood $L(\hat{\zeta}(\mathbf{s}))$, or (b) to minimize a cross-validation sum-of-squares. Details of these approaches are not given here, but they are presented in Laurence and Morgan (1987).

Setting $s_1 = s_2 = s$ and allowing $s \to 0$, the method of equating theoretical and empirical Laplace transforms reduces to one of conditional mean and zero-frequency with, from equation (5.8), the estimating equations

$$\alpha R(\alpha) = \hat{p}_m$$

and

$$\alpha R(\alpha)(1 + \alpha^2) - \alpha^2 = \mu \tilde{t} \qquad (5.10)$$

where

$$\alpha = \frac{\mu}{\sqrt{\gamma d}}$$

$R(\alpha) = \dfrac{1 - \phi(\alpha)}{\Phi(\alpha)}$ is the Mills ratio, $(\phi(\alpha) = \Phi'(\alpha))$

$\hat{p}_m$ is the observed proportion of nonresponders by time t_m and $\tilde{t}$ is an estimate of the mean time to response for the responders, given by:

$$\tilde{t} = \sum_{j=1}^{m-1} c_j(\hat{p}_j - \hat{p}_{j-1})$$

Equations (5.10) do not have an explicit solution, but they can be readily solved using the graph of Figure 5.4.

The summary statistics $\hat{p}_m$ and a conditional mean time are also advocated by Jarrett (1984) for summarizing and comparing sets of time-to-response quantal assay data without assuming any particular model. The MICE index is a function of the product of such terms (Exercise 5.8).

The parameter estimates obtained when this model is fitted to each sex-dose combination using the method of conditional mean and zero-frequency are given in Table 5.5, together with the maximum-likelihood estimates obtained by Diggle and Gratton. The overall agreement between both these parameter estimates and the measures of error can be seen to be very good.

It is shown in Laurence and Morgan (1990a) that there is no serious loss of efficiency in using the method of conditional mean and zero-frequency, and this paper also examines the bias of the method (Exercise 5.9).

It is natural to ask whether one might expect to take $s_1 = s_2 = 0$, to give the method of conditional mean and zero-frequency, in all

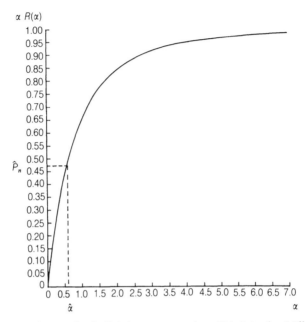

Figure 5.4 *The graph of* $\alpha R(\alpha)$ *versus* α, *where* $R(\alpha)$ *is the Mills ratio,* $R(\alpha) = \{1 - \Phi(\alpha)\}/\phi(\alpha)$.

cases of quantal response data collected over time. In fact when the transform estimation procedure is applied to the data of Table 5.1 we find that the best **s** does not correspond to **s = 0**. For further discussion of this case, see Exercise 5.13.

5.3.5   The case $\beta \neq 0$

Here we shall for simplicity just consider the case $\gamma = 0$. The likelihood function remains difficult to calculate in this case. For example, Puri and Senturia (1972) show that if $\beta\mu/\delta$ is not an integer, which is all we require for numerical analysis, the survivor function $S(t)$ for an initial dose d is given by,

$$S(t) = \sum_{k=0}^{\infty} \frac{1}{k!} \left\{ A_k\left(\frac{\beta\mu}{\delta} - 1\right) \left(\frac{\delta d}{\mu + \delta d}\right)^k \exp(-\beta d) I_{k,\mu + \delta d}(t) \right\}$$

$$+ \left\{ A_k\left(\frac{\beta\mu}{\delta}\right)\left(\frac{\delta}{\beta}\right)^k I_{k+1,\beta}(d)t^k \exp[-(\mu + \delta d)t] \right\}$$

where $A_k(x) = x(x + 1)(x + 2) \cdots (x + k - 1)$, $k \geqslant 1$, $A_0(x) = 1$, and for $\zeta > 0$, $I_{k,\zeta}(x)$ is the incomplete gamma integral,

$$I_{k,\zeta}(x) = \int_0^x \frac{\zeta^k y^{k-1}}{\Gamma(k)} \exp(-\zeta y) dy, \quad \text{for } k \geqslant 1, \text{ and } I_{0,\zeta}(x) = 1$$

For computation it is more efficient to generate the components of $S(t)$ by recursion formulae, together with a suitable procedure for truncating the summation. These are described in Exercise 5.10.

The properties of the estimators $\hat{\beta}$, $\hat{\mu}$ and $\hat{\delta}$ have been examined using simulated data. The variability of $\hat{\beta}$ was so great that numerical overflow problems were sometimes encountered, due to large estimates of β, when the simulations were based on 100 insects exposed at each dose level. Hence the results below are based on 1000 insects at each dose, and 100 replications.

As an illustration, consider the case:

$$\beta = 5.0, \ \mu = 0.25, \ \delta = 0.7, \ d = 0.5, \ t_i = 2i, \ 1 \leqslant i \leqslant 10$$

From the simulations we obtain

$$V(\hat{\beta}) = 50.03, \quad V(\hat{\mu}) = 0.09, \quad V(\hat{\delta}) = 0.0009$$
$$\text{Corr}(\hat{\beta}, \hat{\mu}) = 0.996, \quad \text{Corr}(\hat{\beta}, \hat{\delta}) = -0.257, \quad \text{Corr}(\hat{\mu}, \hat{\delta}) = -0.240$$

Thus the same high correlation involving $\hat{\beta}$ is encountered as in the analysis of section 5.3.1, which was for $t = \infty$ and was taken over all doses. If we reparameterize in terms of (β, τ, δ), with $\tau = \beta\delta/\mu$ as before, then

$$\text{Corr}(\hat{\beta}, \hat{\tau}) = 0.465, \ \text{Corr}(\hat{\tau}, \hat{\delta}) = 0.301$$

which is a more satisfactory result than that obtained above. However, $\hat{\beta}$ is still relatively imprecisely estimated, and with a positively skewed distribution, as in the study of section 5.3.1.

When fitting their full, four-parameter model to the flour-beetle data, Diggle and Gratton found that they could take $\beta = 0$. This was not possible for the $t = \infty$ analyses of section 5.3.1, involving all doses simultaneously, as it resulted in just a one-parameter model for quantal assay data. Likelihood ratio tests of the hypothesis that $\beta = 0$ for the three parameter model with $\gamma = 0$ are inappropriate when applied to the flour-beetle data, since this model does not provide a good description of those data. Consequently the tests were applied to simulated data of the same structure as above, but for a variety of values for β. We find, for example, that for $\beta = 20$, the hypothesis that $\beta = 0$ was not rejected 9 times out of 10, but for $\beta = 40$, the hypothesis was rejected 9 times out of 10.

We may conclude, therefore, that for analyses for each dose separately, and for time-evolving data, the original Puri–Senturia model could well be over-parameterized, and a much simpler model with $\beta = 0$ could then be used to describe the data. If it is necessary to take $\beta \neq 0$, then precision on $\hat{\beta}$ is likely to be low. The simulations used here involved substantially more individuals per dose than is realistic in practice, and it therefore seems unlikely that a model with $\beta \neq 0$ would be needed for an analysis of time-evolving data.

The Puri–Senturia model and its extension make use of the terms of survival analysis, through the hazard and the survivor function. However, the model provides a very specific model for survival analysis based on an assumed mechanism. In section 5.5 we shall employ a more standard application of the methods of survival analysis.

5.4   Multi-event models

Multistage and multi-hit models for quantal assay data were introduced in Exercises 1.17 and 1.19, and discussed also in section 4.6. The multistage model has been generalized to include time by Hartley and Sielken (1977a, b, 1978) to give an approximation to the probability of response to dose d by time t in the proportional hazards form:

$$P(t|d) = 1 - \exp\left\{ -\left(\sum_{i=0}^{s} \zeta_i d^i \right)\left(\sum_{j=1}^{l} \beta_j t^j \right) \right\}$$

In which the parameters, $\{\zeta_i\}$ and $\{\beta_j\}$ are all non-negative. Exact solution for the multi-hit model is straightforward, and so we concentrate on that model here.

As in Exercise 1.17, a dose d of substance is assumed to result in a Poisson process of 'hits' of intensity ζd, and an integer number of h hits are now deemed necessary to produce response. Directly, we can write

$$P(t|d) = \int_0^{\zeta t d} \frac{y^{h-1} e^{-y}}{(h-1)!} \, dy$$

and so the probability of response in (t_{j-1}, t_j) following dose d_i is then:

$$p_{ij} = \int_{\zeta t_{j-1} d_i}^{\zeta t_j d_i} \frac{y^{h-1} e^{-y}}{(h-1)!} dy, \quad 1 \leqslant i \leqslant k, \quad 1 \leqslant j \leqslant m+1$$

where $t_0 = 0$ and $t_{m+1} = \infty$.

Now let $n_{ij} = (r_{ij} - r_{i,j-1})$, denote the number of individuals given dose d_i which respond in (t_{j-1}, t_j).

The likelihood is then

$$L(h, \zeta) \propto \prod_{i=1}^{k} \prod_{j=1}^{m+1} p_{ij}^{n_{ij}}$$

Expressions needed for iterative maximum-likelihood maximization

Table 5.7(a) *Percentage of times the hypothesis: $h = 1$ is rejected in 1000 simulations, using a likelihood-ratio test. The case of five doses: $1, 2, \cdots, 5$ and one sampling time: 2. Values in parentheses correspond to a nominal 5% level. In some cases, for large ζ values, it was not possible to analyse the data due to large numbers of individuals responding*

No. of individuals per dose	h	ζ/h				
		0.1	0.2	0.3	0.5	0.75
5	1	(3.9)	(3.6)	(5.1)		
	2	15.6	17.6	21.9		
	3	19.9	35.2	39.0		
10	1	(4.0)	(4.8)	(3.6)	(4.6)	
	2	18.6	25.6	19.6		
	3	38.8	53.9	46.4		
20	1	(2.9)	(3.1)	(3.6)	(4.1)	(4.1)
	2	30.9	38.3	34.6	24.7	
	3	63.2	82.3	70.6	43.7	
30	1	(2.4)	(3.1)	(3.2)	(3.8)	(3.0)
	2	43.6	55.4	48.2	26.9	
	3	80.9	94.5	83.0	48.2	
60	1	(1.3)	(1.9)	(2.1)	(2.2)	(2.7)
	2	69.8	81.6	78.1	46.5	20.2
	3	97.6	99.7	99.1	82.4	23.2

and for error estimation are:

$$\frac{\partial p_{ij}}{\partial \zeta} = \frac{(\zeta t_j d_i)^h e^{-\zeta t_j d_i}}{(h-1)!\zeta} - \frac{(\zeta t_{j-1} d_i)^h e^{-\zeta t_{j-1} d_i}}{(h-1)!\zeta}$$

and

$$\frac{\partial p_{ij}}{\partial h} = \int_{\zeta t_{j-1} d_i}^{\zeta t_j d_i} \frac{y^{h-1} e^{-y} \log y}{(h-1)!} dy - \psi(h) p_{ij}$$

Discussion is given in Laurence and Morgan (1990b) and Exercise 5.11. It was noted in section 4.6 that for many sets of data taken at just one time-point, it might be difficult to distinguish between a one-hit and a multi-hit model. This becomes less difficult if repeated time samples are taken. An illustration of this is given

Table 5.7(b) *Percentage of times the hypothesis: $h = 1$ is rejected in 1000 simulations, using a likelihood-ratio test. The case of five doses: $1, 2, \cdots, 5$ and three sampling times: 0.25, 1.0, 2.0. Values in parentheses correspond to a nominal 5% level*

No. of individuals per dose	h	ζ/h					
		0.1	0.2	0.3	0.5	0.75	1
5	1	(3.7)	(3.2)	(3.6)	(3.9)	(3.1)	(3.8)
	2	35.1	50.9	52.9	54.5	43.2	35.3
	3	59.3	86.3	89.7	87.3	77.6	66.9
10	1	(2.9)	(3.3)	(2.5)	(3.1)	(4.6)	(3.7)
	2	61.1	77.8	80.4	76.7	69.0	60.0
	3	89.7	98.9	99.6	99.6	97.7	93.1
20	1	(1.5)	(1.0)	(1.4)	(1.6)	(1.3)	(2.2)
	2	82.8	93.7	94.7	92.9	90.0	82.7
	3	99.1	100.0	100.0	100.0	100.0	99.7
30	1	(0.2)	(0.1)	(0.2)	(0.3)	(0.4)	(1.1)
	2	91.4	98.2	98.9	98.2	94.5	91.1
	3	100.0	100.0	100.0	100.0	100.0	100.0
60	1	(0.1)	(0.0)[a]	(0.1)	(0.0)[a]	(0.1)	(0.0)
	2	99.4	99.8	100.0	99.8	99.5	99.1
	3	100.0	100.0	100.0	100.0	100.0	100.0

[a] In all cases $h = 1$

in Table 5.7 which gives the results of a test of $h = 1$ versus h is a positive integer > 1. Making use of the information from all three sampling times enables one to be reasonably confident of detecting whether or not $h > 1$ in an experiment with as few as 100 animals overall, and five dose levels. With only one sampling time a much larger experiment is needed. A similar point is made in Exercise 5.15, in the context of the data of Table 1.5. Related discussion is given by Armitage (1982); see also Exercises 5.13 and 5.14.

5.5 Survival analysis

The work of this section has its origins in the papers by Sampford (1952a, b, 1954) and White and Graca (1958), and makes use of the considerable development in the theory and application of methods of survival analysis since the early 1970s. As we are more interested in derived quantities, such as the ED_{50}, than evaluating the effect of covariates, we shall use parametric models rather than the semi-parametric approach of Cox (1972).

5.5.1 A mixture model for the flour-beetle data

Let us suppose that an individual given dose d_i of the substance under test has probability of response in the time interval (t_{j-1}, t_j) given by

$$p_{ij} = H(t_j; d_i) - H(t_{j-1}; d_i) \qquad (5.11)$$

for a suitable monotonic increasing function of t, $H(t; d_i)$, and $H(0; d_i) = 0$. If $H(\infty; d_i) = 1$ then we would model $H(t; d)$ with a cumulative distribution function, and that is done in section 5.5.2. However, in general, not all individuals may respond to any dose d_i, as appears to be the case for the flour-beetle data of Table 1.5, for which it seems appropriate to take $1 - H(\infty; d_i) > 0$ as the probability of surviving a dose d_i.

Pack (1986a) investigated taking

$$H(t_j; d_i) = \frac{a_i}{(1 + e^{-\eta_{ij}})}$$

where a_i is just a function of the dose level, d_i, and η_{ij} is a function of dose and time, and then fitting the model of equation (5.11) to the flour-beetle data.

Experimentation with fitting a variety of different models revealed that it was only necessary for η_{ij} to be a function of time alone, and a model that was fitted separately to the male and female beetle data had the form:

$$H(t_j; d_i)^{-1} = (1 + e^{-(\alpha_1 + \alpha_2 \log d_i)})(1 + \beta t_j^{-\phi})$$

(see also Morgan *et al.*, 1987). This model has a conveniently simple interpretation, which ties in with the classical threshold model for quantal assay data as follows: at dose level d_i, a proportion, $(1 + e^{-(\alpha_1 + \alpha_2 \log d_i)})^{-1}$ of individuals responds to the test substance. These susceptible individuals will respond at various times, and the survival-time-distribution has cumulative distribution function given by

$$F(t) = (1 + \beta t^{-\phi})^{-1}, \quad \text{for } t \geqslant 0, \text{ and parameters } \beta, \phi > 0 \quad (5.12)$$

irrespective of the dose. This is the same qualitative conclusion that was reached by Hewlett (1974) in his original analysis of the flour-beetle data. We recognize the cumulative distribution function of equation (5.12) as being that of a log–logistic distribution, i.e. the distribution of a random variable W, whose natural logarithm, $Z = \log_e W$ has a logistic distribution. An **assumption** of a time-to-response distribution which does not involve the dose was made by McLeish and Tosh (1990), to illustrate methodology described in section 8.5.

The Nelder–Mead simplex method was used to fit the model of equations (5.11) and (5.12) to the flour-beetle data, separately for the males and for the females. The maximum-likelihood parameter estimates, with estimated asymptotic standard errors (in parentheses) and correlation matrices, are shown below.

			Males			
Parameter	*Estimate*		*Estimated errors (in parentheses) and correlations*			
			$\hat{\alpha}_1$	$\hat{\alpha}_2$	$\hat{\phi}$	$\hat{\beta}$
α_1	4.63	$\hat{\alpha}_1$	(0.46)	–	–	–
α_2	3.37	$\hat{\alpha}_2$	0.976	(0.33)	–	–
ϕ	2.70	$\hat{\phi}$	−0.171	−0.152	(0.14)	–
β	14.28	$\hat{\beta}$	−0.105	−0.093	0.834	(2.38)

		Females			
Parameter	Estimate	Estimated errors (in parentheses) and correlations			
		$\hat{\alpha}_1$	$\hat{\alpha}_2$	$\hat{\phi}$	$\hat{\beta}$
α_1	2.61	$\hat{\alpha}_1$ (0.27)	–	–	–
α_2	2.61	$\hat{\alpha}_2$ 0.937	(0.22)	–	–
ϕ	3.48	$\hat{\phi}$ − 0.047	− 0.031	(0.20)	–
β	57.05	$\hat{\beta}$ − 0.035	− 0.023	0.906	(14.83)

Fitted values are given in Table 5.8. The agreement appears to be good, especially as the model contains only four parameters for each sex of beetle. The residual deviances and corresponding degrees of freedom are, for males, 48.39 (48), and for females 49.94 (48), again indicative of a good fit. The asymptotic chi-square approximation for the deviance was checked using simulation, and found to be satisfactory.

Fitting the same model to both male and female data sets produced a deviance of 139.91 on 100 degrees of freedom, and so the likelihood ratio test statistic of the hypothesis that $(\alpha_1, \alpha_2, \phi, \beta)$ does not vary with sex is 41.58, extremely significant when referred to chi-square tables on four degrees of freedom.

One way of investigating this difference further is as follows. We can assume that only (α_1, α_2) does not vary with sex. The residual deviance is then 128.38 on 98 degrees of freedom, highly indicative of a significant effect of sex on (α_1, α_2), suggesting in fact that males are more susceptible than females. Conversely, if we assume that only (ϕ, β) does not vary with sex, the residual deviance is then 110.70 on 98 degrees of freedom, suggesting that the distribution of time-to-death differs between sexes, males having a smaller mean time-to-death than females. Both of these conclusions were reached also by Diggle and Gratton (1984), after the much more complicated analysis of section 5.3.2. In many insect species there is a sex difference in susceptibility to insecticides. Females tend to be less susceptible than males, but this is not a general rule. In the case of *T. castaneum* beetles, females are slightly heavier than males, and this may account for the difference in susceptibility (Hewlett, P. S., personal communication).

Table 5.8 *The results of fitting the model of (5,11) and (5.12) to the flour-beetle data of Table 1.3, separately for each sex. O denotes observed values and F denotes fitted values*

(a)   *Male beetles*

Dose	0.20		0.32		0.50		0.80	
Group size	144		69		54		50	
Time (days)	O	F	O	F	O	F	O	F
1	3	3.0	7	3.1	5	3.2	4	3.2
2	14	14.1	17	14.9	13	15.4	14	15.3
3	24	16.1	28	27.5	24	28.3	22	28.3
4	31	33.8	44	35.6	39	36.7	36	36.6
5	35	38.1	47	40.2	43	41.4	44	41.4
6	38	40.6	49	42.8	45	44.1	46	44.0
7	40	42.1	50	44.3	46	45.7	47	45.6
8	41	43.0	50	45.2	47	46.7	47	46.6
9	41	43.5	50	45.9	47	47.3	47	47.2
10	41	43.9	50	46.3	47	47.7	48	47.6
11	41	44.2	50	46.6	47	48.0	48	47.9
12	42	44.4	50	46.8	47	48.2	48	48.2
13	43	44.6	50	46.9	47	48.4	48	48.3

(b)   *Female beetles*

Dose	0.20		0.32		0.50		0.80	
Group size	152		81		44		47	
Time (days)	O	F	O	F	O	F	O	F
1	0	0.4	1	0.6	0	0.5	2	0.7
2	2	4.2	6	5.4	4	5.0	9	6.8
3	6	11.5	17	14.8	10	13.5	24	18.4
4	14	17.7	27	22.8	16	20.8	33	28.4
5	23	21.3	32	27.4	19	25.0	36	34.3
6	26	23.2	33	29.9	20	27.3	40	37.3
7	26	24.2	33	31.2	21	28.5	41	39.0
8	26	24.8	34	31.9	25	29.2	42	39.9
9	26	25.1	34	32.4	25	29.6	42	40.4
10	26	25.3	34	32.6	25	29.8	43	40.7
11	26	25.5	34	32.8	25	30.0	43	41.0
12	26	25.6	34	32.9	26	30.1	43	41.1
13	26	25.6	34	33.0	27	30.1	43	41.2

The model considered here was extended by Pack (1986a) by embedding the log–logistic distribution for the survival-time distribution within a wider parametric family indexed by a third shape parameter. There are therefore clear similarities between this work and that of Chapter 4. The form adopted by Pack was

$$F(t; d_i) = 1 - [1 + \lambda \exp\{\eta(t, d_i)\}]^{1/\lambda}, \ \lambda \neq 0$$

$$F(t; d_i) = 1 - \exp[-\exp\{\eta(t, d_i)\}], \ \lambda = 0 \qquad (5.13)$$

where, initially, $\eta(t, d_i) = \psi \log(t) - \beta_i$. Note that the parameter λ is not completely unrestricted since we require $1 + \lambda \exp(\eta) > 0$. This permitted separate β_i values to be estimated for each dose level in the first instance.

The extended form of equation (5.13) has been discussed by Farewell and Prentice (1977) and Pettitt (1984). A more general form, giving a generalized Pareto model, has been considered by Clayton and Cuzick (1985) and has also been investigated by Bennett (1986). If $\lambda = 1$ then the cumulative distribution function of equation (5.13) reduces to the log–logistic form considered already, while if $\lambda = 0$ it becomes the Weibull distribution. In this family we therefore have representatives of proportional odds ($\lambda = 1$) and proportional hazards ($\lambda = 0$) families of models. Analysis based on the extended model of equation (5.13) is given in Pack and Morgan (1990a) – see Exercise 5.15 for extended discussion. Pack and Morgan (1990a) also experimented with extending the logit form adopted for the $\{a_i\}$, but found this unnecessary (Exercise 5.16).

5.5.2 The case of no long-term survivors

In the flour-beetle example considered above, the assumption of long-term survivors at each dose seemed reasonable. The model used extends that proposed by Farewell (1982b), who employed a logistic mixing proportion and a Weibull survival time distribution. In many cases, and the data of Tables 5.1 and 5.2 provide suitable illustrations, it would be unrealistic to assume that there are long-term survivors. In such an event we would model the probability of response by time t_j given dose d_i simply and directly by

$$p_{ij} = F(t_j; d_i) - F(t_{j-1}; d_i)$$

in which $F(t_j; d_i)$ is a suitable cumulative distribution function. Two examples of this will now be provided.

Example 5.1   Kooijman's Daphnia data

Pack (1986a) used the model of equation (5.13) to describe the data set of Table 5.1. Using maximum-likelihood estimation and the Nelder-Mead (1965) simplex method, he obtained a residual deviance of 71.22 on 54 degrees of freedom, resulting in a fit which is just acceptable at the 5% level. When the $\{\hat{\beta}_i\}$ were plotted against the doses $\{d_i\}$ a definite relationship was evident, and from the simpler model

$$\eta(t, d_i) = \psi \log t - \{\alpha_1 + \alpha_2 \exp(-\alpha_3 d_i)\} \qquad (5.14)$$

a residual deviance of 75.27 on 58 degrees of freedom was obtained.

Table 5.9   *Details of fitting the model of equation (5.14) to the data of Table 5.1 (Pack, 1986a)*

Parameter	Maximum likelihood estimate		Estimated asymptotic errors (in parentheses) and correlations				
			$\hat{\lambda}$	$\hat{\psi}$	$\hat{\alpha}_1$	$\hat{\alpha}_2$	$\hat{\alpha}_3$
λ	0.559	$\hat{\lambda}$	(0.285)	–	–	–	–
ψ	2.643	$\hat{\psi}$	0.904	(0.313)	–	–	–
α_1	−0.156	$\hat{\alpha}_1$	−0.469	−0.489	(0.675)	–	–
α_2	12.311	$\hat{\alpha}_2$	0.875	0.950	−0.695	(1.458)	–
α_3	0.031	$\hat{\alpha}_3$	−0.304	−0.387	0.945	−0.559	(0.004)

Dose (μg/l)	Group size	Fitted values Time (days)							
		2	4	7	11	14	16	18	21
0.0	50	0.0	0.0	0.0	0.1	0.3	0.4	0.5	0.8
3.2	50	0.0	0.0	0.1	0.5	0.9	1.3	1.7	2.5
5.6	49	0.0	0.1	0.6	1.0	1.9	2.7	3.6	5.3
10.0	50	0.0	0.3	2.3	3.8	6.8	9.3	12.0	16.4
18.0	53	0.3	2.1	14.4	21.2	30.8	36.0	40.2	44.7
32.0	50	3.7	17.6	44.2	47.2	48.9	49.4	49.6	49.8
56.0	59	29.7	54.0	58.8	58.9	59.0	59.0	59.0	59.0

This model describes death due both to natural causes and exposure to cadmium chloride. The resulting parameter estimates and fitted values are displayed in Table 5.9. For discussion of the fit see Exercise 5.18.

Example 5.2   Fish data

For the data of Table 5.2, Petkau and Sitter (1989) fit the Weibull model:

$$F_l(t_j; x_i) = 1 - \exp\{-t_j^{\gamma_l} \exp(\alpha_l + \beta_l \log x_i)\}$$

for the probability of death by time t_j at concentration x_i, in block l. It was found possible to fit the same model to the two blocks of data – cf. the polynomial growth curve analysis of these data described in section 5.2.2. A good fit, $X^2 = 14.76$ referred to χ^2_{39} tables, results from maximum likelihood parameter estimates of $\hat{\alpha} = -2.398$ (0.332), $\hat{\beta} = 0.889$ (0.137), $\hat{\gamma} = 1.630$ (0.249). For further discussion of this fit, see Exercise 5.17. We may observe that Ostwald's equation, mentioned in section 5.1, is appropriate for this model. A similar model is considered in Preisler and Robertson (1989).

5.6   Non-monotonic response

As we have already seen from the data of Table 1.6, responses over time are not always monotonic: affected individuals may recover, given sufficient time. The data of Table 1.6 are part of a larger set of data, involving four replicates. For illustration, just one of these is shown in Table 5.10, taken from Pack (1986a).

The model proposed by Pack (1986a) for these data has clear similarities with the mixture model of section 5.5.1, and makes the following assumptions:

1. Only a fraction of the group of flies are susceptible at concentration x. The probability that a fly will be knocked down at some time is given by $p_s(x) = [1 + \exp\{-\beta(\log x - \gamma)\}]^{-1}$.
2. A susceptible fly is knocked down at time T_K, which is a random variable from a distribution whose location, and possibly scale, depend on the concentration.
3. The duration of knock-down is T_L. This will also be a random variable whose mean and scale depend on concentration. T_K and

Table 5.10 *Data (and fitted values of the model of this section) describing numbers of flies (Musca domestica) flying at various times following exposure to aerosol fly-sprays filled with Bioallethrin*

Time (min)	Concentration (μg/l)				
	0.3	0.5	0.75	1.0	1.5
1	20	20	20	22	17
	20.0	19.8	19.3	20.6	17.8
2	20	19	19	17	10
	20.0	19.2	17.4	17.3	11.5
4	20	15	18	10	0
	19.8	16.3	12.3	8.8	4.0
6	19	15	17	2	0
	19.6	12.8	12.8	6.3	0.0
10	18	13	16	0	0
	18.5	12.9	12.6	1.5	0.0
15	18	15	14	1	0
	18.4	13.8	14.6	0.5	0.1
30	19	18	19	2	0
	19.6	18.2	15.7	4.3	0.6
60	20	20	19	10	3
	19.9	19.7	19.6	12.0	3.3
90	20	20	20	16	6
	20.0	20.0	19.6	16.3	7.5
150	20	20	20	20	17
	20.0	20.0	20.0	20.1	13.3
180	20	20	20	19	18
	20.0	20.0	20.0	20.7	17.9
Group size	20	20	20	22	20

T_L have independent log–logistic distributions. In detail we set:

$$F_{T_K}(t) = (1 + e^{-\eta_K})^{-1}$$

$$F_{T_L}(t) = (1 + e^{-\eta_L})^{-1}$$

where $\eta_K = \psi_K \log t - \alpha_K - \beta_K/x$

and $\eta_L = \psi_L \log t - \alpha_L - \beta_L \log x$

where $\psi_K, \psi_L > 0$.

(The forms of dependence of η_K and η_L on x follow from fitting location parameters for each dose and plotting these versus concentration. We would expect, as here, large concentration values to correspond to short times to knock-down and long periods knocked-down.)

4. At time $T_K + T_L$ a susceptible fly recovers and flies off.
5. We also assume that flies behave independently of each other.

Assumption 1 is regarded as reasonable by entomologists. The independence of T_K and T_L is questionable, and may be relaxed in elaborations of the model.

Model-fitting by maximum-likelihood to aggregate data, in which the movement of particular individuals is unknown, can be extremely difficult. Consequently, the above model was fitted by the weighted least squares procedure of Kalbfleisch *et al.* (1983), which proceeds as follows.

There are only two possible states in our example, namely flying and being knocked down. We assume that counts are taken at m time points, $t_j, j = 1, 2, \ldots, m$. Let

$$\mathbf{N}_j = (n_{F_j} \, n_{K_j})' \quad j = 1, 2, \cdots, m$$

where n_{F_j} is the number observed flying at time t_j and n_{K_j} is the number knocked down. The number of flies initially exposed is $n = n_{F_j} + n_{K_j}$, for any j.

The transition matrix $\mathbf{P}_j$ contains the probabilities of moving between the states between times t_{j-1} and t_j:

$$\mathbf{P}_j = \begin{pmatrix} \Pr(F_j|F_{j-1}), \Pr(K_j|F_{j-1}) \\ \Pr(F_j|K_{j-1}), \Pr(K_j|K_{j-1}) \end{pmatrix}$$

where, for example, $\Pr(F_j|K_{j-1})$ denotes the probability of being observed flying at time t_j conditional on having been observed knocked down at time t_{j-1}. As we only have two states, we can write

$$Pr(K_j|F_{j-1}) = 1 - Pr(F_j|F_{j-1})$$
$$Pr(K_j|K_{j-1}) = 1 - Pr(F_j|K_{j-1})$$

We also set

$$\mathbf{P}_j^* = \begin{pmatrix} Pr(F_j|F_{j-1}) \\ Pr(F_j|K_{j-1}) \end{pmatrix}$$

The means of $\mathbf{N}_j$ and n_{F_j}, conditional on $\mathbf{N}_{j-1}$ are

$$E(\mathbf{N}_j|\mathbf{N}_{j-1}) = \mathbf{P}_j'\mathbf{N}_{j-1}$$
$$E(n_{F_j}|\mathbf{N}_{j-1}) = \mathbf{P}_j^{*'}\mathbf{N}_{j-1}$$

The conditional variance of n_{F_j} is

$$V(n_{F_j}|\mathbf{N}_{j-1}) = \text{diag}(\mathbf{P}_j^{*'}\mathbf{N}_{j-1}) - \mathbf{P}_j'\,\text{diag}(\mathbf{N}_{j-1})\mathbf{P}_j = \Sigma_j^*$$

where diag(.) is a diagonal matrix with the elements of the vector argument on the diagonal.

Minimum chi-square estimates are then obtained by minimizing

$$S = \sum_{j=1}^{m} (n_{F_j} - \mathbf{P}_j^{*'}\mathbf{N}_{j-1})' \Sigma_j^{*-1}(n_{F_j} - \mathbf{P}_j^{*'}\mathbf{N}_{j-1})$$

This reduces to

$$S = \sum_{j=1}^{m} \frac{\zeta_j^2}{v_j}$$

where

$$\zeta_j = n_{F_j} - \{n_{F_{j-1}} Pr(F_j|F_{j-1}) + (n - n_{F_{j-1}}) Pr(F_j|K_{j-1})\}$$

and

$$v_j = n_{F_{j-1}} Pr(F_j|F_{j-1})\{1 - Pr(F_j|F_{j-1})\}$$
$$+ (n - n_{F_{j-1}}) Pr(F_j|K_{j-1})\{1 - Pr(F_j|K_{j-1})\}$$

We now need to derive the required transition probabilities for our model. First, we have the probability of observing an individual given concentration x flying at time t_j is

$$Pr(F_j) = 1 - p_s + p_s Pr(T_K > t_j) + p_s Pr(T_K + T_L < t_j)$$

where we write $p_s = p_s(x)$.

In the same way, the probability of being observed flying at both t_{j-1} and t_j is

$$\Pr(F_j, F_{j-1}) = 1 - p_s + p_s \Pr(T_K > t_j) + p_s \Pr(T_K + T_L < t_{j-1})$$
$$+ p_s \Pr(t_{j-1} < T_K < t_j, T_K + T_L < t_j)$$

We combine these probabilities to give

$$\Pr(F_j | F_{j-1}) = \frac{\Pr(F_j, F_{j-1})}{\Pr(F_{j-1})}$$

In a similar way, we also have the probability of being knocked-down at time t_j is

$$\Pr(K_j) = p_s \Pr(T_K < t_j, T_K + T_L > t_j)$$
$$= p_s[\Pr(T_K < t_j) - \Pr(T_K + T_L < t_j)]$$
$$= 1 - \Pr(F_j)$$

Now,

$$\Pr(F_j, K_{j-1}) = \Pr(K_{j-1}) - \Pr(K_j, K_{j-1})$$

and

$$\Pr(K_j, K_{j-1}) = p_s \Pr(T_K < t_{j-1}, T_K + T_L < t_j)$$
$$= p_s \{\Pr(T_K < t_{j-1}) - \Pr(T_K + T_L < t_j)$$
$$+ \Pr(t_{j-1} < T_K < t_j, T_K + T_L < t_j)\}$$

so that

$$\Pr(F_j, K_{j-1}) = p_s \{\Pr(T_K + T_L < t_j) - \Pr(T_K + T_L < t_{j-1})$$
$$- \Pr(t_{j-1} < T_K < t_j, T_K + T_L < t_j)\}$$

and we can then form

$$\Pr(F_j | K_{j-1}) = \frac{\Pr(F_j, K_{j-1})}{\Pr(K_{j-1})}$$

In principle, fitting this model is straightforward once we have chosen the distribution of time to, and duration of, knock-down. First we note that both $\Pr(T_K + T_L < t_j)$ and $\Pr(t_{j-1} < T_K < t_j, T_K + T_L < t_j)$ are convolution integrals. If f_K and f_L are the density functions for time to, and duration of, knock-down respectively, and F_L is the cumulative distribution function for the duration of knock-down, then

$$\Pr(T_K + T_L < t_j) = \int_0^{t_j} \int_0^{t_j - t} f_K(t) f_L(l) \, dl \, dt$$

$$= \int_0^{t_j} f_K(t) F_L(t_j - t) \, dt$$

and

$$\Pr(t_{j-1} < T_K < t_j, T_K + T_L < t_j) = \int_{t_{j-1}}^{t_j} f_K(t) F_L(t_j - t) \, dt$$

The chi-square criterion S was minimized using the Nelder and Mead (1965) simplex method, resulting, for the data of Table 5.10, in a minimized value for S of 67.93(d.f. = 47) and the estimates:

Parameter	Estimate		$\hat{\beta}$	$\hat{\gamma}$	$\hat{\psi}_K$	$\hat{\alpha}_K$	$\hat{\beta}_K$	$\hat{\psi}_L$	$\hat{\alpha}_L$	$\hat{\beta}_L$
					Estimated asymptotic errors (in parentheses) and correlation matrix					
β	2.701	$\hat{\beta}$	(0.512)	–	–	–	–	–	–	–
γ	−0.728	$\hat{\alpha}$	0.72	(0.118)	–	–	–	–	–	–
ψ_K	2.464	$\hat{\psi}_K$	0.06	0.21	(0.216)	–	–	–	–	–
α_K	1.084	$\hat{\alpha}_K$	0.30	0.38	0.40	(0.332)	–	–	–	–
β_K	1.466	$\hat{\beta}_K$	−0.43	−0.42	0.34	−0.60	(0.252)	–	–	–
ψ_L	2.471	$\hat{\psi}_L$	0.02	0.02	−0.07	−0.07	0.01	(0.233)	–	–
α_L	9.579	$\hat{\alpha}_K$	0.08	0.09	−0.05	−0.04	−0.03	0.98	(0.963)	–
β_L	4.524	$\hat{\beta}_L$	−0.46	−0.52	−0.15	−0.30	0.33	0.56	0.51	(0.652)

The fitted values are shown in Table 5.10. An overall reduction in the number of parameters results from fitting this model to the four sets of data, with a common value for ψ_K and a common value for ψ_L. The model may now be used to summarize salient features of the data, such as the EC_{50}, estimated as 0.483 (0.06). For further discussion, see Pack and Morgan (1990b).

5.7 Discussion

For the repeated measures type of time-to-response data considered in this chapter, the most suitable form of analysis is through survival analysis. In this chapter we have considered both mixture models and more conventional models, of proportional hazards and proportional odds forms.

Additional work in the area is described by Finkelstein (1986) who provides full details of a direct analysis of interval-censored failure time data using a proportional hazards model, and in a pair of similar papers by Aranda-Ordaz (1983) and Tibshirani and Ciampi (1983). The last two papers provide the detail of a model-fitting procedure using GLIM, for grouped data, which enables one to try to distinguish between proportional and additive hazards models, and also to estimate time-dependent baseline hazards. It is suggested in the paper by Pierce *et al.* (1979) that investigation of the effect of covariates may be easier within the context of such models, as opposed to mixture models, though Farewell's (1982b) results contradict this for the particular data set considered there. Williams (1986c) fitted a log-normal survival-time distribution to a further set of data from aquatic toxicology.

Unconventional survival analysis models, as in sections 5.3 and 5.6, can be very difficult to analyse and fit to data. Frequently, however, mechanistic models will have a clear justification, and estimates are required of the model parameters. This relates to the discussion of section 4.6, and also the pragmatic approch discussed in sections 3.10 and 4.7. An instance of a further set of data which may be described by a mechanistic model is provided in Exercise 5.21. Other examples of mechanistic models in this area are provided by Gart (1963), Finkelstein and Ryan (1987) and Chew and Hamilton (1985). In the last case a compartment model is used for time-to-response data for fish swimming in tanks contaminated with a toxicant.

Observations of insects passing through 'stages' of development, as in Exercise 5.22, result in ordered categorical data, of a kind which motivated the paper of Aitchison and Silvey (1957), resulting in the analyses described as section 3.5.

We note finally that Petkau and Sitter (1989) investigated whether it was necessary to add over-dispersion to the fitted model described in Example 5.2, and concluded that this was not, in fact, necessary (McCullagh and Nelder, 1989, p. 175). They used the Dirichlet-multinomial distribution. O'Hara Hines and Lawless (1992) suggest that this is not realistic for data sampled over time, and present alternative approaches, involving additive errors (see the work of section 6.6.4). Models are fitted using quasi-likelihood as described in section 6.4. They emphasize that failure to model over-dispersion will overestimate precision. It is over-dispersion which is the topic of the next chapter.

5.8 Exercises and complements

The exercises vary in complexity. It is particularly desirable to attempt the five exercises marked with a †. Exercises marked with a * are generally more difficult or speculative.

5.1 Boyce and Williams (1967) estimated the values of the concentration × time product to give 90% mortality in a study. Their results are illustrated in Figure 5.5. Discuss the possible implications of such a figure.

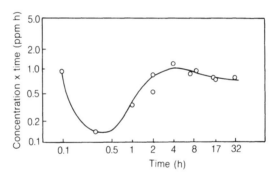

Figure 5.5 *(Boyce and Williams, 1967)* Product of concentration of N-tritylmorpholine and exposure time required to give 90% mortality of the snail, Australorbis glabratus, *plotted versus time. Reproduced from the Annals of Tropical Medicine and Parasitology, with permission.*

5.2† Calculate the MICE index for the four concentrations applied to the male flour-beetles of Table 1.5, and comment on the values which result.

5.3 Discuss why the multivariate normal distribution assumption of equation (5.1) is unlikely to be satisfied for the particular application of section 5.2.2.

5.4 In his probit analysis of the endpoint mortality data of

Table 1.5, Hewlett (1974) obtained the following results:

$$\text{Males Log}_{10} LD_{50} = -0.589 \, (0.021)$$
$$\text{Females Log}_{10} LD_{50} = -0.436 \, (0.021)$$
$$\hat{\beta} = 3.807 \, (0.298)$$
$$X^2 = 5.53, \, p = 0.35 \, (\text{d.f.} = 5)$$
$$\text{Test statistic for parallelism} = 0.80, \, p = 0.37$$
$$\text{Susceptibility ratio, male/females} = 1.42$$

Explain these results, and compare these with the results of a logit analysis which produced

$$\text{Males Log}_e LD_{50} = -1.361 \, (0.044)$$
$$\text{Females Log}_e LD_{50} = -0.992 \, (0.056)$$

$5.5^\dagger$ Verify the form of equation (5.3), and show how Newton–Raphson may be used to solve for the ED_{50}.

5.6 Provide the Newton–Raphson iteration for fitting the model of equation (5.2) to endpoint mortality data.

5.7 Consider how the results of Table 5.5 might lead to a reduction in the number of parameters in the Diggle and Gratton (1984) description of the data of Table 1.5.

5.8 Show that the MICE index may be regarded as a function of the product of a conditional mean time and a proportion not responding.

5.9* Consider how to derive measures of error for the pair of parameters, (μ, γ), estimated by the method of conditional mean and zero-frequency of section 5.3.3.

5.10* Consider how the components of $S(t)$ in section 5.3.5 may be generated by recursion formulae, and suggest a possible truncation rule.

$5.11^\dagger$ Provide the Newton–Raphson iteration for fitting the multi-hit model to time-dependent data, using the information outlined in section 5.4.

*5.12** (Laurence and Morgan, 1990a.) For the Kooijman data of Table 5.1, when fitting the model of equation (5.5) using a transform approach, methods of selecting the transform parameters produced a number of non-zero values of s, as shown below for a constrained optimization approach. Discuss these results.

Dose	$\hat{\gamma}$	$\hat{\mu}$	s
3.2	0.0001	0.0530	0.2298
5.6	0.0002	0.0218	0.2087
10.0	0.0009	0.1329	0.0000
18.0	0.0028	0.042	0.0000

5.13 Given below are data corresponding to Table 5.7, but with a geometric progression of dose levels. Discuss the results, and make a comparison with those in Table 5.7.

(a) Percentage of times the hypothesis: $h = 1$ is rejected in 1000 simulations, using a likelihood-ratio test. The case of five doses: 5/16, 5/8, 5/4, 5/2, 5, and three sampling times: 0.25, 1.0, 2.0.

No. of individuals per dose	h	ζ/h						
		0.1	0.2	0.3	0.5	0.75	1
5	1	(4.8)	(5.1)	(5.8)	(4.2)	(3.8)	(4.2)
	2	32.5	43.9	51.2	50.0	46.2	44.6
	3	–	81.1	86.6	87.2	81.7	80.7
10	1	(3.6)	(3.4)	(4.0)	(2.5)	(3.1)	(2.7)
	2	51.5	70.8	76.6	78.2	74.0	70.5
	3	83.6	98.4	99.2	99.1	99.5	98.4
20	1	(1.9)	(1.2)	(1.3)	(0.7)	(1.2)	(1.4)
	2	81.9	81.6	93.2	92.7	93.0	90.7
	3	98.6	99.9	100.0	100.0	100.0	100.0
30	1	(1.0)	(0.5)	(0.2)	(0.1)	(0.8)	(0.8)
	2	90.1	96.2	97.3	97.6	97.0	95.9
	3	99.8	100.0	100.0	100.0	100.0	100.0
60	1	(0.1)	(0.1)	(0.0)	(0.0)	(0.1)	(0.1)
	2	98.0	100.0	99.8	99.9	99.9	100.0
	3	100.0	100.0	100.0	100.0	100.0	100.0

Values in parentheses correspond to a nominal 5% level. In one case it was not possible to analyse the data, due to small numbers of individuals responding.

(b) Percentage of times the hypothesis $h = 1$ is rejected in 1000 simulations, using a likelihood-ratio test. The case of five doses: 5/16, 5/8, 5/4, 5/2, 5, and one sampling time: 2. Values in parentheses correspond to a nominal 5% level. In two cases it was not possible to analyse the data, due to large/small numbers of individuals responding.

No. of individuals per dose	h	ζ/h					
		0.1	0.2	0.3	0.5	0.75	1
5	1	(6.8)	(3.7)	(5.3)	(4.6)	(4.9)	–
	2	25.4	28.7	31.3	22.1	16.6	13.1
	3	–	52.8	54.1	40.2	26.8	22.5
10	1	(4.3)	(1.9)	(4.7)	(4.4)	(5.3)	(5.7)
	2	29.9	39.1	39.0	36.3	25.9	23.7
	3	55.5	76.0	77.5	65.4	52.7	40.3
20	1	(3.6)	(2.8)	(3.3)	(3.1)	(4.3)	(3.2)
	2	52.7	66.9	66.2	55.9	38.7	29.9
	3	84.0	97.4	97.5	91.7	76.8	61.9
30	1	(2.6)	(2.2)	(2.8)	(3.1)	(3.8)	(3.4)
	2	67.2	81.0	80.8	70.9	52.3	40.5
	3	95.8	99.6	99.7	98.8	91.6	78.2
60	1	(1.0)	(0.7)	(1.0)	(1.7)	(2.4)	(2.4)
	2	88.9	94.1	95.5	90.8	79.3	66.0
	3	99.9	100.0	100.0	99.9	99.3	97.3

5.14   In Figure 5.6 we present graphs of the ratio of the asymptotic variance of $\hat{h}$ in the multi-hit model when there are three sampling times in arithmetic progression as in Table 5.7, to the same variance when only the data at the last time-point are utilized. Explain the non-monotonic nature of the graphs.

5.15[†]   (Pack and Morgan, 1990a.)

(i) Write down the form of the likelihood when the model (Model a) based on the extended cumulative distribution function of

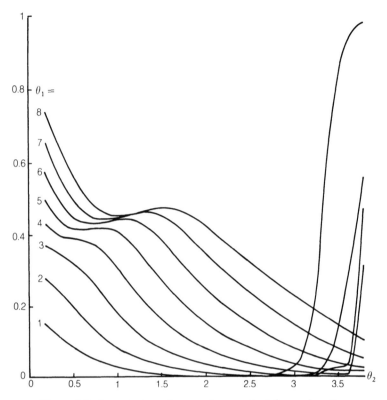

Figure 5.6 *Asymptotic variance ratio – see text for explanation*

equation (5.13) is to be fitted to the flour-beetle data of Table 1.5.
The residual deviances for the fitted models are 37.75 and 41.95
for males and females respectively, each being on 44 degrees of
freedom. Maximum-likelihood estimates of parameters, together
with estimated asymptotic errors, are shown below. Discuss
whether the log–logistic survival time distribution provides a
suitable description of the data.

(ii) Two further models were fitted to the data: Model b, with
$\beta_1 = \beta_2 = \beta_3 = \beta$, and Model c, with α_2 the same for males and
females. The resulting point estimates are shown below. For
Model c, the deviance is 92.03 on 95 degrees of freedom, while
for Model b we have:

Deviances	Males	Females	d.f.
Model b	40.85	49.93	47

Model a

	Males		Females	
Parameter	m.l.e.	s.e.	m.l.e.	s.e.
α_1	4.286	0.384	2.596	0.274
α_2	3.148	0.275	2.598	0.219
λ	0.244	0.113	0.922	0.297
ψ	2.124	0.159	3.503	0.404
β_1	2.635	0.197	4.590	0.529
β_2	2.168	0.183	3.782	0.404
β_3	2.356	0.195	4.687	0.432
β_4	2.478	0.199	3.729	0.355

Model b

	Males		Females	
Parameter	m.l.e.	s.e.	m.l.e.	s.e.
α_1	4.302	0.388	2.603	0.275
α_2	3.158	0.277	2.603	0.220
λ	0.327	0.112	0.953	0.267
ψ	2.193	0.157	3.425	0.355
β	2.427	0.158	4.006	0.339

Model c

	Males		Females	
Parameter	m.l.e.	s.e.	m.l.e.	s.e.
α_1	3.870	0.249	2.876	0.226
α_2	2.836	0.172	2.836	0.172
λ	0.321	0.110	0.972	0.276
ψ	2.189	0.156	3.438	0.359
β	2.426	0.157	4.014	0.341

Discuss these results.

(iii) Estimates of the $\log(LD_{50})$s for Model c for day 13 are:

Males:	-1.365	(0.034)
Females:	-1.014	(0.036)

Verify that when a logit model with parallelism is fitted to the day 13 data above, the corresponding values are as given in Exercise 5.4. Compare these two sets of results. Cf. also section 5.4.

(iv) In Figure 5.7 we present contours corresponding to $F(t; d) =$ constant. Discuss the shape of these contours.

5.16 (Continuation) Although Model c provides a satisfactory fit to the data it does not model the endpoint mortalities very closely. The mortality counts in most of the cells are zero and this may have some effect on the test for goodness-of-fit. Therefore, a small-scale simulation study was undertaken to check the fit of the model. One hundred sets of data were simulated from Model c using the maximum-likelihood estimates of the parameters and the same group sizes. The range of deviances was 75.76 to 129.92 with a mean of

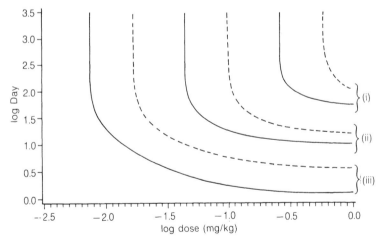

Figure 5.7 *Contours corresponding to* F(t; d) = p *for Model* c *with (i)* p = 0.9, (ii) p = 0.5, (iii) p = 0.1. *Reproduced from Pack and Morgan (1990a) with permission from the Biometric Society.*
Key: −−−−− F
　　　 ——— M

90.12 and a sample variance of 150.75. The fitted model is therefore clearly consistent with the simulated data.

One way in which we might try to improve the model is to use a different function for the mixing proportion as this will determine the endpoint mortalities. One such function which contains the logistic function as a special case is

$$a(d_i) = 1 - [1 + \lambda_2 \exp\{\eta(d_i)\}]^{-1/\lambda_2} \quad \lambda_2 \neq 0$$

$$a(d_i) = 1 - \exp[-\exp\{\eta(d_i)\}] \qquad \lambda_2 = 0$$

where

$$\eta(d)_i = \alpha_1 + \alpha_2 \log(d_i)$$

which we have already been using for the survival time cumulative distribution function. We compare this model with Model b by fitting it separately to the two sexes. The deviances for this model (Model d) are 39.35 and 49.22 respectively for the males and females. These have to be compared to the deviance values for Model b, given in Exercise 5.15(ii). The parameter estimates are shown below:

Maximum likelihood parameter estimates and estimated asymptotic standard errors from fitting Model d

Parameter	Males m.l.e.	Males s.e.	Females m.l.e.	Females s.e.
α_1	20.854	10.922	1.478	0.734
α_2	12.760	6.297	1.909	0.476
λ_2	6.428	3.360	0.121	0.566
λ	0.322	0.110	0.965	0.273
ψ	2.190	0.156	3.433	0.358
β	2.426	0.157	4.011	0.340

Discuss whether extending the description of $a(d)$ in this way is worthwhile.

5.17 Petkau and Sitter (1989) provide the comparisons shown below between the fit of their model, described in Example 5.2, and that of the Carter and Hubert (1984) model of section 5.2.2. Graph and discuss these results.

Point estimates and confidence intervals for LD_{50} for the data of Table 5.2:

	Model of Example 5.2		Model of section 5.2.2	
Time (days)	Estimate	95% CI	Estimate	95% CI
2.0	2.76	(1.82, 4.88)	5.60	(3.99, 8.76)
2.5	1.83	(1.32, 2.77)	1.83	(1.61, 2.11)
3.0	1.31	(0.98, 1.81)	1.02	(0.93, 1.38)
3.5	0.99	(0.73, 1.32)	0.84	(0.68, 1.07)
4.0	0.77	(0.54, 1.04)	0.70	(0.56, 0.91)
6.0	0.37	(0.20, 0.57)	0.49	(0.37, 0.66)

5.18 Discuss the fit of the model of equation (5.14) to the data of Table 5.1. Does Ostwald's equation hold? Might a mixture model, such as those of section 5.5.1, be appropriate?

5.19 Using standard notation, why can we describe the model:

$$\text{logit}\,(p_{ij}) = \lambda \log t_j - \alpha_1 - \alpha_2 \log d_i$$

as a proportional odds model? Discuss how you would fit this model to data.

5.20 Consider how you would analyse the data of Table 1.8.

5.21[†] The data below have been presented by Fenlon (1988), and represent the cumulative mortality of larvae of the diamond-backed moth (*P. xylostella*) at different dose levels of *P.x. granulosis* virus.

Dose level ($\times 10^5$)	No. of larvae	Percentage cumulative mortality by day					
		3	4	5	6	7	8
1	32	0	0	9	25	25	50
3	28	0	0	4	32	36	39
6	35	0	0	20	34	34	69
10	28	0	0	25	64	71	89
30	27	0	0	33	63	66	74
60	34	0	0	38	88	88	91
100	37	0	22	62	95	95	97
300	37	0	16	89	89	97	100
600	34	0	3	65	100	100	100

Plot the data, and discuss any unusual features.

5.22 An insect passes through k stages before it becomes adult. Denote the time in the ith stage by T_i, and let $T = \Sigma_{i=1}^k T_i$. Suppose that the T_i are independent, identically distributed exponential random variables, with respective means, λ_i^{-1}. For the cases $k = 2, 3$, evaluate the probability that the adult form is reached by time t, and provide approximations for the case small λ_i. How do these values compare with the multi-stage expression of section 5.4?

5.23 Pool the data of Table 5.2 over the two blocks. Compare and contrast the ED_{50} values, and the confidence-intervals for these, resulting from separate logit models at each of the time points: 2, 3 and 4 days, with the values given below, resulting from the survival-analysis approach of Petkau and Sitter (1989), described in Example 5.2.

Time (days)	ED_{50}	95% confidence-interval
2	2.76	(1.82, 4.88)
3	1.31	(0.98, 1.81)
4	0.77	(0.54, 1.04)

The computer program for obtaining the trimmed Spearman–Kärber estimate of the ED_{50} (section 7.4.1 and Appendix E) appears to treat data from successive time points as independent.

CHAPTER 6

Over-dispersion

In the work considered so far in this book we have encountered models based on simple distributions such as the binomial or the Poisson. In this chapter we shall present ways of relaxing the basic assumptions which result in these simple models. The main emphasis will be placed on extensions of the binomial distribution, and so the work will be particularly relevant to experiments in which it is litters, rather than individual animals, which are to be regarded as the experimental units.

Williams (1988a) listed five different types of experiment which could result in the kind of data that we shall consider in this chapter. These are as follows:

1. Teratology experiments, where pregnant females are treated and counts are made of dead or deformed foetuses in litters. Deformities may be of several different kinds (e.g. Eriksson and Styrud, 1985), but this multiple-response feature will not be considered here (cf. section 3.8).
2. Dominant lethal assays result when it is the **male** animals which are treated. These are then mated to one or more females. Here again response may be measured in terms of features of foetuses, or in some cases also in terms of **implants**.
3. Certain reproduction studies span several generations and may result in indices measuring the viability of surviving individuals. An interesting example is provided by Vince's work on the effect of auditory stimulation of quail eggs (e.g. Vince, 1979; Vince and Chinn, 1971; Morgan and North, 1985).
4. In *in vivo* cytogenetic assays, samples of cells are taken from animals and examined for chromosomal aberration. An example of the kind of data which may result is given in Exercise 6.8.
5. In chronic toxicity trails animals receiving the same treatment may be caged to ensure that they do not suffer adversely from

being in isolation. (For treatment of related problems, see Huster *et al.*, 1989 and Clayton, 1991.)

Other examples will also be encountered later. We shall find it convenient to use the terminology of 'litters' in the rest of the chapter.

6.1 Extra-binomial variation

Tables 1.10a and 1.10b present data describing foetal deaths in mice. Simple inspection of these tables is sufficient for us to doubt the adequacy of a binomial model to describe all of the data. As it is now the litters, rather than the individuals, which are the experimental units, then on biological grounds we would expect differences between litters.

To start we shall just consider ways of modelling variation **in excess** of what we might expect from simple models such as the binomial, without reference to making comparisons, assessing treatment effects and incorporating dose-related features; these will come later. The problem of simple models being too simple in this context is not a new one, and was recognized over 40 years ago by writers such as Cochran (1943) and Finney (1947, section 18).

The notation we adopt is as follows:

We have k treatment groups, the ith with m_i litters. The different groups may or may not be associated with different dose levels of some substance. The jth litter out of the m_i has size n_{ij}, and out of the n_{ij} individuals, x_{ij} respond.

It may be useful to refer to the diagram below for a quick check of the notation:

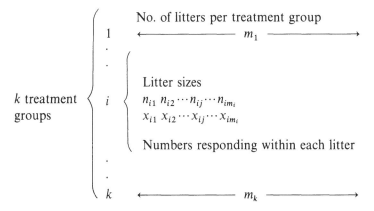

For the moment we shall assume that the probability of response is independent of litter size, though later we shall consider models which incorporate litter size. We shall also usually base our inferences on proportions responding within litters, **conditional** upon observed litter sizes. Clear biological arguments have been advanced to support the hypothesis that litter size should not be treatment-dependent in certain special cases, as, for example, in the case of dominant lethal assays (Pack, 1986a, p. 2; UKEMS report, 1983), or if dose levels are all at a somewhat low level. However, in other cases some useful information may be lost if this assumption is made – for further discussion see Rai and Van Ryzin (1985), Williams (1987b, 1988a), and section 6.6.5.

The basic binomial model assumes a constant probability p_i of response for all individuals in all litters in the ith treatment group. Conditional upon the litter size, the number, X_{ij}, of foetuses responding out of a total of n_{ij} in a litter then has a binomial distribution, with probability function.

$$\Pr(X_{ij} = x_{ij} \mid n_{ij}) = \binom{n_{ij}}{x_{ij}} p_i^{x_{ij}} (1 - p_i)^{n_{ij} - x_{ij}}$$

$$0 \leqslant p_i \leqslant 1$$
$$0 \leqslant x_{ij} \leqslant n_{ij}$$
$$1 \leqslant j \leqslant m_i$$
$$1 \leqslant i \leqslant k$$

The maximum likelihood estimate of p_i is then,

$$\hat{p}_i = \sum_{j=1}^{m_i} x_{ij} \left/ \sum_{j=1}^{m_i} n_{ij} \right.$$

Many authors have found this to be an unsatisfactory model for their data (e.g. Haseman and Soares, 1976; McCaughran and Arnold, 1976; Aeschbacher et al., 1977; James and Smith, 1982). As an illustration, when the binomial model is fitted to the data of Tables 1.10a and 1.10b then we obtain the fitted values of Tables 6.1a and 6.1b. In each case there is just the one group being considered here, so that $k = 1$. The fits are clearly poor (see also Exercise 6.10). They were obtained by Pack (1986a), who also formally examined the goodness-of-fit using a Monte Carlo approach: 499 sets of binomial data were simulated, using the fitted probabilities of response, and simulating litter sizes from the empirical distribution of litter sizes. For each data set the binomial model was fitted and the log-likelihood was noted. It was found that for the data of Table 6.1a the maximum log-likelihood was -842.61, compared with a range of $(-753.37, -657.44)$ for the simulated data, and for the data of Table 6.1b the corresponding values were -765.06 and $(-684.94, -594.35)$; these results confirm the poor binomial fits.

When the binomial model is inappropriate, this is usually due to variation in the data which is **greater** than expected under the binomial assumptions, and the term 'extra-binomial variation' has come to be used to describe this case. Later we shall encounter examples in which there is **less** variation than expected, which are less frequently considered in the literature (Exercises 3.10 and 3.11).

6.1.1 The beta-binomial model

The binomial model assumes that the n_{ij} individuals in the jth litter of the ith treatment group respond independently to treatment, each with some fixed probability p_i of responding, **which is the same for all litters**. This assumption may be relaxed by supposing that p_i varies between litters. A convenient distributional form to adopt for p_i is the beta distribution, with probability density function given by:

$$f(p_i) = \frac{p_i^{\alpha_i - 1}(1 - p_i)^{\beta_i - 1}}{B(\alpha_i, \beta_i)}, \quad 0 \leqslant p_i \leqslant 1, \quad \alpha_i, \beta_i > 0$$

and where $B(\alpha_i, \beta_i)$ is the beta function. In general, we expect α_i and β_i to vary between treatment groups. The mean and variance of this distribution are given respectively by:

$$E[p_i] = \left(\frac{\alpha_i}{\alpha_i + \beta_i}\right)$$

$$V(p_i) = \frac{\alpha_i \beta_i}{(\alpha_i + \beta_i)^2 (\alpha_i + \beta_i + 1)}, \quad 1 \leqslant i \leqslant k$$

If the random variable X_{ij} denotes the number of individuals responding out of n_{ij}, then as shown above, conditional upon p_i, and litter size n_{ij},

$$\Pr(X_{ij} = x_{ij} | n_{ij}, p_i) = \binom{n_{ij}}{x_{ij}} p_i^{x_{ij}}(1 - p_i)^{n_{ij} - x_{ij}}, \quad 0 \leqslant x_{ij} \leqslant n_{ij}$$

Table 6.1a *Observed frequency distribution of foetal death and fitted frequencies under the binomial model. Data taken from Haseman and Soares (1976) – data set no. 1. Results from Pack (1986a)*

Litter size (n)	Number of dead foetuses (x)													
	0	1	2	3	4	5	6	7	8	9	10	11	12	13
1	2	–	–	–	–	–	–	–	–	–	–	–	–	–
	1.8	0.2	–	–	–	–	–	–	–	–	–	–	–	–
2	2	–	–	–	–	–	–	–	–	–	–	–	–	–
	1.7	0.3	–	–	–	–	–	–	–	–	–	–	–	–
3	3	–	–	–	–	–	–	–	–	–	–	–	–	–
	2.3	0.7	0.1	–	–	–	–	–	–	–	–	–	–	–
4	5	1	1	–	–	–	–	–	–	–	–	–	–	–
	4.8	1.9	0.3	–	–	–	–	–	–	–	–	–	–	–
5	2	2	–	–	–	–	–	–	–	–	–	–	–	–
	2.5	1.2	0.2	–	–	–	–	–	–	–	–	–	–	–
6	2	2	–	–	–	–	–	–	–	–	–	–	–	–
	2.3	1.3	0.3	–	–	–	–	–	–	–	–	–	–	–
7	2	2	2	1	–	–	–	–	–	–	–	–	–	–
	3.6	2.5	0.7	0.1	–	–	–	–	–	–	–	–	–	–
8	6	1	–	1	1	–	–	–	–	–	–	–	–	–
	4.3	3.3	1.1	0.2	–	–	–	–	–	–	–	–	–	–
9	2	3	1	–	–	–	–	–	–	–	–	–	–	–
	2.6	2.3	0.9	0.2	–	–	–	–	–	–	–	–	–	–
10	2	4	2	–	2	–	–	–	–	–	–	–	–	–
	3.9	3.8	1.7	0.4	0.1	–	–	–	–	–	–	–	–	–
11	19	11	3	3	–	–	–	–	–	–	–	–	–	–
	12.9	13.9	6.8	2.0	0.4	0.1	–	–	–	–	–	–	–	–
12	33	24	11	5	4	4	–	–	–	–	–	–	1	–
	26.7	31.4	16.9	5.5	1.2	0.2	–	–	–	–	–	–	–	–
13	39	27	12	6	5	2	–	–	1	–	–	–	–	–
	27.3	34.8	20.4	7.3	1.8	0.3	–	–	–	–	–	–	–	–
14	34	30	14	6	6	–	–	1	–	–	–	–	–	–
	24.6	33.7	21.5	8.4	2.3	0.4	0.1	–	–	–	–	–	–	–
15	38	22	18	4	2	1	–	–	–	–	–	–	–	–
	20.9	30.7	21.1	9.0	2.6	0.6	0.1	–	–	–	–	–	–	–
16	13	16	14	4	3	1	–	–	–	–	–	–	–	–
	11.4	17.9	13.2	6.0	1.9	0.5	0.1	–	–	–	–	–	–	–
17	8	4	3	3	2	1	–	1	–	–	–	–	–	–
	4.5	7.5	5.9	2.9	1.0	0.3	–	–	–	–	–	–	–	–
18	–	4	2	1	–	–	–	–	–	–	–	–	–	–
	1.3	2.3	1.9	1.0	0.4	0.1	–	–	–	–	–	–	–	–
19	2	1	–	–	–	–	–	–	–	–	–	–	–	–
	0.5	0.9	0.8	0.5	0.2	0.1	–	–	–	–	–	–	–	–
20	–	–	–	–	–	–	–	–	–	–	–	–	–	1
	0.2	0.3	0.3	0.2	0.1	–	–	–	–	–	–	–	–	–

Table 6.1b *Observed frequency distribution of foetal death and fitted frequencies under the binomial model. Data taken from Haseman and Soares (1976) – data set no. 3. Results from Pack (1986a)*

Litter size (n)	Number of dead foetuses (x)									
	0	1	2	3	4	5	6	7	8	9
1	7	–	–	–	–	–	–	–	–	–
	6.5	0.5	–	–	–	–	–	–	–	–
2	7	–	–	–	–	–	–	–	–	–
	6.0	0.9	–	–	–	–	–	–	–	–
3	6	–	–	–	–	–	–	–	–	–
	4.8	1.1	0.1	–	–	–	–	–	–	–
4	5	2	1	–	–	–	–	–	–	–
	5.9	1.8	0.2	–	–	–	–	–	–	–
5	8	2	1	–	1	1	–	–	–	–
	9.0	3.5	0.5	–	–	–	–	–	–	–
6	8	–	–	–	–	–	–	–	–	–
	5.1	2.4	0.5	–	–	–	–	–	–	–
7	4	4	2	1	–	–	–	–	–	–
	6.5	3.5	0.8	0.1	–	–	–	–	–	–
8	7	7	1	–	–	–	–	–	–	–
	8.3	5.1	1.4	0.2	–	–	–	–	–	–
9	8	9	7	1	1	–	–	–	–	–
	13.3	9.3	2.9	0.5	0.1	–	–	–	–	–
10	22	17	2	–	1	–	–	1	1	–
	20.9	16.2	5.6	1.2	0.2	–	–	–	–	–
11	30	18	9	1	2	–	1	–	1	–
	27.3	23.2	9.0	2.1	0.3	–	–	–	–	–
12	54	27	12	2	1	–	2	–	–	–
	40.1	37.2	15.8	4.1	0.7	0.1	–	–	–	–
13	46	30	8	4	1	1	–	1	–	–
	34.5	34.7	16.1	4.6	0.9	0.1	–	–	–	–
14	43	21	13	3	1	–	–	1	–	1
	29.2	31.7	15.9	4.9	1.0	0.2	–	–	–	–
15	22	22	5	2	1	–	–	–	–	–
	17.0	19.7	10.7	3.6	0.8	0.1	–	–	–	–
16	6	6	3	–	1	1	–	–	–	–
	5.2	6.4	3.7	1.3	0.3	0.1	–	–	–	–
17	–	–	–	–	–	–	–	–	–	–
18	3	–	2	1	–	–	–	–	–	–
	1.6	2.2	1.4	0.6	0.2	–	–	–	–	–

and now after unconditioning with respect to the distribution of p_i we have:

$$
\begin{aligned}
\Pr(X_{ij} = x_{ij} | n_{ij}) &= \int_0^1 \Pr(X_{ij} = x_{ij} | n_{ij}, p_i) f(p_i) dp_i \\
&= \binom{n_{ij}}{x_{ij}} \frac{B(\alpha_i + x_{ij}, n_{ij} + \beta_i - x_{ij})}{B(\alpha_i, \beta_i)}, \quad \text{for } 0 \leqslant x_{ij} \leqslant n_{ij} \quad (6.1)
\end{aligned}
$$

The distribution of equation (6.1) is termed the **beta-binomial** distribution. It has been employed in modelling consumer purchasing behaviour (Chatfield and Goodhart, 1970), in dental studies of caries in children (Weil, 1970), and in describing disease incidence in households (Griffiths, 1973). It was suggested for toxicological data by Williams (1975). Earlier references are given by McNally (1990). An alternative parameterization is in terms of (μ_i, θ_i), where

$$
\mu_i = \left(\frac{\alpha_i}{\alpha_i + \beta_i} \right)
$$

$$
\theta_i = \frac{1}{(\alpha_i + \beta_i)}, \quad 1 \leqslant i \leqslant k
$$

We now have $0 < \mu_i < 1$ and $\theta_i > 0$. In this parameterization we have, from the above results,

$$
\begin{aligned}
\mathrm{E}[p_i] &= \mu_i \\
\mathrm{V}(p_i) &= \mu_i(1 - \mu_i)\theta_i/(1 + \theta_i)
\end{aligned}
$$

We see that as $\theta_i \to 0$, $\mathrm{V}(p_i) \to 0$ and so the beta-binomial distribution reduces to the binomial form as the variance of p_i decreases. This new parameterization is therefore preferable in terms of interpretation of the parameters, and it is also more **stable** (Ross, 1975, 1990, and the discussion of section 2.2), resulting in more circular contours of the likelihood surface, which might facilitate iterative maximum-likelihood estimation. In the rest of section 6.1 it is convenient to drop the i-suffix, denoting treatment group.

In terms of the (μ, θ) parameterization we can rewrite the beta-binomial probabilities of equation (6.1) as follows:

$$\Pr(X_j = x_j \mid n_j) = \frac{\binom{n_j}{x_j} \prod_{r=0}^{x_j-1} (\mu + r\theta) \prod_{r=0}^{n_j-x_j-1} (1 - \mu + r\theta)}{\prod_{r=0}^{n_j-1} (1 + r\theta)},$$

for $0 \leqslant x_j \leqslant n_j, \quad 1 \leqslant j \leqslant m$ \hfill (6.2)

where $\prod_{r=0}^{-1}$ is interpreted as unity (Exercise 6.1).

We can now see clearly that the standard binomial distribution results when $\theta = 0$. With this parameterization, the mean and variance of X_j are given by:

$$E[X_j] = n_j \mu$$

$$V(X_j) = n_j \mu (1 - \mu) \left\{ 1 + \frac{\theta}{1 + \theta} (n_j - 1) \right\} \hfill (6.3)$$

Although we require the parameter $\theta > 0$ in the beta-binomial distribution, the $(n + 1)$ terms of equation (6.2) sum to unity for any real μ and θ (Exercise 6.2), although individual terms may lie outside the range $[0, 1]$. In fact, if

$$\theta \geqslant \max \left(\frac{\mu}{(1 - n_j)}, \frac{(1 - \mu)}{(1 - n_j)} \right)$$

then as pointed out by Prentice (1986), the terms of equation (6.2) form a valid probability distribution, which we may call an **extended** beta-binomial distribution, though it no longer results from a beta-mixture of binomial distributions, as before. This is a useful result, since it means that the binomial distribution, when $\theta = 0$, does not occur at an endpoint to the allowable range for θ, and consequently it is possible to perform a valid likelihood-ratio test of the null hypothesis that $\theta = 0$. We shall see examples of this later. When $\theta < 0$ then the extended beta-binomial model in fact is **under-dispersed**, in the sense that the $V(X_j)$ is now **less** than the corresponding binomial variance. However, in practice, when we might expect to encounter 'large' values for n_j, there will be little potential for modelling under-dispersal with this model, due to the above bound on θ.

An interesting feature of the beta-binomial model is that it induces a correlation, ρ, between the responses of individuals in the same litter. This feature was first appreciated by Altham (1978) and it is not difficult to show that $\rho = \theta/(1 + \theta)$ (Exercise 6.3).

6.1.2 Fitting the beta-binomial model

For simplicity we shall continue with the case of just one treatment group. Under the beta-binomial model we have the following expression for the log-likelihood:

$$l(\mu, \theta) = \sum_{j=1}^{m} \left\{ \sum_{r=0}^{x_j - 1} \log(\mu + r\theta) + \sum_{r=0}^{n_j - x_j - 1} \log(1 - \mu + r\theta) \right.$$
$$\left. - \sum_{r=0}^{n_j - 1} (1 + r\theta) \right\} + \text{constant term} \qquad (6.4)$$

First and second derivatives with respect to μ and θ are

$$\frac{\partial l}{\partial \mu} = \sum_{j=1}^{m} \left\{ \sum_{r=0}^{x_j - 1} \frac{1}{(\mu + r\theta)} - \sum_{r=0}^{n_j - x_j - 1} \frac{1}{(1 - \mu + r\theta)} \right\}$$

$$\frac{\partial l}{\partial \theta} = \sum_{j=1}^{m} \left\{ \sum_{r=0}^{x_j - 1} \frac{r}{(\mu + r\theta)} + \sum_{r=0}^{n_j - x_j - 1} \frac{r}{(1 - \mu + r\theta)} - \sum_{r=0}^{n_j - 1} \frac{r}{(1 + r\theta)} \right\}$$

$$\frac{\partial^2 l}{\partial \mu^2} = \sum_{j=1}^{m} \left\{ -\sum_{r=0}^{x_j - 1} \frac{1}{(\mu + r\theta)^2} - \sum_{r=0}^{n_j - x_j - 1} \frac{1}{(1 - \mu + r\theta)^2} \right\}$$

$$\frac{\partial^2 l}{\partial \theta^2} = \sum_{j=1}^{m} \left\{ -\sum_{r=0}^{x_j - 1} \frac{r^2}{(\mu + r\theta)^2} - \sum_{r=0}^{n_j - x_j - 1} \frac{r^2}{(1 - \mu + r\theta)^2} + \sum_{r=0}^{n_j - 1} \frac{r^2}{(1 + r\theta)^2} \right\}$$

$$\frac{\partial^2 l}{\partial \mu \partial \theta} = \sum_{j=1}^{m} \left\{ -\sum_{r=0}^{x_j - 1} \frac{r}{(\mu + r\theta)^2} + \sum_{r=0}^{n_j - x_j - 1} \frac{r}{(1 - \mu + r\theta)^2} \right\}$$

Maximum-likelihood estimates therefore require numerical iteration, as provided, for example, by the Newton–Raphson method. An efficient FORTRAN algorithm is given by Smith (1983), while Pack (1986a) used the NAG (1982) library routine E04LAF, which employs a modified Newton–Raphson algorithm but also copes with constraints on parameters; this is relevant here in view of the bounds on θ. Paul (1982, 1985) used the Nelder–Mead simplex method, as did Segreti and Munson (1981), though in the latter case they were modelling dose-response data, and set μ to be a logistic function of dose. We shall consider their model in detail later. The method-of-moments estimates are given below (Kleinman, 1973; Exercises 6.4 and 6.15)

$$\hat{\mu} = \frac{x_{\cdot}}{n_{\cdot}}$$

$$\hat{\theta} = \frac{\dfrac{m-1}{m} \sum_{j=1}^{m} n_j \left(\dfrac{x_j}{n_j} - \hat{\mu}\right)^2 - \hat{\mu}(1-\hat{\mu}) \sum_{j=1}^{m} \left(1 - \dfrac{n_j}{n_{\cdot}}\right)}{\hat{\mu}(1-\hat{\mu}) \sum_{j=1}^{m} (n_j - 1)\left(1 - \dfrac{n_j}{n_{\cdot}}\right)} \qquad (6.5)$$

where $x_{\cdot} = \sum_{j=1}^{m} x_j$ and $n_{\cdot} = \sum_{j=1}^{m} n_j$.

The need to use specialist programs is avoided if the approximate procedure proposed by Brooks (1984) is adopted, since this is based on GLIM. For further discussion, see Exercise 6.6 and section 6.4. Chatfield and Goodhart (1970) fitted the beta-binomial model by the method of mean and zero-frequency, which was discussed also in section 5.3.4 (Exercise 6.5). As an illustration we give, in Table 6.2, the point estimates from fitting binomial and beta-binomial models to the data of Tables 1.10a and 1.10b, using maximum-likelihood.

It is clear that for these data it is far too restrictive to set $\theta = 0$, and, as suggested earlier, extra-binomial variation needs to be incorporated in the model. The fitted values from the beta-binomial

Table 6.2 *A summary of the maximum-likelihood binomial and beta-binomial fits to the data of Table 1.10a and 1.10b, taken from Haseman and Soares (1976). These results were obtained by Pack (1986a)*

Table	Binomial fit		Beta-binomial fit		
	p	$-$ Max. log. lik.	μ	θ	$-$ Max. log. lik.
1.10a	0.089		0.090	0.073	
		842.61	(0.005)	(0.011)	777.79
	$(0.003)^a$		$[0.005]^b$	[0.015]	
1.10b	0.072		0.074	0.081	
		765.06	(0.004)	(0.012)	701.33
	(0.003)		[0.005]	[0.019]	

[a]Estimates in parentheses: ( ) provide asymptotic estimates of standard error, from the inverse Hessian evaluated at the maximum likelihood estimate.
[b]Estimate in brackets: [ ] provide estimates of standard error based on 1000 simulations: simulated data were obtained using the maximum likelihood estimates of the parameters and by matching the observed litter sizes.

Table 6.3a *Observed frequency distribution of foetal death and fitted frequencies under the beta-binomial model. Data taken from Haseman and Soares (1976) – data set no. 1. Results from Pack (1986a)*

Litter size (n)	Number of dead foetuses (x)													
	0	1	2	3	4	5	6	7	8	9	10	11	12	13
1	2	–	–	–	–	–	–	–	–	–	–	–	–	–
	1.8	0.2	–	–	–	–	–	–	–	–	–	–	–	–
2	2	–	–	–	–	–	–	–	–	–	–	–	–	–
	1.7	0.3	–	–	–	–	–	–	–	–	–	–	–	–
3	3	–	–	–	–	–	–	–	–	–	–	–	–	–
	2.3	0.6	0.1	–	–	–	–	–	–	–	–	–	–	–
4	5	1	1	–	–	–	–	–	–	–	–	–	–	–
	5.0	1.6	0.4	1.0	–	–	–	–	–	–	–	–	–	–
5	2	2	–	–	–	–	–	–	–	–	–	–	–	–
	2.6	1.0	0.3	0.1	–	–	–	–	–	–	–	–	–	–
6	2	2	–	–	–	–	–	–	–	–	–	–	–	–
	2.5	1.0	0.4	0.1	–	–	–	–	–	–	–	–	–	–
7	2	2	2	1	–	–	–	–	–	–	–	–	–	–
	4.1	1.9	0.7	0.2	0.1	–	–	–	–	–	–	–	–	–
8	6	1	–	1	1	–	–	–	–	–	–	–	–	–
	4.9	2.5	1.1	0.4	0.1	–	–	–	–	–	–	–	–	–
9	2	3	1	–	–	–	–	–	–	–	–	–	–	–
	3.1	1.7	0.8	0.3	0.1	–	–	–	–	–	–	–	–	–
10	2	4	2	–	2	–	–	–	–	–	–	–	–	–
	4.9	2.8	1.4	0.6	0.2	0.1	–	–	–	–	–	–	–	–
11	19	11	3	3	–	–	–	–	–	–	–	–	–	–
	16.6	10.8	5.2	2.5	1.1	0.4	0.2	–	–	–	–	–	–	–
12	33	24	11	5	4	4	–	–	–	–	–	–	1	–
	35.9	22.6	12.4	6.2	2.9	1.2	0.5	0.2	0.1	–	–	–	–	–
13	39	27	12	6	5	2	–	–	1	–	–	–	–	–
	38.3	25.1	14.4	7.6	3.7	1.7	0.7	0.3	0.1	–	–	–	–	–
14	34	30	14	6	6	–	–	1	–	–	–	–	–	–
	36.2	24.5	14.6	8.0	4.2	2.0	0.9	0.4	0.2	0.1	–	–	–	–
15	38	22	18	4	2	1	–	–	–	–	–	–	–	–
	32.3	22.6	13.9	7.9	4.3	2.2	1.1	0.5	0.2	0.1	–	–	–	–
16	13	16	14	4	3	1	–	–	–	–	–	–	–	–
	18.5	13.3	8.4	5.0	2.8	1.5	0.8	0.4	0.2	0.1	–	–	–	–
17	8	4	3	3	2	1	–	1	–	–	–	–	–	–
	7.7	5.6	3.7	2.2	1.3	0.7	0.4	0.2	0.1	–	–	–	–	–
18	–	4	2	1	–	–	–	–	–	–	–	–	–	–
	2.3	1.8	1.2	0.7	0.4	0.3	0.1	0.1	–	–	–	–	–	–
19	2	1	–	–	–	–	–	–	–	–	–	–	–	–
	1.0	0.7	0.5	0.3	0.2	0.1	0.1	–	–	–	–	–	–	–
20	–	–	–	–	–	–	–	–	–	–	–	–	–	1
	0.3	0.2	0.2	0.1	0.1	–	–	–	–	–	–	–	–	–

Table 6.3b *Observed frequency distribution of foetal death and fitted frequencies under the beta-binomial model. Data taken from Haseman and Soares (1976) – data set no. 3. Results from Pack (1986a)*

Litter size (n)	Number of dead foetuses (x)									
	0	1	2	3	4	5	6	7	8	9
1	7	–	–	–	–	–	–	–	–	–
	6.5	0.5	–	–	–	–	–	–	–	–
2	7	–	–	–	–	–	–	–	–	–
	6.0	0.9	0.1	–	–	–	–	–	–	–
3	6	–	–	–	–	–	–	–	–	–
	4.8	1.0	0.2	–	–	–	–	–	–	–
4	5	2	1	–	–	–	–	–	–	–
	6.1	1.5	0.3	0.1	–	–	–	–	–	–
5	8	2	1	–	1	1	–	–	–	–
	9.3	2.8	0.7	0.2	–	–	–	–	–	–
6	8	–	–	–	–	–	–	–	–	–
	5.4	1.8	0.6	0.2	–	–	–	–	–	–
7	4	4	2	1	–	–	–	–	–	–
	7.1	2.6	0.9	0.3	0.1	–	–	–	–	–
8	7	7	1	–	–	–	–	–	–	–
	9.2	3.7	1.4	0.5	0.2	–	–	–	–	–
9	8	9	7	1	1	–	–	–	–	–
	15.3	6.5	2.7	1.0	0.4	0.1	–	–	–	–
10	22	17	2	–	1	–	–	1	1	–
	24.8	11.0	4.9	2.1	0.8	0.3	0.1	–	–	–
11	30	18	9	1	2	–	1	–	1	–
	33.5	15.7	7.3	3.3	1.4	0.6	0.2	0.1	–	–
12	54	27	12	2	1	–	2	–	–	–
	50.9	24.8	12.2	5.8	2.6	1.1	0.4	0.2	0.1	–
13	46	30	8	4	1	1	–	1	–	–
	45.5	23.0	11.7	5.9	2.8	1.3	0.5	0.2	0.1	–
14	43	21	13	3	1	–	–	1	–	1
	40.0	20.9	11.1	5.7	2.9	1.4	0.6	0.3	0.1	–
15	22	22	5	2	1	–	–	–	–	–
	24.2	13.0	7.1	3.8	2.0	1.0	0.5	0.2	0.1	–
16	6	6	3	–	1	1	–	–	–	–
	7.6	4.2	2.4	1.3	0.7	0.4	0.2	0.1	–	–
17	–	–	–	–	–	–	–	–	–	–
18	3	–	2	1	–	–	–	–	–	–
	2.5	1.5	0.9	0.5	0.3	0.2	0.1	–	–	–

fits are given in Tables 6.3a and 6.3b. The results of Pack's (1986a)
Monte Carlo tests of goodness-of-fit of the beta binomial model are
given below:

Values of maximum log-likelihood for beta-binomial model

Table	Actual data	Range of values from 99 matched simulations	Rank of max., log-likelihood from actual data in the range of 100 values available
1.10a	-777.79	$(-815.25, -730.07)$	33
1.10b	-701.33	$(-746.55, -634.42)$	46

So, on the basis of these results alone, we may conclude that the beta-binomial model is providing a reasonable description of the data. We do note, however, the possibility of a number of extreme, or outlying litters exhibiting high-mortality, and this is a point to which we shall return in section 6.6.2. There is further discussion of the beta-binomial fit, of Monte Carlo tests and additional examples, in Exercise 6.7 and its solution.

Haseman and Kupper (1979) took random samples of pairs of 20 litters from the data of Tables 1.10a and 1.10b and compared the pairs for differences using likelihood ratio tests based on the beta-binomial distribution. They concluded that the Type I errors were inflated. Pack (1986a) repeated their comparison, but for forty litters in each case, and also found inflated Type I errors. He did not feel that this finding was sufficient to reject the beta-binomial as a suitable model for these data.

6.1.3   Tarone's test

An alternative test to the likelihood-ratio test of the hypothesis that $\theta = 0$ was proposed by Tarone (1979). This was a $C(\alpha)$ test (Neyman, 1959), which does not require the fitting of the beta-binomial model. It is the locally most powerful test of $\theta = 0$, and is asymptotically optimal against beta-binomial alternatives. It is in fact a score test (Pack 1986; Prentice, 1986; Exercise 6.10 and Appendix D) and the test statistic, which is asymptotically normally distributed, is given by:

$$Z = \frac{\sum_{j=1}^{m} \dfrac{(x_j - n_j \hat{p})^2}{\hat{p}(1 - \hat{p})} - \sum_{j=1}^{m} n_j}{\left\{ 2 \sum_{j=1}^{m} n_j(n_j - 1) \right\}^{1/2}}$$

where $\hat{p} = x_{\cdot}/n_{\cdot}$.

Basing his approach on the ten litter treatment group in Kupper and Haseman (1978), Tarone conducted a small Monte Carlo study of the size of the test, and found it to be conservative.

6.1.4 The possibility of bias

In a paper prompted by the simulation work of Kupper *et al.* (1986), Williams (1988c) presented the results of Table 6.4. We can see that unless $\rho = 0$, corresponding to ignoring over-dispersion, or ρ is set approximately equal to its true value, appreciable bias may result in the estimate of μ. This result is discussed further in Exercise 6.11. While this finding may appear slightly strange in comparison with linear modelling, when estimates of mean value remain unbiased if variances are mis-specified, what is their relevance here? We would

Table 6.4 *(Williams, 1988c) The effect of estimating μ in the beta-binomial model when the correlation parameter $\rho = \theta/(1 + \theta)$ is fixed at an assumed value. Results of 1000 simulations: sample mean values for μ, with sample standard errors in parentheses*

	True parameter values	
Assumed value of ρ	$\mu = 0.125$ $\rho = 0.200$	$\mu = 0.250$ $\rho = 0.333$
0.00	0.126 (0.001)	0.253 (0.002)
0.05	0.112 (0.001)	0.231 (0.002)
0.10	0.113 (0.001)	0.227 (0.002)
0.20	0.127 (0.001)	0.237 (0.002)
0.30	0.146 (0.001)	0.249 (0.002)
0.40	0.167 (0.001)	0.269 (0.002)
0.50	0.191 (0.001)	0.290 (0.002)

not normally contemplate fixing ρ in order to estimate μ. The answer lies in the hidden dangers this effect may reveal when the beta-binomial distribution is used for making **comparisons** between treatments. In such cases it may seem attractive to model ρ or θ as a function of dose level, say, or of mean response (e.g. Moore, 1987). Mis-specifications in such cases might then result in biased comparisons. Fears of problems of this nature lead naturally to the consideration of more robust procedures as the basis of inference, and in particular to the technique of quasi-likelihood. These are matters to which we shall return later in the chapter, and which lead naturally into the following section.

Williams (1988a) has reported serious biases resulting in the estimation of θ when the beta-binomial model is fitted to data simulated from the logistic-normal-binomial distribution, and mixtures of binomial distributions (see sections 6.6.3 and 6.6.4 for relevant discussion of these two cases).

There is more to be said on the subject of modelling extra-binomial variation and we shall continue the discussion in sections 6.5 and 6.6.

6.2 Making comparisions in the presence of extra-binomial variation

6.2.1 An example involving treatment versus control

We shall now return to the example of Exercise 1.6, involving teratology data. It appears from the control group that a binomial distribution might suffice to describe the variation in the data. However, the proportions surviving in the treated group range from 0% to 100%, and it seems unlikely that a binomial distribution will suffice to describe these data. The data were analysed by Williams (1975), assuming a beta-binomial model. The maximum-likelihood results are given below. Here subscripts of C and T refer to control and treated groups respectively, and estimated asymptotic standard errors are given in parentheses:

$$\hat{\mu}_C = 0.898\,(0.026),\ \hat{\theta}_C = 0.021\,(0.048);$$
$$\max l_C = -51.69$$
$$\hat{\mu}_T = 0.740\,(0.069),\ \hat{\theta}_T = 0.465\,(0.234);$$
$$\max l_T = -64.99$$

When the binomial model is fitted separately to each group the values of the maximum log-likelihood are:

$$\max l_C = -51.80; \max l_T = -77.77$$

As anticipated, therefore, for the control group there is virtually no improvement in fit from including the between-litter variation, but there is a significant improvement for the treated group.

When a single beta-binomial distribution is fitted to both the control and treated groups simultaneously we obtain:

$$\hat{\mu} = 0.818 \ (0.038), \hat{\theta} = 0.271 \ (0.117); \max l = -120.54$$

Consequently a likelihood ratio test of the hypothesis:

$$H_0: \mu_C = \mu_T; \theta_C = \theta_T$$

versus

$$H_1: \mu_C \neq \mu_T; \theta_C \neq \theta_T$$

results in the likelihood-ratio (LR) statistic:

$$2(120.54 - 51.69 - 64.99) = 7.72$$

which is significant at the 2.5% level when referred to χ_2^2 tables. We shall refer to this test as $H1$.

Several other tests are possible, in each case with LR statistics referred to χ_1^2 tables:

$$H2 \begin{Bmatrix} H_0: \mu_C = \mu_T | \theta_C = \theta_T \\ H_1: \mu_C \neq \mu_T | \theta_C = \theta_T \end{Bmatrix} \quad \text{LR statistic: } 2.07$$

$$H3 \begin{Bmatrix} H_0: \theta_C = \theta_T; \mu_C = \mu_T \\ H_1: \theta_C \neq \theta_T; \mu_C = \mu_T \end{Bmatrix} \quad \text{LR statistic: } 1.95$$

$$H4 \begin{Bmatrix} H_0: \theta_C = \theta_T; \mu_C \neq \mu_T \\ H_1: \theta_C \neq \theta_T; \mu_C \neq \mu_T \end{Bmatrix} \quad \text{LR statistic: } 5.65$$

$$H5 \begin{Bmatrix} H_0: \mu_C = \mu_T; \theta_C \neq \theta_T \\ H_1: \mu_C \neq \mu_T; \theta_C \neq \theta_T \end{Bmatrix} \quad \text{LR statistic: } 5.77$$

The $H1$–$H5$ notation is due to Aeschbacher *et al.* (1977). We see from $H4$ that there is significant evidence that $\theta_C \neq \theta_T$, and then from $H5$ that there is significant evidence that $\mu_C \neq \mu_T$. We conclude that the treatment has affected both the variation and the mean response.

There is more discussion of this example in Exercises 6.12–6.14. Note also the comparisons which may be drawn with the approach detailed in the solution to Exercise 1.6. In that case a non-significant result was obtained. See also Exercise 6.16.

The comparison of this section is readily extended to the case of comparing k treatments. If the ith treatment corresponds to beta-binomial parameters, (μ_i, θ_i), $1 \leqslant i \leqslant k$, then the log-likelihood is given by:

$$l = \sum_{i=1}^{k} \sum_{j=1}^{m_i} \left\{ \sum_{r=0}^{x_{ij}-1} \log(\mu_i + r\theta_i) + \sum_{r=0}^{n_{ij}-x_{ij}-1} \log(1 - \mu_i + r\theta_i) \right.$$
$$\left. - \sum_{r=0}^{n_{ij}-1} \log(1 + r\theta_i) \right\} + \text{constant} \qquad (6.6)$$

It is possible that treatment differences may correspond to differences between the $\{\mu_i\}$, but not between the $\{\theta_i\}$, and we shall encounter examples of this in sections 6.2.2 and 6.6.5. However, incorrect specification of the $\{\theta_i\}$ (e.g. $\theta_i = \theta$, for all i) may give rise to errors in estimation of treatment differences, due to the possibility of bias mentioned in section 6.1.4. As pointed out by Williams (1988c), if one had assumed $\theta_C = \theta_T$ for the example of this section, then the estimate of the difference: $(\mu_T - \mu_C)$ would be reduced from 0.158 to 0.098.

6.2.2 Comparing alternative test procedures

In this section we continue the one treatment versus control comparison of the last section. There are many *ad hoc* ways of comparing control and treatment groups. For instance, as discussed already in the solution to Exercise 1.6, in each group we may form proportions responding in each litter, and then compare groups by means of a Mann–Whitney U test applied to these proportions. If we prefer to use a t-test, rather than a non-parametric test then we might well first of all transform the proportions by means of the variance-stabilizing arcsine square-root transformation; see also Brooks (1983) and Chanter (1975). A popular modification of this transformation which has seen much use in teratology is the 'Freeman–Tukey Binomial' transformation, given by:

$$y = \frac{1}{2} \left\{ \sin^{-1} \left(\frac{x}{n+1} \right)^{1/2} + \sin^{-1} \left(\frac{x+1}{n+1} \right)^{1/2} \right\}$$

corresponding to x animals responding out of n (Bishop *et al.*, 1975, p. 367).

It is interesting to consider which approach to use in practice. If data do in fact conform to the beta-binomial model, we might expect likelihood-ratio procedures based on the beta-binomial model to be more powerful than simpler approaches. However, this expectation would inevitably also be tempered by the realization that the likelihood-ratio tests are asymptotic, and so may not perform particularly well for small samples.

A number of authors have been interested in the small-sample properties of likelihood-ratio tests and in comparing alternative test procedures (Haseman and Kupper, 1979; Vuataz and Sotek, 1978; Gladen, 1979; Haseman and Soares, 1976; Shirley and Hickling, 1981; Pack, 1986b). Comparisons for small samples need to be based on simulation studies, and the picture is complicated by the range of possible tests, $H1-H5$ above, to be considered. While some of the early work in this area did not find likelihood-ratio methods to be more powerful than tests using simple transformations of proportions, Pack (1986b) has argued that in various cases this work did not match sufficiently well the kind of configurations that are most likely to be encountered in practice. His conclusions are, fortunately, quite simple to state, though inevitably they are qualified by the particular configurations of his simulations. Being compared were likelihood-ratio tests of the three comparisons: $H1$, $H2$ and $H5$, the t-test following a Freeman–Tukey Binomial transformation, and a t-test based on a weighted estimator due to Kleinman (1973). This last test has been shown by Pack (1986a) to be equivalent to a test resulting from a quasi-likelihood approach (Exercise 6.15). Pack (1986b) concluded that if treatment effects are small (for example, $\mu_T - \mu_C \leqslant 0.2$) then no test has a clear advantage. Otherwise, however, under the assumed beta-binomial model, a likelihood-ratio test of $H5$ (which places no restrictions on the θ parameters) was recommended, as it had acceptable error-rates, and was the most powerful of the likelihood-ratio tests in a wide range of instances. By contrast the two t-tests were found to be appreciably less powerful in many cases. From a computational point of view, the Wald test (Appendix D) is preferable to the likelihood-ratio test for $H5$, and has equivalent operating characteristics. As has been pointed out by Williams (1988c), the greater power of the likelihood-ratio test of $H5$, when compared with that of $H2$ (which is when the constraint of $\theta_1 = \theta_2$ is imposed), can be partially explained in terms of the bias in estimation of μ which may arise when θ is mis-specified (section 6.1.4).

The conclusion that a parametric procedure is best if the assumed model holds should come as no great surprise. Also unsurprising is the greater sensitivity of this procedure to departures from assumptions. The likelihood-ratio tests have been found to have inflated error rates when applied to real data, and they may then lose power superiority in comparison with the more robust and simpler t-tests (Pack 1986a). A study to investigate this comparison fully is long overdue.

An important comparison between treatments occurs when different treatments correspond to different doses of some substance, and this is the topic of the next section. We now therefore return to the main dose-response topic of this book, having prepared the ground, in sections 6.1 and 6.2 for dealing with extra-binomial variation.

6.3 Dose-response with extra-binomial variation

The data of Table 6.5 result from an investigation into neonatal acute toxicity to trichloromethane, a common contaminant of drinking water. The doses were administered (by oral gavage) to

Table 6.5 *Neonatal acute toxicity to trichloromethane. Data taken from Segreti and Munson (1981). There is no explanation of the common litter size (8) in all cases, but this is likely to result from using litters of size $\geqslant 8$ and then, as a consequence of experimental constraints, taking a sample of 8 mice when necessary. Five litters are exposed to each treatment and they are arranged below in order, from left to right, of increasing response*

Dosage (mg/kg)	Number dead per litter					Total dead
Control	0	0	0	2	2	4
250	0	0	1	3	6	10
300	0	0	0	1	8	9
350	0	2	2	5	8	17
400	1	2	4	6	7	20
450	1	4	5	6	8	24
500	1	7	8	8	8	32

mice seven days after birth, and toxicity was recorded by the number dead within 14 days of treatment.

The variation in response between litters within doses is clearly greater than could be explained by simple binomial response, and we shall now consider how to incorporate extra-binomial variation into the dose-response framework, with particular reference to the results of Table 6.5.

6.3.1   The basic beta-binomial model

As in section 6.1.1, we assume that for the ith treatment, p_i has a beta-distribution with parameters μ_i and θ_i. There is control mortality present in Table 6.5, and we shall use Abbott's formula to model the control mortality (section 3.2). We also take, as the simplest possibility, a logistic dose-response relationship to describe the dose-dependence of μ_i. Thus, following Segreti and Munson (1981), we set: $\mu_i = \lambda + (1 - \lambda)/(1 + e^{-(\alpha + \beta z_i)})$ where z_i is log (dosage). The regression of μ_i on continuous variables was first suggested by Crowder (1978) in a context to be described in section 6.6.5. We know from the work of Chapter 4 that many other possibilities may be considered as alternatives to, and possibly in preference to, the logistic link function (Exercise 6.17), and we return to this point in section 6.5.

When this model is fitted to the data of Table 6.5, using the Nelder–Mead simplex method, we obtain the maximum-likelihood estimates and estimated asymptotic correlation matrix given below. Estimated asymptotic standard errors are given on the diagonal.

Parameter	Estimate		$\hat{\lambda}$	$\hat{\alpha}$	$\hat{\beta}$	$\hat{\theta}$
			\multicolumn{4}{c}{Estimated errors (in parentheses) and correlations}			
λ	0.161	$\hat{\lambda}$	(0.087)	–	–	–
α	36.810	$\hat{\alpha}$	−0.534	(18.222)	–	–
β	−14.030	$\hat{\beta}$	−0.524	−0.999	(6.891)	–
θ	0.681	$\hat{\theta}$	0.150	−0.095	0.098	(0.230)

It was found necessary to centre the data (apart from the control case) by subtracting the mean log dosage from each log dosage (see here

the discussion of section 2.1). In all, five models were fitted, and the
details are shown below.

Model	$-$ Max. log likelihood	Number of parameters in the model
$\{\mu_i\}, \theta_i = 0$	158.76	7
$\{\mu_i\}, \{\theta_i\}$	122.43	14
$\{\mu_i\}, \theta_i = \theta$	125.76	8
$\mu_i = \mu, \theta_i = \theta$	133.84	2
$\alpha, \beta, \theta, \lambda$	126.43	4

We may conclude that there is significant over-dispersion
$(2 \times (158.76 - 122.43) = 72.66$, referred to $\chi_7^2)$, that we may take $\theta_i = \theta$
for all i, since $2 \times (125.76 - 122.43) = 6.66$, referred to χ_6^2, and that
the logit model provides a satisfactory fit $(2 \times (126.43 - 125.76) =$
1.34, referred to $\chi_4^2)$. The LD_{50} is estimated by $10^{-\hat{\alpha}/\hat{\beta}}$, in the original
parameterization, before log dosages were centred, to give 420.4 mg/kg,
with a 95% Fieller interval of (105.8, 530.6). This is to be contrasted
with the results from a standard logit analysis, corrected for over-
dispersion by a heterogeneity factor (section 2.5), which estimates
the LD_{50} as 413 mg/kg, with a 95% Fieller interval of $(2.5 \times 10^{-7}$,
637.5). Thus in this example the confidence-interval using the
beta-binomial distribution is somewhat narrower than if the over-
dispersion is corrected by using the heterogeneity factor. For further
discussion of this example, see Exercise 6.17 and section 6.4. An
overall assessment of the fit of the beta-binomial model is given
below.

Dosage (mg/kg)	Average number responding	Beta-binomial fit
Control	0.8	1.29
250	2.0	1.56
300	1.8	2.05
350	3.4	2.94
400	4.0	4.14
450	4.8	5.33
500	6.4	6.27

There is no suggestion of a systematic lack of fit, which might have encouraged an investigation of alternative link functions to the logistic.

6.3.2   Describing variation through the parameter θ (or ρ)

In the example of the last section, and also in the illustration of section 6.6.5 to be considered later, it was found possible to set $\theta_i = \theta$, and this clearly simplifies the analysis. In general we should not expect such a neat result, as the example of section 6.2.1 has already shown. In the light of the discussion of section 6.1.4, we might also be wary of biases which may result from constraining the $\{\theta_i\}$. The remedy of Kupper *et al.* (1986) is not to make any assumption regarding the $\{\theta_i\}$, but an alternative approach suggested in Prentice (1986) and supported by Williams (1988a) is to suppose that there may be systematic variation of θ_i (or equivalently ρ_i) with, for instance, μ_i, litter size, or dose, d_i, and to express this variation in a parametric form, with parameters to be estimated from the data. The form presented by Moore (1987) has:

$$\rho_i = \rho^* \{\mu_i(1 - \mu_i)\}^{\zeta - 1} \tag{6.7}$$

Thus here $\zeta = 1$ reduces to the ordinary beta-binomial case, and when $\zeta = 2$ we obtain approximately constant variance on the logit scale (section 6.6.4 and Exercise 6.18). Williams (1988a) considers that the model of equation (6.7) may be more useful than one in which ρ_i is solely an explicit function of dose. We shall illustrate the performance of this model by reference to the example presented by Moore (1987).

Example 6.1   Chromosome aberration data (Moore, 1987)

Blood samples were taken from 648 survivors of the bombing of Hiroshima, 20 years after the event. From each individual a number (in the range 30–100) of lymphocytes were examined, and the number with chromosome observations was recorded. Here each individual plays the role of the litter in earlier work. Natural variation between individuals leads us to expect extra binomial variation, but a further contributing factor is likely to be the uncertainty in measurement of the dose (cf. section 3.9). Figure 6.1 presents a plot of proportions of lymphocytes in samples which have aberrant chomosomes, against

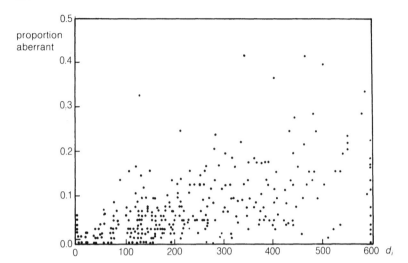

Figure 6.1 *Plot of proportions of samples of lymphocytes with aberrant chromosomes, plotted against estimated dose of radiation for survivors of the Hiroshima bombing. Reproduced from Moore (1987), with permission from the Royal Statistical Society.*

estimated radiation doses (d_i). For full details, see Moore (1987) and Awa *et al.* (1978). Note in particular that there are inevitable imprecisions in the dose estimation (cf. section 3.9) and doses exceeding 600 rads have been reduced to 600 rads.

Moore considered a beta-binomial model, with a linear, rather than logistic, link: $\mu_i = \alpha + \beta d_i$, and also set $\rho_i = \rho^*\{\mu_i(1 - \mu_i)\}^{\zeta-1}$, where i now indexes the individual subjects. The model was fitted, using maximum-likelihood and a Gauss–Newton method, and the results are given below, together with estimated asymptotic standard errors:

$$\hat{\alpha} = 0.008328 \ (0.000681)$$
$$\hat{\beta} = 0.0002707 \ (0.0000266)$$
$$\hat{\rho}^* = 0.772 \,(0.312)$$
$$\hat{\zeta} = 2.144 \,(0.143).$$

The maximum log-likelihood is $l = -9397.453$. If we set $\zeta = 1$, we obtain the corresponding figure of $l = -9445.779$, so there is substantial evidence that for these data we need to take $\zeta > 1$. Note that the resulting estimates of μ_i are for $\hat{\mu}_i < 0.5$, and so taking $\zeta > 1$ in this case results in a ρ_i parameter which increases with increasing μ_i. Incorrectly setting $\zeta = 1$, corresponding to the standard beta-binomial model, results in substantial bias in the other estimators and their errors, which is the same finding as in section 6.1.4. A limited examination of the goodness-of-fit of this model was made by the addition of a quadratic component to the expression for μ_i, which was not found to be significant.

In addition to fitting the beta-binomial model by maximum likelihood, Moore (1987) also used the approach of quasi-likelihood. We shall now explain this procedure in section 6.4, and illustrate its use.

6.4 The quasi-likelihood approach

Suppose the random variable X has the binomial $\text{Bin}(n, P)$ distribution, and that P itself is distributed over the range $[0,1]$, with mean μ and variance $\mu(1 - \mu)\rho$. If we make no particular assumptions regarding the distribution of P, then

$$
\begin{aligned}
E[X] &= E_P[E[X \mid P = p]] = n\mu \\
V(X) &= V_P(E[X \mid P = p]) + E_P[V(X \mid P = p)] \\
&= n\mu(1 - \mu)\{1 + \rho(n - 1)\}
\end{aligned}
\tag{6.8}
$$

which is the same mean and variance as have been obtained for the beta-binomial distribution. A more robust procedure results if we fit the data using the expressions for unconditional mean and variance only. This may be done by the method of quasi-likelihood (Wedderburn, 1974; McCullagh and Nelder, 1989, p. 323). Early proponents of this approach are Cochran (1943) and Kleinman (1973) (Exercise 6.15). Conveniently, for the problem under discussion, the method of quasi-likelihood may be performed by iterated weighted regression and is readily carried out by GLIM. The basic iteration here for general logistic regression has been given in equation (2.19). It is simply modified, for fixed ρ, to carry out the method of quasi-likelihood based on equation (6.8), by simply replacing the diagonal matrix of weights, V, by WV, where W is the diagonal matrix,

$$
W = \text{diag}\left[\{1 + \rho(n_i - 1)\}^{-1}\right]
$$

(Williams, 1982a; Moore, 1987).

We can see that if all of the n_i are equal ($= n$) then the quasi-likelihood fit will be **identical** to the binomial maximum-likelihood fit, but the dispersion matrix of $\hat{\boldsymbol{\beta}}$ has to be scaled up by the heterogenity factor, $[1 + \rho(n-1)]$, which in this case is precisely the procedure recommended by Finney (section 2.5). We would not expect small differences between the $\{n_i\}$ to result in greatly different point estimates of $\boldsymbol{\beta}$, whether obtained by quasi-likelihood or from an assumption of a binomial distribution and fitting by maximum likelihood.

We see, therefore, that for any ρ, $\hat{\boldsymbol{\beta}}$ can be obtained by a standard iterated weighted regression procedure. In practice we have also to estimate ρ. One approach is to equate the Pearson generalized chi-square goodness-of-fit statistic, X^2, to its expected value, which is approximately $k - v$, the degrees of freedom, where v is the number of parameters estimated and k is the number of dose levels; another approach is to equate X^2 to k. Both methods are considered by Moore (1986), who proves consistency and asymptotic normality for the resulting parameter estimates. We have:

$$X^2 = \sum_{i=1}^{k} \frac{(X_i - n_i\hat{\mu}_i)^2}{[n_i\hat{\mu}_i(1 - \hat{\mu}_i)\{1 + \rho(n_i - 1)\}]} \tag{6.9}$$

Approximately we have (Williams, 1982a):

$$E[X^2] = k - v + \rho \sum_{i=1}^{k} \{(n_i - 1)(1 - v_iq_i)\}, \qquad \text{if } \mathbf{W} = \mathbf{I}$$

$$E[X^2] = \sum_{i=1}^{k} [w_i(1 - w_iv_iq_i)\{1 + \rho(n_i - 1)\}] \quad \text{if } \mathbf{W} \neq \mathbf{I} \tag{6.10}$$

where q_i is the ith diagonal element of $\mathbf{X}(\mathbf{X'WVX})^{-1}\mathbf{X'}$, v_i is the ith diagonal element of $\mathbf{V}$, and w_i is the ith diagonal element of $\mathbf{W}$.

For any fitted values of $\boldsymbol{\beta}$, we can use equation (6.10) to update $\hat{\rho}$, and continue to iterate in this 'see-saw' manner until X^2 is sufficiently close to $(k - v)$. See Exercise 6.15 and its solution, which also gives the relationship with Kleinman's (1973) weighted estimator.

An equivalent approach, presented by Breslow (1984) for the related case of extra-Poisson variation, is to equate $X^2 = (k - v)$ in equation (6.9). Then solve for ρ by writing equation (6.9) in the form:

$$\rho = \frac{1}{(k - v)} \sum_{i=1}^{k} \frac{(X_i - n_i\hat{\mu}_i)^2}{\{n_i\hat{\mu}_i(1 - \hat{\mu}_i)(\rho^{-1} + n_i - 1)\}} \tag{6.11}$$

and iterate on ρ (Moore, 1987).

It is important to distinguish between two cases: in one we have replication within doses, and in the other, as in Example 6.2 below, we do not. In the former case it is possible to fit a separate parameter to represent the effect of each dose level, and it is in the context of this model that ρ would be estimated. Any sub-models, such as a logistic regression on dose, would then make use of the ρ which results from fitting the saturated model. In that case the chi-square goodness-of-fit statistics for sub-models would not equal their degrees of freedom. An illustration is provided later in Example 6.3.

A third suggestion is that of Brooks (1984, see Exercise 6.6), who combines ideas of maximum likelihood and maximum quasi-likelihood. Having estimated $\hat{\beta}$ by using quasi-likelihood for fixed ρ, he evaluates $\hat{\rho}$ by assuming a beta-binomial distribution and then maximizing a profile 'likelihood' with respect to ρ.

We shall now illustrate the use of equations (6.10) for incorporating over-dispersion using quasi-likelihood by applying them to the data of Table 3.9.

Example 6.2   An illustration of using quasi-likelihood for fitting a dose-response relationship incorporating over-dispersion

The data of Table 3.9 are not well-fitted by a logit model and binomial variation. It was shown in Chapter 4 how the fit could be improved, either by a suitable dose transformation, or by changing the link function, but in each case preserving the binomial error structure. We now consider the alternative approach of adding extra-binomial variation.

First of all we note that the Finney heterogeneity factor is given by the goodness-of-fit X^2 value from the logit fit divided by the appropriate degrees of freedom (section 2.5). Thus, from Table 4.2, the heterogeneity factor is $13.375/4 = 3.344$. If we have good reason to suspect over-dispersion we may now scale up all standard errors by $\sqrt{3.344}$, to give

$$\hat{\alpha} = -4.451 \quad \text{(s.e.} = 1.116)$$
$$\hat{\beta} = 4.460 \quad \text{(s.e.} = 0.942)$$

We may note that this is done **routinely** by the SAS statistical package with procedure PROBIT. Users are asked to check that a large X^2 goodness-of-fit statistic is not caused by systematic departure from the model. (Note here the critical comments of Sanathanan *et al.*

1987, the cautionary remarks of Finney, 1971, p. 72, and the material of section 6.5.1.) In order to use GLIM to perform a quasi-likelihood analysis based on equation (6.9) and (6.10) we need to extract the system vector $\%VL$, which holds the $\{d_i\}$ values and mimic in GLIM the algebra of equations (6.9) and (6.10) when appropriate. The initial fit is simply binomial, and then successive fits iterate on the value of the over-dispersion parameter ρ, until a good enough fit is obtained. The details can be appreciated from the code given below, together with the output, for three cycles on ρ, which can be seen to be adequate in this case.

```
$UNITS 6 $DATA R N W X1 $READ
16  49  1  0.71
18  48  1  1.00
34  48  1  1.31
47  49  1  1.48
47  50  1  1.61
48  48  1  1.70
$YVAR R $ERROR B N $WEIGHT W
$FIT X1 $
$DIS E C R $
$EXTRACT %VL  $CAL   WVQ=%PW*%WT*%VL
   :  %P=(%X2-%CU(%PW*(1-WVQ)))/(%CU((%BD-1)*%PW*(1-WVQ)))
   :  W=1/(1+%P*(%BD-1))
$PRINT '%P=' %P
$FIT X1 $
$DIS E C R $
$EXTRACT %VL .$CAL   WVQ=%PW*%WT*%VL
   :  %P=(%X2-%CU(%PW*(1-WVQ)))/(%CU((%BD-1)*%PW*(1-WVQ)))
   :  W=1/(1+%P*(%BD-1))
$PRINT  '%P=' %P
$FIT X1 $
$DIS E C R $
$EXTRACT %VL $CAL WVQ=%PW*%WT*%VL
   :  %P=(%X2-%CU(%PW*(1-WVQ)))/(%CU((%BD-1)*%PW*(1-WVQ)))
   :  W=1/(1+%P*(%BD-1))
$PRINT '%P=' %P


GLIM 3.77 update 3 (copyright)1985 Royal Statistical Society, London

scaled deviance = 15.412 at cycle  4
             d.f. =  4

          estimate        s.e.      parameter
      1     -4.451       0.6102      1
      2      4.460       0.5152      X1
      scale parameter taken as  1.000

Correlations of parameter estimates
  1   1.0000
  2  -0.9634    1.0000
         1          2

  unit   observed    out of    fitted   residual
     1         16        49     10.63      1.863
     2         18        48     24.11     -1.764
     3         34        48     38.44     -1.606
```

```
        4           47          49         43.89        1.454
        5           47          50         46.94        0.036
        6           48          48         45.99        1.448
-- change to data affects model
%P=  0.0491
scaled deviance = 4.6348 at cycle  4
          d.f. = 4

              estimate         s.e.       parameter
        1       -4.460         1.117        1
        2        4.465         0.9434       X1
        scale parameter taken as  1.000

Correlations of parameter estimates
    1    1.0000
    2   -0.9636    1.0000
          1         2

  unit   observed    out of     fitted     residual
    1       16         49        10.58        1.027
    2       18         48        24.06       -0.962
    3       34         48        38.42       -0.879
    4       47         49        43.88        0.795
    5       47         50        46.94        0.021
    6       48         48        45.99        0.796

-- change to data affects model
%P=  0.0493
scaled deviance = 4.6180 at cycle  4
          d.f. = 4

              estimate         s.e.       parameter
        1       -4.460         1.119        1
        2        4.465         0.9451       X1
        scale parameter taken as  1.000

Correlations of parameter estimates
    1    1.0000
    2   -0.9636    1.0000
          1         2

  unit   observed    out of     fitted     residual
    1       16         49        10.58        1.025
    2       18         48        24.06       -0.961
    3       34         48        38.42       -0.877
    4       47         49        43.88        0.794
    5       47         50        46.94        0.021
    6       48         48        45.99        0.795

-- change to data affects model
%P=  0.0493
```

We see that there is very little change in the point estimates and the errors are well-approximated by those based on the Finney heterogeneity factor. This was anticipated since in this case there was little variation in the $\{n_i\}$. (Cf. Exercise 6.20.)

From fitting the logit model with binomial error to log doses the estimated errors for α and β are, in order, (0.610, 0.516). From fitting this model to untransformed doses with binomial error the estimated

errors are, in order, (0.293, 0.016). From fitting the Aranda-Ordaz
asymmetric model to log doses with binomial error, we obtain the
error estimates of Table 4.3. It is clear, therefore, that different assumptions can result in quite different models producing good descriptions
of the data, but differing in the precision of their predictions. In
practice the decisions regarding choice of link, scale of measurement
for dose and the inclusion of over-dispersion are likely to be made
on the basis of experience with an experimental procedure, or an
observational study, which may typically be expected to be
producing more than a single set of data for consideration. We return
to this point in section 6.5.   □

A further example is given in section 6.7.3. The quasi-likelihood
approach, combined with estimation of ρ, may be extended to the
case in which the over-dispersion parameter ρ is a function of the
mean or additional covariates. One approach is described in Moore
(1987) for the illustration of Example 6.1. Further discussion may
be found in Nelder (1988), Williams (1988a) and Nelder and Pregibon
(1987). Changing the expression for the variance function can greatly
affect estimates of standard errors. A correction, resulting in a robust
approach, is described by Moore and Tsiatis (1991).

When we gain in robustness we expect to lose in efficiency. Various
studies have suggested that if the amount of over-dispersion is
moderate then efficiency loss is low, and restricted to the estimation
of ρ. See here the papers by Cox (1983), Firth (1987), Crowder (1987),
Kleinman (1973) and Moore (1986). It is not customary to estimate
the variance of ρ when it has been estimated by a moment
approach—see, for example, Moore (1986), and the discussion of
Moore (1987) and McCullagh and Nelder (1989, p. 128).

6.5 Over-dispersion versus choice of link function

We have seen in section 6.4 that the data of Table 3.9 may be
described by changing the link function, or by adding a term to
absorb over-dispersion about an original link function. For examples
involving substantial numbers of doses it may well be easier to detect
systematic departures from an assumed link function. Replicated
observations at individual doses may indicate over-dispersion. In
many cases experiments will be part of an extended series and prior
knowledge may support the presence of over-dispersion, or the
selection of a particular link function or scale of measurement for

the dose variable. Further problems of choice are illustrated in section 6.5.1.

6.5.1 Binomial examples

The examples of this section were kindly provided by Martin Ridout. Dose-response experiments were simulated with the total number of individuals as in Table 3.9. Expected values were those resulting from the logit fit of Figure 3.9, but instead of being binomially distributed, the data were distributed according to a beta-binomial distribution, with the amount of over-dispersion estimated from the lack of fit of the logit model in Figure 4.1. Five examples of the resulting numbers responding are given below:

	Examples				
Log_{10} dose	I	II	III	IV	V
0.71	28	19	14	26	11
1.00	28	21	26	26	36
1.31	32	37	25	44	44
1.48	43	47	40	42	46
1.61	49	49	48	49	44
1.70	47	47	47	48	48

In each case there is a poor logit fit when dose is measured on the logarithmic scale, but an acceptable logit fit can be found, wihout assuming any over-dispersion, when a different scale of measurement is adopted. The details are given in Figure 6.2.

An additional example is given in section 6.6.5.

6.5.2 Wadley's problem with over-dispersion

In Wadley's problem, introduced in section 3.3, dose-related quantal responses are taken to have a Poisson distribution (Wadley, 1949). Over-dispersion was first discussed in this context by Anscombe (1949). A GLIM macro, which allows for a control group, a number of treated groups, a range of link functions and a negative-binomial

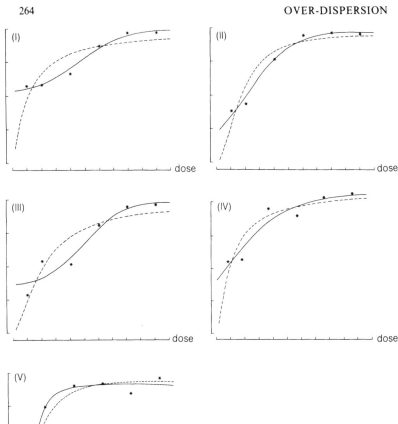

Figure 6.2   I:------logit model for log x:$X^2 = 15.67$ ⎫
                    ————logit model for x²:$X^2 = 2.33$ ⎬

                 II:------logit model for log x:$X^2 = 11.02$ ⎫
                     ————logit model for x:$X^2 = 3.40$ ⎬

                III:------logit model for log x:$X^2 = 17.18$ ⎫
                     ————logit model for x²:$X^2 = 6.66$ ⎬

                IV:------logit model for log x:$X^2 = 13.64$ ⎫
                    ————logit model for x:$X^2 = 8.67$ ⎬

                 V:------logit model for log x:$X^2 = 16.89$ ⎫
                    ————logit model for x⁻¹:$X^2 = 6.58$ ⎬

error distribution is provided by Smith and Morgan (1989). The over-dispersion parameter is estimated using the approach of equation (6.11). We shall now illustrate the use of this macro in analysing the colony count data of Table 1.12. For further discussion, see Morgan and Smith (1992).

Example 6.3   Analysis of the data of Example 1.9, taken from Trajstman (1989)

With just one exception, for this example, sample means are less than sample variances. The recommended approach (e.g. Anderson, 1988; Brooks, 1984; Williams 1982a) is to fit a saturated model, with a separate factor for each dose level, and thus use only the information **within** doses to estimate dispersion. Setting the goodness-of-fit X^2 statistic equal to the degrees of freedom (McCullagh and Nelder, 1989, p. 127) provides an over-dispersion factor of 0.0558 for the negative-binomial model. The residual deviance is a little higher than one might expect, at 130.42 on 116 degrees of freedom. There is a suggestion of variation in heterogeneity with respect to dose. However, there is no pattern to this and so no attempt was made to model it – cf. section 6.3.2.

From fitting separate regressions and assuming a logit link the residual deviance is 148.65 on 123 degrees of freedom. There is therefore a significant lack of fit, which persists when alternative link functions are used. This can be seen to arise from the turn-down of mean counts at the lowest HPC dose. Indeed an explanation of why this may be anticipated for low doses is given by Trajstman (1989) in his paper. Simpson and Margolin (1986) discuss this kind of non-monotonic dose-response relationship.   □

Using this approach it is sometimes possible to distinguish between competing link functions in the presence of over-dispersion, and an example of this is described in Exercise 6.22. A GENSTAT procedure of the GLIM macro of Smith and Morgan (1989) is given in Smith (1991). This procedure incorporates a scaled Poisson and a negative binomial mean-variance relationship, with estimation by quasi-likelihood, as well as the exact negative-binomial distribution.

Tests for over-dispersion relative to the Poisson model can be obtained as score tests within mixed Poisson models. For details and comparisons, see Dean and Lawless (1989). O'Hara Hines *et al.* (1992) provide diagnostic tools for detecting whether the link function is mis-specified. This is done in the context of models for data collected over time (Chapter 5), and extend to the multinomial case the ideas of influence and diagnostics described in section 3.4.

6.6 Additional models

In this section we focus again on extra-binomial variation, and comment on a variety of models which have been proposed to explain and account for the over-dispersion. In sections 6.6.1–6.6.3 we concentrate on different models for over-dispersion relative to the binomial distribution, separate from considerations of dose-dependence. We return to dose-dependence in sections 6.6.4 and 6.6.5.

6.6.1 The correlated-binomial model

The correlated-binomial model was proposed by Kupper and Haseman (1978), and independently, as an 'additive binomial' generalization of the binomial distribution, by Altham (1978). In this model, for all litters within a treatment group there is a constant probability, p, of death, but an explicit *intra*-litter correlation is introduced between the foetuses within each litter. We have already seen that such a correlation arises as a by-product of beta-binomial modelling, but unless an extended beta-binomial is used the correlation is always positive in that case. The results of Bahadur (1961) show that when mutual independence does not hold, then the usual binomial form for $\Pr(X_j = x_j = |n_j)$ is multiplied by a function of 2nd, 3rd, ..., n_jth order correlations between the littermate responses. (As usual, j references the jth litter.) If we assume that all 3rd and higher order correlations are zero then the correlated-binomial distribution results, and is given by:

$$\Pr(X_j = x_j | n_j) = \binom{n_j}{x_j} p^{x_j}(1-p)^{n_j - x_j}$$

$$\times \left[1 + \frac{\rho}{2p(1-p)} \{(x_j - n_j p)^2 + x_j(2p-1) - n_j p^2\} \right]$$

for $0 \leqslant p \leqslant 1$, $\quad 0 \leqslant x_j \leqslant n_j$, $\quad 1 \leqslant j \leqslant m$,

where ρ is the correlation between the responses of any two littermates. The mean and variance are given by:

$$E[X_j | n_j] = n_j p$$
$$V(X_j | n_j) = n_j p(1-p)\{1 + \rho(n_j - 1)\}$$

Hence we have the same mean/variance relationship as with the beta-binomial model, and the same fit would result from a quasi-likelihood approach.

We see immediately from the expression for the variance that we must have

$$\rho \geqslant -1/(n_j - 1), \quad \text{for all } j \text{ for which } n_j > 1$$

and so the largest litter provides the most restrictive bound on ρ. In fact, as Kupper and Haseman (1978) have shown, there are additional bounds on ρ, given by:

$$-\frac{2}{n_j(n_j - 1)}\min\left(\frac{p}{1-p}, \frac{1-p}{p}\right) \leqslant \rho \leqslant \frac{2p(1-p)}{(n_j - 1)p(1-p) + 0.25 - \gamma_0},$$

for all j, where $\gamma_0 = \min_{x_j}\left[\{x_j - (n_j - 1)p - 0.5\}^2\right]$.

We reproduce below a table from Kupper and Haseman (1978) which quantifies these bounds for a range of values for n_j and p.

	p		
n_j	0.10	0.30	0.50
2	$(-0.111, 1.000)$	$(-0.429, 1.000)$	$(-1.000, 1.000)$
3	$(-0.037, 0.529)$	$(-0.143, 0.636)$	$(-0.333, 1.000)$
5	$(-0.011, 0.300)$	$(-0.043, 0.420)$	$(-0.100, 0.500)$
7	$(-0.005, 0.231)$	$(-0.020, 0.296)$	$(-0.048, 0.333)$
10	$(-0.002, 0.200)$	$(-0.010, 0.200)$	$(-0.022, 0.200)$
15	$(-0.001, 0.120)$	$(-0.004, 0.135)$	$(-0.010, 0.143)$
20	$(-0.001, 0.100)$	$(-0.002, 0.100)$	$(-0.005, 0.100)$

In many practical examples we might anticipate a combination of 'large' $\{n_j\}$ and 'small' p, in which case the correlation ρ is prevented from taking negative values of ρ appreciably different from 0. Consequently the correlated binomial model does not offer a real alternative to the beta-binomial, especially since model-fitting by maximum-likelihood is more complex due to the data-dependent bounds on ρ. Pack (1986a) used the simplex method for model-fitting by maximum-likelihood, setting the likelihood to a very small value

Table 6.6 *Parameter estimates for the maximum-likelihood correlated-binomial fit to the data of Table 1.10a and 1.10b, taken from Haseman and Soares (1976). Results are taken from Pack (1986a)*

Table	Correlated-binomial fit		
	p	ρ	Max. log likelihood
1.10a	0.0930	0.0436	801.68
	$(0.0044)^a$	(0.0058)	
	$[0.0057]^b$	[0.0064]	
1.10b	0.0761	0.0389	732.19
	(0.0042)	(0.0062)	
	[0.0063]	[0.0080]	

[a] Estimates in parentheses: ( ) provide asymptotic estimates of standard error, from the inverse Hessian evaluated at the maximum likelihood estimate.
[b] Estimates in brackets: [ ] provide estimates of standard error based on 1000 simulations: simulated data were obtained based on the observed litter sizes and the maximum likelihood estimates of the parameters.

whenever the simplex strayed into an inadmissible region. The results in Table 6.6 provide the parameter estimates from the maximum-likelihood fit to the data of Tables 1.10a and 1.10b; cf. Table 6.2.

Paul (1985) and Pack (1986a) also investigate a beta-correlated-binomial model which includes both the beta-binomial and correlated binomial models as special cases (Exercise 6.24). We continue discussion of the correlated-binomial model in section 6.6.3.

6.6.2 *Mixtures of binomials; outliers and influence*

The data of Tables 1.10a and 1.10b exhibit a small number of litters with high mortality. It is natural to wonder to what extent the presence of such litters might influence the fit of models such as the beta-binomial and correlated-binomial. It is unusual to encounter studies as large as these, which are the result of pooling information from a number of experiments. One may also, therefore, question whether there may be present a small number of litters with abnormally high response rates, which may even be responsible for a beta-binomial model (say) being used in preference to an inadequate binomial model. A mixture of two binomials was included in the study by Smith and James (1984), which involved fitting the mixture, the beta-binomial and the correlated-binomial models to 48 sets of data from dominant-lethal assays. Two- and three-component mixture models have been simulated by Williams (1988a) in order to investigate the performance of the beta-binomial model when fitted to such data.

A general c-component mixture has probability function given by:

$$\Pr(X_j = x_j | n_j) = \sum_{r=1}^{c} \gamma_r \binom{n_j}{x_j} p_r^{x_j} (1 - p_r)^{n_j - x_j}$$

where

$$0 \leqslant x_j \leqslant n_j, \quad 1 \leqslant j \leqslant m$$
$$0 \leqslant p_r \leqslant 1, \quad 1 \leqslant r \leqslant k$$
$$\sum_{r=1}^{k} \gamma_r = 1, \quad 1 \leqslant r \leqslant k \tag{6.11}$$

For simplicity we shall write equation (6.11) as:

$$P_j = \sum_{r=1}^{c} \gamma_r P_{jr}, \quad 1 \leqslant j \leqslant m$$

We may then write the log-likelihood as

$$l = \sum_{j=1}^{m} \log(P_j) - \lambda \left(\sum_{r=1}^{c} \gamma_r - 1 \right) \tag{6.12}$$

where λ is a Lagrange multiplier. We can now use equation (6.12) as the basis for an iterative procedure for obtaining maximum-likelihood estimates as follows:

$$\frac{\partial l}{\partial \gamma_r} = \sum_{j=1}^{m} \frac{P_{jr}}{P_j} - \lambda = 0, \quad 1 \leqslant r \leqslant c \tag{6.13}$$

$$\frac{\partial l}{\partial p_r} = \sum_{j=1}^{m} \frac{1}{P_j} \frac{\partial P_j}{\partial p_r} = \sum_{j=1}^{m} \frac{\gamma_r}{P_{jr}} \frac{\partial P_{jr}}{\partial p_r} = 0, \quad 1 \leqslant r \leqslant (c-1) \tag{6.14}$$

If we now multiply equation (6.13) by γ_r and sum over r, we obtain

$$\sum_{r=1}^{c} \gamma_r \sum_{j=1}^{m} \frac{P_{jr}}{P_j} - \lambda \sum_{r=1}^{c} \gamma_r = \sum_{j=1}^{m} \left(\sum_{r=1}^{c} \frac{\gamma_r P_{jr}}{P_j} \right) - \lambda = 0$$

resulting in the maximum likelihood estimate $\hat{\lambda} = m$.

Returning now to equation (6.13), substituting for $\hat{\lambda}$ gives, trivially,

$$\hat{\gamma}_r m = \sum_{j=1}^{m} \frac{\hat{\gamma}_r \hat{P}_{jr}}{\hat{P}_j},$$

which we may write in the form:

$$\hat{\gamma}_r = \frac{1}{m} \sum_{j=1}^{m} \hat{\mathrm{Pr}}(r|x_j)$$

where $\mathrm{Pr}(r|x_j)$ is the probability, using the rth component, that x_j foetuses have responded, conditional upon x_j foetuses having responded.

Now since $\dfrac{\partial P_{jr}}{\partial p_r} = \dfrac{(x_j - n_j p_r)}{p_r(1 - p_r)P_{jr}}$

(cf. section 2.2), then from equation (6.14) we have

$$\sum_{j=1}^{m} \frac{\gamma_r P_{jr}(x_j - n_j p_r)}{P_j p_j (1 - p_j)} = 0$$

which implies that

$$\hat{p}_r = \sum_{j=1}^{m} x_j \frac{\gamma_r P_{jr}}{P_j} \Big/ \sum_{j=1}^{m} n_j \frac{\gamma_r P_{jr}}{P_j}$$

$$= \frac{\displaystyle\sum_{j=1}^{m} x_j \mathrm{Pr}(r|x_j)}{\displaystyle\sum_{j=1}^{m} n_j \mathrm{Pr}(r|x_j)}$$

These equations suggest the following iterative procedure: let $\hat{\zeta}^{(t)}$ denote the tth iterate of any parameter ζ, then we form

$$\hat{\mathrm{Pr}}^{(t+1)}(r|x_j) = \frac{\hat{\gamma}_r^{(t)} \hat{P}_{jr}^{(t)}}{\hat{P}_j^{(t)}} \tag{6.15}$$

where $\hat{P}_{jr}^{(t)}$ and $\hat{P}_j^{(t)}$ are the estimates of P_{jr} and P_j evaluated at $\{\hat{\gamma}_r^{(t)}\}$, $\{\hat{p}_r^{(t)}\}$. We then set:

$$\hat{p}_r^{(t+1)} = \frac{\displaystyle\sum_{j=1}^{m} x_j \hat{\mathrm{Pr}}^{(t+1)}(r|x_j)}{\displaystyle\sum_{j=1}^{m} n_j \hat{\mathrm{Pr}}^{(t+1)}(r|x_j)}$$

$$\hat{\gamma}_r^{(t+1)} = \frac{1}{m} \sum_{j=1}^{m} \hat{\mathrm{Pr}}^{(t+1)}(r|x_j) \tag{6.16}$$

Table 6.7 (Pack, 1986a) *The maximum-likelihood estimates from fitting two- and three-component binomial mixture models to the data of Table 1.10a and 1.10b. Estimated asymptotic standard errors are given in parentheses*

(i) The two-component case

Table	Parameter			Max. log-likelihood
	p_1	p_2	γ	
1.10a	0.061	0.284	0.872	782.44
	(0.005)	(0.033)	(0.035)	
1.10b	0.056	0.471	0.959	686.24
	(0.003)	(0.052)	(0.011)	

(ii) The three-component case

Table	Parameter					Max. log-likelihood
	p_1	p_2	p_3	γ_1	γ_2	
1.10a	0.051	0.199	0.721	0.763	0.231	770.71
	(0.007)	(0.029)	(0.084)	(0.073)	(0.073)	
1.10b	0.045	0.145	0.568	0.821	0.153	677.85
	(0.012)	(0.079)	(0.065)	(0.200)	(0.197)	

and so the iteration continues, cycling backwards and forwards between equations (6.15) and (6.16) until a termination criterion is satisfied. The convergence of such algorithms is discussed in Titterington *et al.* (1985, section 4.3), where the above is shown to result from the EM algorithm (section 3.2). An ALGOL program is provided by Agha and Ibrahim (1984). If initial estimates satisfy, $0 \leqslant \hat{p}_r^{(0)} \leqslant 1$ and $\sum_{r=1}^{c} \hat{\gamma}_r^{(0)} = 1$ then the final estimates will also satisfy these restrictions (Everitt and Hand, 1981, p. 96). Pack (1986a) found this algorithm slow to converge, and preferred the Nelder–Mead simplex method. We now give, in Table 6.7, his results from fitting mixtures involving two and three components to the data of Tables 1.10a and 1.10b.

As an illustration, the estimated asymptotic correlation matrix for parameters when the three-component mixture is fitted to the data of Table 1.10a, is shown below.

	$\hat{p}_1$	$\hat{p}_2$	$\hat{p}_3$	$\hat{\gamma}_1$	$\hat{\gamma}_2$
$\hat{p}_1$	1	–	–	–	–
$\hat{p}_2$	0.75	1	–	–	–
$\hat{p}_3$	0.11	0.19	1	–	–
$\hat{\gamma}_1$	0.84	0.88	0.14	1	–
$\hat{\gamma}_2$	−0.84	−0.88	−0.13	−1.00	1

An interesting comparison may be drawn with the corresponding binomial and beta-binomial fits summarized in Table 6.2. We shall present a more detailed comparison in section 6.6.3. We note here that the three-component mixture fits appear to be worth the cost of the additional two parameters. However, formal comparison by means of a likelihood ratio test is not possible here since the two-parameter mixture model lies on the boundary of the parameter-space for the three-parameter mixture model. For related discussion, see Titterington *et al.* (1985, Chapter 5) and Everitt and Hand (1981, Chapter 5). An investigation based on profile likelihoods is given in Ridout *et al.* (1992). As an illustration, in Table 6.8 we provide the fit of the two-component mixture to the data of Table 1.10b.

In comparison with Table 6.3, we see that the mixture models provide a better fit than the beta-binomial model to the litters with large mortality, at the expense of the quality of the fit to litters with small mortality. This suggests that to have the best of both worlds one should try fitting a mixture of a binomial model with a beta-binomial model (Healy, M.J.R., pers. comm.), or possibly even a mixture of two beta-binomial models. This idea is explored by Ridout *et al.* (1992) and, in a slightly different context, to which we return in section 6.7.2, by Maruani and Schwartz (1983). For data such as that of Table 1.10 it has been found that significant improvements to the fit can be obtained from fitting the mixture of a beta-binomial and a binomial distribution. In several cases the binomial term improves the fit by accommodating outlying litters with high mortality. However, there may also be alternative solutions due to multiple maxima. For a full discussion see Exercise 6.25 which also addresses the influence of outlying litters.

Table 6.8 (Pack. 1986a) *The two-component binomial mixture model fitted to the data of Table 1.10b: observed and fitted frequencies*

Litter size (n)	Number of dead foetuses (x)									
	0	1	2	3	4	5	6	7	8	9
1	7	–	–	–	–	–	–	–	–	–
	6.5	0.5	–	–	–	–	–	–	–	–
2	7	–	–	–	–	–	–	–	–	–
	6.1	0.9	0.1	–	–	–	–	–	–	–
3	6	–	–	–	–	–	–	–	–	–
	4.9	1.0	0.1	–	–	–	–	–	–	–
4	5	2	1	–	–	–	–	–	–	–
	6.1	1.5	0.3	0.1	–	–	–	–	–	–
5	8	2	1	–	1	1	–	–	–	–
	9.4	2.9	0.5	0.2	0.1	–	–	–	–	–
6	8	–	–	–	–	–	–	–	–	–
	5.4	2.0	0.4	0.1	0.1	–	–	–	–	–
7	4	4	2	1	–	–	–	–	–	–
	7.0	3.0	0.6	0.2	0.1	0.1	–	–	–	–
8	7	7	1	–	–	–	–	–	–	–
	9.1	4.3	1.0	0.3	0.2	0.1	0.1	–	–	–
9	8	9	7	1	1	–	–	–	–	–
	14.8	8.0	2.0	0.5	0.3	0.2	0.1	0.1	–	–
10	22	17	2	–	1	–	–	1	1	–
	23.6	14.1	3.9	0.9	0.5	0.4	0.3	0.2	0.1	–
11	30	18	9	1	2	–	1	–	1	–
	31.4	20.6	6.2	1.4	0.6	0.6	0.5	0.3	0.2	–
12	54	27	12	2	1	–	2	–	–	–
	46.9	33.6	11.1	2.5	0.9	0.9	0.9	0.7	0.4	0.2
13	46	30	8	4	1	1	–	1	–	–
	41.1	31.9	11.5	2.7	0.8	0.7	0.8	0.7	0.5	0.2
14	43	21	13	3	1	–	–	1	–	1
	35.4	29.5	11.5	2.8	0.7	0.6	0.7	0.7	0.5	0.3
15	22	22	5	2	1	–	–	–	–	–
	20.9	18.7	7.8	2.1	0.5	0.3	0.4	0.4	0.4	0.3
16	6	6	3	–	1	1	–	–	–	–
	6.5	6.2	2.8	0.8	0.2	0.1	0.1	0.1	0.1	0.1
17	–	–	–	–	–	–	–	–	–	–
18	3	–	2	1	–	–	–	–	–	–
	2.0	2.2	1.1	0.4	0.1	–	–	–	–	–

6.6.3   Comparison of models

It is clearly of interest to compare the beta-binomial, correlated-binomial and (two-point) mixture distributions. From a large number of empirical studies, Pack (1986a) concluded that the correlated-binomial model was out-performed by the beta-binomial model, in the sense that in all cases the beta-binomial maximum likelihood was greater than that of the correlated-binomial model. Of course, here we are comparing non-nested models with the same number of parameters, but both models are special cases of the correlated-beta-binomial model described in Exercise 6.26, which does therefore provide an umbrella under which comparisons may take place. As the beta-binomial is easier to fit than the correlated-binomial there seems little point in considering the correlated-binomial distribution any further. The models are indistinguishable if fitted by quasi-likelihood, as we have seen.

For the data of Table 1.10a, the beta-binomial model has greater likelihood than the mixture model and has one fewer parameter, but experience suggests that this is an unusual occurrence. For the data of Table 1.10b the beta-correlated-binomial model provides a significantly better fit than the beta-binomial model, but in this case the mixture model fits with a greater maximum likelihood than the beta-correlated-binomial model, and these last two models have the same number of parameters. Detailed comparisons may be carried out by Monte Carlo procedures, and examples are shown in Exercise 6.26 and 6.27. Pack's (1986a) conclusion, from analysing a wide range of data sets, was that both the beta-binomial model and the mixture-of-binomials model are worth considering. Undoubtedly the beta-binomial model is the easier to use, as it extends readily, as we have seen, to the case of treatment comparisons and the description of dose effects. Mixture models may be of greater use in identifying litters with high mortality, which may possibly be linked to atypical conditions. In such a case exclusion of such atypical litters might allow analysis to proceed without recourse to methods to accommodate extra-binomial variation. A mixture model has already been discussed in Example 3.3, but in that case the experimental unit was a single egg, rather than a litter as has been the case here.

The range of possible extensions to the binomial model is much larger than those already covered. Examples are provided in

Exercises 6.28–6.31. All were considered by Pack (1986a) and not found to be generally useful. Rudolfer (1990) has studied the properties of a Markov chain model of extrabinomial variation, which could be particularly attractive if the practical context involved some form of temporal or spatial dependence (as in the dental illustration of Weil, 1970). We now consider two further models which have been found to have particular advantages in certain applications.

6.6.4 Logistic-normal-binomial and probit-normal-binomial models

In a survey on consumer purchasing, carried out by Social and Community Planning Research, there were 1265 respondents, spread over 32 different areas, with two different interviewers being responsible for each area, giving 64 interviewers in all. Some of the results of this survey are analysed by Anderson and Aitkin (1985). The differential effects of different interviewers are well known, and one might therefore model the probability of a positive response by the jth interviewee of the ith interviewer by:

$$p_{ij} = F(\boldsymbol{\beta}' \boldsymbol{x}_{ij} + \zeta a_i)$$

where

ζ is a scalar parameter, $\boldsymbol{\beta}$ is a vector of parameters,

$\boldsymbol{x}_{ij}$ is a vector of covariates describing the interviewee (e.g. age, sex, marital state, etc.),

a_i is an additive interviewer effect, $1 \leqslant i \leqslant 64$ here,

and F is some suitable cumulative distribution function.

If the number of interviewers is small, we may take the interviewer effect as a fixed factor. Otherwise it is more sensible to model the $\{a_i\}$ as random effects, and the natural distribution to adopt is $a_i \sim N(0, 1)$. Anderson and Aitkin (1985) adopt a logistic form for $F(\ )$, and so the complete model has:

$$p_{ij} = 1/\{1 + \exp - (\boldsymbol{\beta}' \boldsymbol{x}_{ij} + \zeta a_i)\}. \tag{6.17}$$

In general, if we have a total of I interviewers, the ith with n_i individuals interviewed, then the log-likelihood has the rather complicated form:

$$l(\boldsymbol{\beta}, \zeta) = \sum_{i=1}^{I} \log \left(\int_{-\infty}^{\infty} \left[\prod_{j=1}^{n_i} \{\delta_{ij} p_{ij} + (1 - \delta_{ij})(1 - p_{ij})\} \right] \phi(a_i) da_i \right)$$

$$\tag{6.18}$$

where

$$\delta_{ij} = 1 \text{ if the } (i, j)\text{th response is positive}$$
$$= 0 \text{ if the } (i, j)\text{th response is negative.}$$

The integration in equation (6.18) is necessary since the form for p_{ij} in equation (6.17) is conditional upon the ith interviewer having effect a_i.

It is shown, by Anderson and Aitkin (1985), how the EM algorithm (section 3.2) may be used to maximize equation (6.18) and so obtain maximum-likelihood estimates of $\boldsymbol{\beta}$ and ζ. However, convergence is no longer guaranteed. In this example the 'experimental unit' is the respondent, on whom covariate information is available. One can envisage other examples where the appropriate unit would be the interviewer. If we look more closely at the expression of equation (6.17) then we see that

$$\text{logit} \, (p_{ij}) = \boldsymbol{\beta}' \boldsymbol{x}_{ij} + \zeta a_i \qquad (6.19)$$

where $a_i \sim N(0, 1)$.

Thus $\text{logit}(p_{ij}) \sim N(\boldsymbol{\beta}' \boldsymbol{x}_{ij}, \zeta)$, and we say that p_{ij} has a **logistic-normal** distribution (e.g. Johnson, 1965). Had the normal distribution also been used for $F(\ )$, then the resulting distribution for p_{ij} would have been the **probit-normal** distribution.

In applications such as the survey illustration of this section, these models are particularly useful as they allow a mixture of random and fixed effects to appear together on the same logistic scale. The model above is easily extended to allow for more than one level of nesting, so that if there may be area effects to be considered we simply replace equation (6.19) by, in an obvious notation,

$$\text{logit} \, (p_{ijk}) = \boldsymbol{\beta}' \boldsymbol{x}_{ijk} + \zeta_1 a_i + \zeta_2 b_j$$

where b_j is a random area effect, which may be assumed to be independent of a_i, and also $N(0, 1)$. Thus the general form for the distribution of the logit of the probability of response is that of a univariate normal distribution, with mean $\boldsymbol{\beta}' \boldsymbol{x}$, for an individual with covariate vector $\boldsymbol{x}$, and variance which is a linear function of variance components. An agricultural application is provided by Jansen (1990), with a GENSTAT program described in Jansen (1988). See also Preisler (1988, 1989b), who provides GLIM macros.

The logistic-normal and probit-normal distributions may also be employed when there is only one random effect, in competition with the beta-binomial distribution. The comparison is then as follows:

if p denotes the probability of response, and X is the random variable denoting the number of individuals responding out of a litter (say) of size n, then if p varies according to a beta distribution, unconditionally, X has a beta-binomial distribution. If logit p (probit p) has a normal distribution, then we say that X has a logistic-normal-binomial (probit-normal-binomial) distribution. The logistic-normal-binomial distribution was proposed by Pierce and Sands (1975), and the probit-normal-binomial distribution results from the correlated-probit distribution derived by Ochi and Prentice (1984). A comparison between the beta-binomial and logistic-normal-binomial distributions is given in Table 6.9.

In this example the minor differences which exist for small litter size can be seen to increase with increasing litter size. Williams (1988a) suggests that the possibility of bias explained in section 6.1.4 might also arise for these other distributional forms. He also emphasized the importance of an adequate quadrature approximation to the integration in equation (6.18), an approximation which preceded the maximum likelihood estimation of Anderson and Aitkin (1985). The quasi-likelihood approach to fitting the logistic-normal-binomial and the probit-normal-binomial distributions is discussed by Williams (1988a); see also Gilmour *et al.* (1985).

Table 6.9. (Williams 1988a) *Comparison of the beta-binomial (BB) and logistic-normal-binomial (LNB) distributions for* $\mu = 0.1$, $\rho = 0.2$, *where* μ *and* $\rho\mu(1 - \mu)$ *are the mean and variance of* p: n *denotes litter size, and* r *denotes the number responding*

| | $n = 5$ | | $n = 10$ | | $n = 15$ | | $n = 20$ | |
	BB	LNB	BB	LNB	BB	LNB	BB	LNB
$r = 0$	0.692	0.686	0.573	0.551	0.505	0.466	0.458	0.406
$r = 1$	0.182	0.196	0.182	0.216	0.172	0.217	0.162	0.212
$r = 2$	0.077	0.071	0.098	0.099	0.102	0.111	0.100	0.120
$r = 3$	0.033	0.130	0.060	0.055	0.068	0.064	0.070	0.070
$r = 4$	0.012	0.013	0.037	0.032	0.047	0.043	0.051	0.045
$r = 5$	0.003	0.004	0.023	0.019	0.034	0.032	0.039	0.035
$6 \leqslant r \leqslant 10$	–	–	0.026	0.028	0.042	0.053	0.095	0.083
$11 \leqslant r \leqslant 15$	–	–	–	–	0.007	0.010	0.095	0.083
$16 \leqslant r \leqslant 20$	–	–	–	–	–	–	0.025	0.029

6.6.5 Modelling the effect of litter-size

The data of Table 6.10 are analysed in Paul (1982). We can see that the average proportion of abnormal foetuses drops between the medium and high dose groups, and also the mean litter size at the highest dose is the lowest. It is suggested by Williams (1987b) that this may be due to the high dose being responsible for deaths of foetuses before birth, or a reduction in the number of implants. Such an effect could be responsible simultaneously for smaller litters and also fewer affected born foetuses (since affected foetuses might be more likely to die *in utero*). Note here the link-up with the approach of Simpson and Margolin (1986), involving non-monotonic dose-response relationships, designed to cope with toxicity of substances at high doses, and mentioned earlier in section 4.3.

The work considered so far in this chapter takes no account of information which may be present in litter size, and it is a good idea to consider utilizing litter size as a covariate in teratological work.

In some cases it may be sensible to model the distribution of litter sizes, and base analyses on the joint distribution of litter size and numbers responding. Rai and Van Ryzin (1985) suggest that one might model litter size with a Poisson distribution about a mean that is log-linearly related to the dose level, but Williams (1987b) argued that over-dispersion relative to the Poisson distribution may need to be included – see also McCaughran and Arnold (1976). An interesting analysis of the data of Table 6.10 is given in Williams (1987b), as follows.

Let p_{ij} be the probability that a foetus in the jth litter of the ith treatment group is abnormal. When the model:

$$\text{logit}(p_{ij}) = \tau_i + \gamma s_{ij} \qquad (6.20)$$

where s_{ij} is the size of the jth litter of the ith treatment group, is fitted to the data the regression parameter γ is not significantly different from zero. Setting $\gamma = 0$ and re-estimating the $\{\tau_i\}$ gives:

$$\hat{\tau}_1 - \hat{\tau}_0 = 0.16 \ (0.52)$$
$$\hat{\tau}_2 - \hat{\tau}_0 = 1.17 \ (0.41)$$
$$\hat{\tau}_3 - \hat{\tau}_0 = 0.66 \ (0.46)$$

where τ_0 denotes the control parameter, τ_1 denotes the low dose parameter, etc. A cruder analysis may be based on whether or not

Table 6.10. *Data presented in Paul (1982). Animals treated were rabbits, and response was the presence of skeletal or visual abnormalities Here* n *denotes the number of live foetuses, and* r *indicates the number affected by treatment*

Group																											
Control	r	1	1	4	0	0	0	0	1	0	2	0	5	2	1	2	0	0	1	0	0	0	0	3	2	4	0
	n	12	7	6	7	10	8	7	6	11	7	9	9	7	9	7	11	10	8	4	10	12	8	7	8	–	–
Low dose	r	0	1	1	0	2	0	1	0	1	0	3	0	0	5	1	0	0	3	0	0	3	–	–	–	–	–
	n	5	11	7	9	12	8	6	7	6	4	9	6	6	9	5	9	1	6	7	6	9	–	–	–	–	–
Medium dose	r	2	3	2	1	2	3	0	4	0	4	0	6	6	6	5	4	1	0	3	10	6	–	–	–	–	–
	n	4	4	9	8	9	7	8	6	9	6	7	3	13	6	8	11	7	6	10	6	6	–	–	–	–	–
High dose	r	1	0	1	0	1	0	1	2	0	4	1	4	2	3	1	–	–	–	–	–	–	–	–	–	–	–
	n	9	10	7	5	4	6	3	8	5	4	3	8	6	8	6	–	–	–	–	–	–	–	–	–	–	–

there are any foetuses which are abnormal in a litter. In this case, if we model:

logit Pr(any foetus is abnormal in litter of size s_{ij}) $= \tau_i + \gamma s_{ij}$

then we obtain:

$$\hat{\gamma} = 0.26 \ (0.11)$$
$$\hat{\tau}_1 - \hat{\tau}_0 = 0.21 \ (0.63)$$
$$\hat{\tau}_2 - \hat{\tau}_0 = 1.28 \ (0.66)$$
$$\hat{\tau}_3 - \hat{\tau}_0 = 1.85 \ (0.77)$$

The litter size regression parameter is now significant and treatment effects are now monotonically related to dose level. It is suggested that more robust procedures may result from such an approach if litter sizes are reduced at high dose levels.

An explicit attempt to model the effect of litter size is made in Rai and Van Ryzin (1985), where they propose the following expression, given in standard notation and deriving from the one-hit model (cf. section 5.4).

$$p_{ij} = [1 - \exp\{-(\alpha + \beta d_i)\}] \exp\{-s_{ij}(\zeta_1 + \zeta_2 d_i)\} \quad (6.21)$$

This model is then fitted to data without allowance for extra-binomial variation. Alternative suggestions of Williams (1987b) are

$$\text{logit}(p_{ij}) = \tau_i + \gamma s_{ij} + \delta d_i s_{ij}$$

or

$$\text{logit}(p_{ij}) = \alpha + \beta d_i + \gamma s_{ij} + \delta d_i s_{ij}$$

though of course in any application other link functions may be more appropriate. Williams (1987b) emphasizes that such models may also require the inclusion of over-dispersion. The over-dispersion parameter, ρ, may also be modelled in terms of litter size (section 6.3.2).

Clearly there are many possibilities to be considered. In practice analysis would be greatly helped by prior experience with similar drugs and experiments, and careful interpretation of residual plots. Williams (1988a) provides an analysis which incorporates a quadratic dependence on litter-size in the systematic part of the model, combined with a constant measure of over-dispersion (Exercise 6.32).

6.7 Additional applications

6.7.1 Urn-model representations; ant-lions

The beta-binomial distribution results if balls are drawn in a particular way from an urn (Johnson and Kotz, 1969, Chapter 9): suppose an urn contains r red balls and b black balls originally, and a ball is drawn at random, examined, and then replaced with an extra ball of the same colour as the one drawn. If this procedure is repeated n times, the number of red balls drawn has a beta-binomial distribution (Exercise 6.33).

An alternative rule is to suppose that when a drawn ball is replaced, it is replaced with a ball of the **opposite** colour from the colour of the ball just drawn. In this case, if we write $\Pr(x|n)$ as the probability that n drawings result in x red balls being drawn we can readily verify the following recursion:

$$\Pr(x|n) = \left(\frac{r+n-x}{r+b+n-1} \right) \Pr(x-1|n-1)$$
$$+ \left(\frac{x+b}{r+b+n-1} \right) \Pr(x|n-1) \qquad (6.22)$$

In contrast to the recursion of Exercise 6.33, the recursion of equation (6.22) does not appear to result in a simple explicit form for $\Pr(x|n)$. However, the probabilities are readily generated from equation (6.22), and the model may be fitted to data without restricting r and b to be integers. Urn models of this type are discussed in Friedman (1949), and Morisita (1971) used the model of equation (6.22) to describe the behaviour of ant-lions, faced with the choice of building burrows in fine or coarse sand. Ant-lions introduced into an area containing both fine and coarse sand will prefer to dig burrows in fine sand, but also prefer to avoid other ant-lions. This anti-social feature of their behaviour could therefore result in a preference for coarse sand, if the fine sand is already well-populated by ant-lions. An example of data resulting from an experiment involving ant-lions is shown in Table 6.11.

The maximum-likelihood fitting of the model of equation (6.22) to these data is discussed in Exercise 6.34. This model results in under-dispersion relative to the binomial distribution. Pack (1986a) found little to choose between the model of equation (6.22) and the

Table 6.11. *Data from Morisita (1971), giving the frequency distribution of ant-lions in fine sand. The total number of experiments conducted = 123*

Number of ant-lions introduced (n)	Number settled in fine sand (x)							
	0	1	2	3	4	5	6	7
3	0	7	24	1	–	–	–	–
4	0	3	17	10	–	–	–	–
5	0	0	10	15	4	–	–	–
6	0	2	4	5	9	2	–	–
7	0	0	1	2	4	3	–	–

generalized beta-binomial distribution when both were fitted to small sets of dominant lethal toxicology data which were under-dispersed rather than over-dispersed.

6.7.2   The beta-geometric distribution; fecundability

The data of Table 6.12 describe the times taken by couples who were attempting to conceive, until pregnancy results. The data were obtained **retrospectively**, starting from a pregnancy in each case. Couples are designated as 'smokers' if the female partner smoked. There are clearly problems of definition here and a detailed discussion is provided in Weinberg and Gladen (1986). The model they used for these data was the beta-geometric distribution, with the distributional form:

$$\Pr(X = k) = \frac{\mu \prod_{i=1}^{k-1} \{1 - \mu + (i - 1)\theta\}}{\prod_{i=1}^{k} \{1 + (i - 1)\theta\}}, \, k \geqslant 1 \qquad (6.23)$$

where μ and θ are the parameters of the beta distribution assumed in order to describe the variation between different couples in the probability of conception at each menstrual cycle. Corresponding to the distribution of equation (6.23) being fitted separately to the smokers and the non-smokers, the two resulting beta-distributions are illustrated in Figure 6.3. This provides a quick visual impression of the effect of smoking. For discussion of the fit of this model to the data, and of possible alternative models, see Weinberg and Gladen (1986), Ridout and Morgan (1991) and Exercise 6.35. This

Table 6.12. *Data from Weinberg and Gladen (1986) on the number of menstrual cycles to pregnancy*

Cycles	Smokers	Non-smokers	Cycles	Smokers	Non-smokers
1	29	198	8	5	9
2	16	107	9	1	5
3	17	55	10	1	3
4	4	38	11	1	6
5	3	18	12	3	6
6	9	22	> 12	7	12
7	4	7	–	–	–

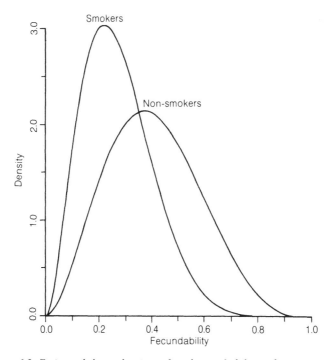

Figure 6.3 *Estimated beta densities for the probability of conception at each cycle, for smokers and non-smokers. Reproduced from Weinberg and Gladen (1986), with permission from the Biometric Society.*

is an application where mixture models may be thought to have some relevance (cf. section 6.6.2), and Maruani and Schwartz (1983) consider a mixture of a beta distribution with a probability of zero success, with the latter included to accommodate possibly sterile couples (Exercise 6.36).

6.7.3   Analysis of variance

In one of the earliest examples of using the beta-binomial distribution to model over-dispersion of proportions, Crowder (1978) presented an analysis of the data of Table 6.13. Here a beta-binomial model was proposed for the variation of probability of germination over groups. It was found possible to take the parameter θ as constant, and then the logit of the beta-binomial mean-value parameter μ was expressed as a function of main effects for seeds and for extracts, and with an interaction term. The results are given in Table 6.14.

It can be seen that the most important effect is due to extracts, but that there is also a suggestion of a significant interaction term. These data have subsequently been re-analysed by Williams (1982a) and by Brooks (1983). In the former case quasi-likelihood was used, resulting in a very similar analysis and virtually identical conclusions. Williams (1982a) also considered analysis using the logistic-normal-binomial model and concluded that for these data, since all the estimated values for μ lie in the range 0.3–0.7, the same conclusions could be expected to result (section 6.6.4).

Table 6.13 *Data presented by Crowder (1978) giving the number (*r*) of seeds germinating out of* n, *in a* 2 × 2 *factorial layout. There are two types of seed, and two root extracts*

Type of seed	0.aegyptiaca 75						0.aegyptiaca 73					
Root extracts	Bean			Cucumber			Bean			Cucumber		
r	n	r/n	r	n	r/n	r	n	r/n	r	n	r/n
10	39	0.26	5	6	0.83	8	16	0.50	3	12	0.25
23	62	0.37	53	74	0.72	10	30	0.33	22	41	0.54
23	81	0.28	55	72	0.76	8	29	0.29	15	30	0.50
26	51	0.51	32	51	0.63	23	45	0.51	32	51	0.63
17	39	0.44	46	79	0.58	0	4	0.00	3	7	0.43
–	–	–	10	13	0.77	–	–	–	–	–	–

Table 6.14 (Crowder, 1978) *Maximum log-likelihoods from fitting the beta-binomial model to the data of Table 6.13*

Factorial effects included	Number of parameters	Maximum log-likelihood
None	2	64.52
Seeds only	3	63.55
Extracts only	3	57.20
Seeds + extracts (main effects)	4	55.83
Full model with interaction	5	53.77

Prior to the introduction of logit models for proportions, a common analysis would have been to transform the proportions by means of an arcsine square-root transformation and then proceed with a standard analysis of variance, based on the usual assumptions of normality. Provided that proportions are not extreme, similar results, for binomially-distributed data, can be expected to be obtained (e.g. Cox and Snell, 1989, p. 22). For binary data with over-dispersion, discrepancies may well arise, as demonstrated by Anderson and Aitkin (1985). Brooks (1983) has shown how to modify the arcsine square-root transformation to accommodate over-dispersion, and hence produce an analysis which may be implemented by standard computer packages for the analysis of variance. He illustrated his approach on the data of Table 6.13.

6.7.4 Incorporating historical control information

The Cochran/Armitage test for trend was introduced in section 2.10. Applied to the data of Table 2.3 a result was obtained which was just significant at the 5% level. In addition to the control data of Table 2.3 there is a substantial amount of **historical** control data, resulting from previous experiments. Corresponding to this particular application, Tarone (1982a) presented the data of Table 6.15.

One way of incorporating this information is to assume that the probability of control response, P, varies over experiments according to a beta-distribution, with parameters (α, β), as in section 6.1.1. If we reparameterize by setting $P = 1/(1 + e^{-a})$, then the corresponding

Table 6.15 (Tarone, 1982a) *Control lung tumour rates (r out of n) from 70 experiments with female F344 rats: historical data for Table 2.3*

r/n	Frequency	r/n	Frequency	r/n	Frequency
0/50	3	0/19	6	1/23	2
0/49	3	0/18	4	1/20	14
0/47	2	0/10	1	1/19	1
0/25	2	1/53	1	1/18	1
0/24	2	1/50	2	2/20	6
0/22	1	1/49	2	2/19	1
0/20	14	1/47	1	2/18	1

density function for a is:

$$f(a) = \frac{\Gamma(\alpha + \beta)}{\Gamma(\alpha)\Gamma(\beta)} \, e^{a\alpha}(1 + e^a)^{-(\alpha + \beta)} \quad -\infty < a < \infty$$

The model is completed by setting the probability of response to dose d as $P(d) = 1/(1 + e^{-(a + bd)})$. The fact that $P = P(0)$ corresponds to the alternative to Abbott's formula discussed in section 3.2, and leading to equation (3.2). The likelihood for (a, b), for data such as that in Table 2.3, is therefore proportional to:

$$\frac{\exp\{a(r_0 + \alpha)\}}{\{1 + \exp(a)\}^{n_0 + \alpha + \beta}} \prod_{i=1}^{k} \frac{\exp\{(a + bd_i)r_i\}}{\{1 + \exp(a + bd_i)\}^{n_i}}$$

where the constant of proportionality does not involve (a, b). In this case, r_0 and n_0 are obtained by pooling over the historical data. In the illustration of Table 6.15, this gives $r_0 = 40$ and $n_0 = 1805$. A score test of the hypothesis that $b = 0$ results in the statistic,

$$\tilde{X}^2 = \frac{\left(\sum_{i=1}^{k} r_i d_i - \tilde{p} \sum_{i=1}^{k} n_i d_i \right)^2}{\left[\tilde{p}(1 - \tilde{p}) \left\{ \sum_{i=1}^{k} n_i d_i^2 - \left(\sum_{i=1}^{k} n_i d_i \right)^2 \Big/ \tilde{n} \right\} \right]}$$

where

$$\tilde{n} = \left(\alpha + \beta + \sum_{i=0}^{k} n_i \right), \text{ and } \tilde{p} = \left(\alpha + \sum_{i=0}^{k} r_i \right) \Big/ \tilde{n} \qquad (6.24)$$

which is referred to χ_1^2 tables. Tarone (1982a) recommends the empirical Bayes approach of substituting for α and β their estimates resulting from fitting a beta-binomial distribution to the data of Table 6.15. The justification for this rests on the assumption that a large amount of historical information is present, resulting in precise estimation of (α, β).

For the particular practical application discussed here, $\hat{\alpha} = 11.52$, $\hat{\beta} = 501.93$, $\tilde{p} = 0.035$, and $\tilde{n} = 623.45$, resulting in $X^2 = 21.96$ and a greatly increased significance level. For further discussion, see Exercise 6.37 and Hoel and Yanagawa (1986). Margolin and Risko (1984) showed that small changes to the historical data can have large influence on $\tilde{X}$. An attempt to avoid this problem is presented in Tamura and Young (1987), who stabilize the estimation of $\eta = (\alpha + \beta)$ in a manner analogous to ridge regression. The corresponding analysis for the Poisson case is given in Tarone (1982b).

6.8 Discussion

In this chapter we have considered over-dispersion relative to the Poisson, geometric and binomial distributions, with the greatest emphasis placed on the last of these. The extension to the multinomial case, based on an underlying Dirichlet distribution, is described by Paul *et al.* (1989), and used also by Petkau and Sitter (1989) for investigating whether over-dispersion was needed for the application of Example 5.2. In many applications we must anticipate over-dispersion, and if tests do not detect it then this probably reflects low power resulting from small sample sizes. Ignoring over-dispersion is unlikely to change point estimates in any radical way, but estimates of standard error will be underestimated, and tests will be in error. For discussion in the context of Poisson regression, see Breslow (1990). However, mis-specification of over-dispersion can result in bias. As seen from the work of section 6.5 and Chapter 4, failure to allow for over-dispersion when it is present can lead to an unproductive exercise in link modification.

Quasi-likelihood clearly provides a powerful robust tool, which is readily available to users of packages such as GLIM and GENSTAT. It has wider regression applications than have been described here (e.g. Breslow, 1990). As emphasized in section 6.5.2, care must be taken with regard to estimation of the over-dispersion parameter. Additionally, as stressed by Williams (1989), tests of sub-models must be based on differences in deviance, and not differences in the Pearson chi-square statistic, X^2.

A large number of possible models for over-dispersion have been presented in this chapter. A further approach is given by Morton (1991), based on conditioning extra-Poisson variables on their sum and then using quasi-likelihood. The logistic-normal-binomial model is useful in allowing random and fixed effects to appear together on the same logistic scale. As discussed in section 6.6.3, the beta-binomial and various mixture models appear to provide adequate tools for modelling over-dispersion relative to the binomial distribution. As emphasized in section 6.6.2, we should also not lose sight of the possibility of over-dispersion being due to (or exaggerated by) the contaminating presence of outliers.

As shown in section 6.7, models for over-dispersion have wide applications. A complex approach is presented by McNally (1990), in which in a study of reproductive performance in cows, ovulation detection rate, pregnancy rate and embryo loss rate are all given beta distributions. Correlation between individuals in survival analysis has been the subject of recent research into *frailty* models (McGilchrist and Aisbett, 1991; Clayton, 1991).

6.9   Exercises and complements

The exercises vary in complexity. It is particularly desirable to attempt the five exercises marked with a †. Exercises marked with a * are generally more difficult or speculative.

6.1[†]   Illustrate graphically different forms for the beta, and beta-binomial distributions. Verify that the beta-binomial probabilities of equation (6.1) can be written in the form given by equation (6.2).

6.2   (Prentice, 1986). Verify that the $(n_j + 1)$ terms of equation (6.2) sum to unity for any real μ and θ.

6.3[†]   Show that, under an assumed beta-binomial model, the correlation, ρ, between the responses of individuals in the same litter is given by $\rho = \theta/(1 + \theta)$, where $\theta^{-1} = (\alpha + \beta)$, and (α, β) are the beta-binomial parameters of equation (6.1). Show that $\rho = (\alpha + \beta + 1)^{-1}$. Provide an intuitive explanation of why $\rho \neq 0$. Inference for ρ, in an ANOVA context, is provided by Crowder (1979).

6.4 Verify the form of the method-of-moments estimators of the beta-binomial parameters (μ, θ), given in section 6.1.2.

6.5 (Chatfield and Goodhart, 1970). Under the beta-binomial model we have, from equation (6.2),

$$\Pr(X_j = 0 | n_j) = \prod_{r=0}^{n_j - 1} \left(1 - \frac{\mu}{(1 + r\theta)} \right)$$

Use this and the fact that $E[X_j] = n_j \mu$ to devise a mean and zero-frequency method of estimating μ and θ.

6.6* Experiment with the GLIM code given below for approximately fitting the beta-binomial model by maximum-likelihood. From Brooks (1984) we have the following GLIM macros reproduced by permission of the Royal Statistical Society.

```
$MAC    FIT
$CAL    W=1(%P*(%BD-1)+1)
$USE    MODEL
$CAL    %C=1/%P-1:C=%C:%Z=1
$VAR    %Z Z $CAL Z=%C $USE LGAM $CAL L=LG $DELETE Z LG
$CAL    %Z=%NU:Z=%BD+C $USE LGAM $CAL L=L-LG
$CAL      PR=%FV/%BD:GAM=PR*C:DEL=C-GAM:Z=Y+GAM $USE LGAM
$CAL    L=L+LG:Z=%BD-%YV+DEL $USE LGAM
$CAL    L=L+LG:Z=GAM $USE LGAM
$CAL      L=L-LG:Z=DEL $USE LGAM
$CAL    L=L-LG:%L=%CU(L)
$DELETE Z LG
$PRINT:'%P='%P'%L='*6%L
$ENDMAC
$MAC    LGAM
$VAR    %Z Z1 Z2 Z3
$CAL    Z=Z+2:Z1=1/Z:Z2=Z1*Z1:Z3=Z2*Z2:
  LG=(Z-0.5)*%LOG(Z)-Z+0.918938533+Z1*(1-Z2/30+Z3/105-Z2*Z3/140)/12:
  LG=LG-%LOG(Z-1)-%LOG(Z-2)
$DELETE Z1 Z2 Z3
$ENDMAC
```

Note that Brooks (1984) found this general approach did not work well for fitting the correlated-binomial distribution. Discuss how you might try to extend this approach to the analysis of variance application of Crowder (1978) (section 6.7.3).

6.7

(i) The table below presents the second data set from Haseman and Soares (1976), together with the observed frequencies under the beta-binomial model, obtained by Pack (1986a). Discuss the adequacy of the fit.

Litter size (n)	Number of dead foetuses (x)										
	0	1	2	3	4	5	6	7	8	9	10
1	15	1	–	–	–	–	–	–	–	–	–
	14.3	1.7	–	–	–	–	–	–	–	–	–
2	6	1	2	–	–	–	–	–	–	–	–
	7.2	1.7	0.1	–	–	–	–	–	–	–	–
3	6	6	–	–	–	–	–	–	–	–	–
	8.6	2.9	0.5	–	–	–	–	–	–	–	–
4	7	2	3	–	2	–	–	–	–	–	–
	9.1	3.9	0.9	0.1	–	–	–	–	–	–	–
5	16	9	3	3	1	–	–	–	–	–	–
	18.9	9.6	2.9	0.6	0.1	–	–	–	–	–	–
6	57	38	17	2	2	–	–	–	–	–	–
	62.4	36.5	13.1	3.4	0.6	0.1	–	–	–	–	–
7	119	81	45	6	1	–	–	1	–	–	–
	124.5	81.6	33.6	10.4	2.5	0.4	0.1	–	–	–	–
8	173	118	57	16	3	–	–	–	1	–	–
	166.1	119.8	55.4	19.7	5.6	1.3	0.2	–	–	–	–
9	136	103	50	13	6	1	1	–	–	–	–
	128.7	100.7	51.3	20.4	6.7	1.8	0.4	0.1	–	–	–
10	54	51	32	5	1	–	–	–	–	–	1
	55.2	46.3	25.6	11.2	4.1	1.3	0.3	0.1	–	–	–
11	13	15	12	3	1	–	–	–	–	–	–
	15.6	13.9	8.2	3.9	1.6	0.5	0.2	–	–	–	–
12	–	4	3	1	–	–	–	–	–	–	–
	2.6	2.5	1.6	0.8	0.3	0.1	–	–	–	–	–
13	–	–	1	–	–	–	–	1	–	–	–
	0.6	0.6	0.4	0.2	0.1	–	–	–	–	–	–

(ii) As for (i), but for a data set presented by Aeschbacher *et al.* (1977).

Litter size (n)	Number of dead foetuses (x)						
	0	1	2	3	4	5	6
1	–	–	–	–	–	–	–
2	1	–	–	–	–	–	–
	0.9	0.1	–	–	–	–	–
3	–	–	–	–	–	–	–
4	2	–	–	–	–	–	–
	1.5	0.4	0.1	–	–	–	–
5	1	–	–	–	–	–	–
	0.7	0.2	0.1	–	–	–	–

(*Cont.*)

Litter size (n)	Number of dead foetuses (x)						
	0	1	2	3	4	5	6
6	1	–	1	–	–	–	–
	1.4	0.5	0.1	–	–	–	–
7	–	–	–	–	–	–	–
8	–	–	–	–	–	–	–
9	–	–	–	1	–	–	–
	0.6	0.3	0.1	–	–	–	–
10	5	–	1	1	–	–	–
	4.0	1.8	0.8	0.3	0.1	–	–
11	3	2	2	1	–	–	–
	4.3	2.1	0.9	0.4	0.1	–	–
12	4	5	2	2	–	–	1
	7.3	3.7	1.7	0.8	0.3	0.1	–
13	8	7	2	2	1	–	–
	10.0	5.3	2.6	1.2	0.5	0.2	0.1
14	11	3	3	–	–	–	–
	8.2	4.5	2.3	1.1	0.5	0.2	0.1
15	13	6	1	1	1	–	–
	10.2	5.9	3.1	1.5	0.7	0.3	0.1
16	10	2	4	1	2	–	–
	8.5	5.0	2.7	1.4	0.7	0.3	0.2
17	3	4	–	–	–	–	–
	3.0	1.8	1.0	0.6	0.3	0.1	0.1
18	–	3	1	–	–	–	–
	1.7	1.0	0.6	0.3	0.2	0.1	–
19	–	–	1	–	–	–	–
	0.4	0.3	0.2	0.1	–	–	–
20	1	–	–	–	–	–	–
	0.4	0.3	0.2	0.1	0.1	–	–

6.8 (Williams, 1988b). In *in vivo* cytogenetic assays, a sample of cells is taken from an animal, and the number of cells with a chromosomal aberration is recorded. An illustration of the kind of data to result is given below. Discuss how you might analyse these data.

Data from an in vivo cytogenetic assay

Treatment	No. of aberrant cells in 50 cells per animal	Proportion aberrant
Negative control	0 4 0 0 4 0 1 1 0 0	0.020
Low dose	1 0 3 0 1 0 3 0 0 1	0.018
Medium dose	6 5 0 3 7 1 1 0 0 0	0.046
High dose	3 2 1 6 4 0 0 0 0 5	0.042

6.9   Consider the reasons for the poor binomial fits presented in Table 6.1.

6.10   Verify that Tarone's test of section 6.1.3 is a score test of the hypothesis that $\theta = 0$ in the extended beta-binomial distribution.

*6.11**   Discuss the results of Table 6.4 in the context of the likely shape of the likelihood surface contours. Derive an expression for the estimated asymptotic correlation: corr$(\hat{\mu}, \hat{\rho})$ under the beta-binomial model.

6.12   Show that beta-distributions are bell-shaped if and only if $\theta < \mu$ and $\theta < (1 - \mu)$. Verify that for the example of section 6.2.1, the beta-distribution fitted to the control data set is bell-shaped, while that fitted to the treated group is J-shaped.

6.13   Investigate the extent to which the conclusion, that we require $\theta > 0$ to describe the treated data set of the Example of section 6.2.1, is influenced by the presence of a single litter with 100% mortality in the treated group. Cf. the discussion of section 6.6.2.

*6.14**   The work of section 6.2.1 ignores an absolute measure of goodness-of-fit. Rectify this omission.

6.15   (Kleinman, 1973; Brooks, 1984; Moore, 1986; Pack, 1986a). Consider a single sample of proportions: $\{x_i/n_i, \ 1 \leqslant i \leqslant m\}$, where the x_i are realizations of binomial Bin(n_i, P) distributions, where P itself is distributed over the range $[0, 1]$, with mean μ and variance $\rho\mu(1 - \mu)$. For a given set of weights $\{w_i\}$, the sample mean and corrected sum of squares are given respectively by:

$$\tilde{p} = \frac{\sum\limits_{i=1}^{m} w_i \left(\dfrac{x_i}{n_i} \right)}{\sum\limits_{i=1}^{m} w_i}$$

$$S = \sum_{i=1}^{m} w_i \left(\frac{x_i}{n_i} - \tilde{p} \right)^2$$

We can equate these terms to their expected values, derived from equation (6.8). Show that this gives:

$$\hat{\mu} = \tilde{p}$$

$$\hat{\rho} = \frac{S - \hat{\mu}(1 - \hat{\mu}) \sum\limits_{i=1}^{m} \left(\dfrac{w_i}{n_i} \right)\left(1 - \dfrac{w_i}{w.} \right)}{\hat{\mu}(1 - \hat{\mu}) \sum\limits_{i=1}^{m} (n_i - 1)\left(\dfrac{w_i}{n_i} \right)\left(1 - \dfrac{w_i}{w.} \right)}$$

where $w. = \Sigma_{i=1}^{m} w_i$.

If we set $w_i = n_i$ we obtain the method-of-moments estimates of section 6.1.2, apart from S being replaced by $(m-1)S/m$, which reduces mean square error (Kleinman, 1973).

If we set $w_i^{-1} = V(x_i/n_i)$, show that we can take:

$$w_i = \frac{n_i}{1 + \rho(n_i - 1)} \tag{6.25}$$

Kleinman (1973) proposed the following two-stage procedure:

1. Set $w_i = n_i$ (i.e. $\hat{\rho} = 0$), form $\hat{\mu}$ and S and then form $\hat{\rho}$. If $\hat{\rho} \leqslant 0$, proceed no further, i.e. accept $\hat{\mu}$, the usual binomial estimate.

2. If $\hat{\rho} > 0$, form a new set of $\{w_i\}$ from equation (6.25) and obtain an updated estimate of μ.
   Show that this corresponds to one-step of the Williams (1982a) approach of section 6.4. (One-step procedures are discussed more generally in Appendix A.)

6.16* Consider how you would make the comparison between control and treated groups of animals of Exercise 1.6, (i) following a Freeman–Tukey Binomial transformation, and (ii) using a permutation test.

*6.17** Consider how to asses the goodness-of-fit of the beta-binomial model to the data of Table 6.5. Consider further whether the fit may be improved by changing the link function.

6.18 Verify that when $\zeta = 2$ in equation (6.7), then approximately $V\{\text{logit}(p_i)\} = \text{constant}$. (See also Exercise 1.2.)

6.19 Moore (1987), in connection with Example 6.1, presented the following standard errors, (i) ignoring the estimation of ζ, and (ii) ignoring the estimation of (ζ, ρ^*). Comment on these results and how they compare with those presented in the text.

| | Estimated asymptotic standard errors | |
Parameter	(i)	(ii)
α	0.000681	0.000680
β	0.0000266	0.0000263
ρ^*	0.076	—

*6.20** Experiment with the comparison of using quasi-likelihood and use of the heterogeneity factor as in Example 6.2, but for cases with greater variation between the $\{n_i\}$.

6.21 Trajstman (1989) presented the data of Table 1.12. In his analysis he both assumed a scaled Poisson distribution, and a negative-binomial error variance with over-dispersion parameter ϕ. In both cases the over-dispersion parameter was chosen, by trial and error, so that the fitted Pearson X^2 is close to the degrees of freedom of the fitted model. Discuss whether this is an acceptable approach to adopt.

6.22 The data of Table 6.16 result from artificially augmenting the data of Table 3.7, so that we now have a comparison between the three treatments, *A*, *B* and *C*. The residual deviance for the saturated model is 56.04 on 57 degrees of freedom, with an over-dispersion parameter of 0.0043, based on a negative binomial error distribution. From fitting separate regressions for each treatment, assuming a negative binomial error distribution, and incorporating the above

Table 6.16 *Data from Baker et al. (1980), artificially augmented. The original data correspond to the first five control values and responses to treatment A*

	Replications					Sample mean	Sample variance	Fitted mean (logit model)
Controls	219	228	202	237	228	216.7	202.0	219.7
	204	217	190	224	218			
Treatments	**Dose**							
A	1	167	167	158	158	175	165.0	51.5	158.4
	5	105	105	123	105	105	108.6	64.8	107.4
	10	88	88	88	61	61	77.2	218.7	84.4
	50	61	44	44	35	35	43.8	112.7	41.3
B	1	166	159	158	181	143	161.4	190.3	154.0
	5	97	103	112	88	120	104.0	156.5	106.9
	10	78	102	80	75	74	81.8	133.2	85.8
	50	49	40	40	57	51	47.4	54.3	45.2
C	1	160	142	143	148	135	145.6	86.3	146.8
	5	101	74	81	82	94	86.4	118.3	81.4
	10	54	63	42	52	48	51.8	60.2	56.5
	50	32	23	15	16	19	21.0	47.5	20.2

over-dispersion parameter, Morgan and Smith (1990) obtain the results below, for a range of link functions (in all cases, degrees of freedom = 63).

Link	Residual deviance
Logit	65.50
Complementary log–log	84.84
Probit	67.93
Cauchit	69.55

Discuss the choice of link function for these data. Cf. also Exercise 6.26. The GLIM output is given in Smith and Morgan (1989).

6.23* (Pack, 1986a). Discuss whether the correlated binomial distribution might in some cases be bi-modal.

6.24* (Paul, 1985; Pack, 1986a). Suppose the probability p in the correlated binomial distribution is given a beta distribution with parameters (α, β). Show that the usual approach of integrating with respect to the beta distribution produces the beta-correlated binomial distribution, with probability function,

$$\Pr(X = x \mid n) = \binom{n}{x} \frac{B(x + \alpha, n - x + \beta)}{B(\alpha, \beta)} \{1 + \frac{\phi}{2} g(x, n, \alpha, \beta)\}$$

where $\phi = \rho p(1 - p)$, and

$$g(x,n,\alpha,\beta) = \frac{x(x - 1) \prod_{i=1}^{4} (\alpha + \beta + n - i)}{\prod_{i=1}^{2} (x + \alpha - i) \prod_{i=1}^{2} (n - x + \beta - i)}$$

$$- \frac{2x(n - 1) \prod_{i=1}^{3} (\alpha + \beta + n - i)}{(x + \alpha - 1) \prod_{i=1}^{2} (n - x + \beta - i)} + \frac{n(n - 1) \prod_{i=1}^{2} (\alpha + \beta + n - i)}{\prod_{i=1}^{2} (n - x + \beta - i)}$$

What is wrong with this approach? Verify that nevertheless a proper distribution results and deduce its mean and variance. How else might this distribution be derived?

6.25[†] (Ridout *et al.* 1992). An alternative to fitting a mixture of two binomial distributions, as in section 6.6.2, is to fit a mixture of a beta-binomial, with parameters (μ, θ), with probability γ, and a binomial distribution, $\text{Bin}(n, v)$, with probability $(1 - \gamma)$. We present below the maximum-likelihood estimates of these parameters for the three data sets of Haseman and Soares (1976), given in Tables 1.10 and Exercise 6.7(i). Estimated asymptotic standard errors are given in parentheses.

Data set	γ	μ	θ	v	Max. log-likelihood
1	0.995	0.0859	0.0539	0.737	771.16
	(0.006)	(0.004)	(0.0117)	(0.137)	
2	0.15	0.05	2.56	0.19	1633.3
	(0.032)	(0.025)	(2.00)	(0.006)	
3	0.97	0.06	0.02	0.56	682.9
	(0.009)	(0.004)	(0.009)	(0.06)	

For data sets 1 and 3, comparisons may be drawn with the binomial mixture results of Table 6.7, and the beta-binomial results of Table 6.3. For data set 2, the corresponding information is given below:

Two-component binomial fit:

$$\hat{p}_1 = 0.103 \, (0.003) \quad \hat{p}_2 = 0.741 \, (0.096) \quad \hat{\gamma} = 0.99 \, (0.004),$$
$$- \text{ Max. log-likelihood} = 1644.40.$$

Three-component binomial fit (no errors available due to boundary estimation of p_3):

$$\hat{p}_1 = 0.082 \quad \hat{p}_2 = 0.205$$
$$\hat{p}_3 = 1.00 \quad \hat{\gamma}_1 = 0.803$$
$$\hat{\gamma}_2 = 0.193, \quad - \text{ Max. log-likelihood} = 1634.4$$

Beta-binomial fit: $\hat{\mu} = 0.109 \, (0.003)$, $\hat{\theta} = 0.045 \, (0.008)$,

$- $ Max. log-likelihood $= - 1657.30.$

Provide a detailed discussion of these results and their possible implications for the three data sets. How might you test whether the beta-binomial fits are significantly improved by mixing with a binomial distribution?

6.26 (Pack, 1986a). Both the beta-binomial (BB) and the correlated binomial (CB) models are particular cases of the beta-correlated-binomial (BCB) model of Exercise 6.24. Show that a conservative test for the BB and CB models results from referring twice the difference between the BB and CB log-likelihood values to χ_1^2 tables. Cf. Exercise 6.22. Apply this result, where appropriate, to the maximum log-likelihood values given below (from Pack, 1986a), and draw conclusions.

Data set	Binomial	CB	BB	BCB	2-element binomial mixture
Haseman and Soares (1976)					
HS1	−842.61	−801.68	−777.79	−776.4	−782.44
HS2	−1686.76	−1669.02	−1657.30	−1652.2	−1644.4
HS3	−765.06	−732.19	−701.33	−695.94	−686.24
Aesbacher et al. (1977)	−179.17	−169.52	−168.93	−168.93	−168.90

Model spans the Binomial, CB, BB, BCB, and 2-element binomial mixture columns.

6.27[†] The comparisons of Exercise 6.26 do not extend to the mixture model. Suppose we want to compare a beta-binomial fit with the two-binomials mixture fit. We may use a Monte Carlo comparison as follows.

Fit each model. From each fitted model simulate 99 sets of data, and to each set fit both models. Calculate the rank of the log-likelihood of each model fitted to the real data in the 100 log-likelihood values available, for each model. Also calculate, for each series of 99 simulated data sets, the difference between the maximum log-likelihoods of the two models fitted to the simulated data, and obtain the rank of the corresponding difference for the real data set in the series of 100 differences available. We illustrate this for both the HS1 and HS2 data sets of Exercise 6.26 (Pack, 1986a).

Data set	Data simulated from	Real data − max. log lik.	Under model	Rank
HS1	Beta-binomial	777.79	Beta-binomial	46
		782.44	Mixture	36
		Real data difference		
		4.66		88
	Binomial mixture	782.44	Mixture	20
		777.79	Beta-binomial	28
		Real data difference		
		−4.66		1
HS2	Beta-binomial	1657.3	Beta-binomial	26
		1644.4	Mixture	37
		Real data difference		
		−12.90		1
	Binomial mixture	1644.4	Mixture	5
		1657.3	Beta-binomial	23
		Real data difference		
		12.90		5

Which model would you choose for each of these data sets? Compare your conclusions with those resulting from the solution to Exercise 6.25.

6.28   (Altham, 1978). The 'multiplicative' extension of the binomial distribution has probability function:

$$\Pr(X = x \mid n) = \frac{\binom{n}{x} p^x (1-p)^{n-x} \zeta^{x(n-x)}}{f(p, \zeta, n)}, \quad 0 \leqslant x \leqslant n, \, 0 < p < 1, \, \zeta > 0$$

where

$$f(p, \zeta, n) = \sum_{s=0}^{n} \binom{n}{s} p^s (1-p)^{n-s} \zeta^{s(n-s)}.$$

Show that this model can accommodate both over- and under-dispersion. Obtain expressions for the mean and variance.

6.29 The probability function for the quasi-binomial distribution is given by:

$$\Pr(X = x|n) = \binom{n}{x} p(1-p)(1 + x\gamma)^{x-1}$$

$$\times \{1 - p + \gamma(n - x)\}^{n-x-1}(1 + n\gamma)^{1-n}$$

where $0 \leqslant p, \gamma \leqslant 1$. The binomial distribution results when $\gamma = 0$.

Derive the mean and variance. Fazal (1976) gives a score test for comparing this distribution with the binomial. A more general form of the distribution is derived by Consul and Mittal (1975), in the context of an urn model.

6.30 Kimball and Friedman (1986) propose a multivariate Bernoulli distribution for correlated binary data. For describing responses within a litter of size n, this reduces to (Pack, 1986a)

$$\Pr(X = x|n) = \binom{n}{x}\left[p^x(1-p)^{n-x} + \rho p(1-p)\left\{ \binom{n}{2} - 2x(n-x) \right\} \right]$$

Derive the mean and variance.

6.31 Bratcher and Bell (1985) extend the beta-binomial distribution to give:

$$f(p) = \frac{p^{\alpha-1}(1-p)^{\beta-1}e^{cp}}{B(\alpha, \beta)M(\alpha, \beta, c)}, \quad 0 < p < 1, \alpha, \beta > 0$$

where $M(\alpha, \beta, c)$ is the confluent hypergeometric function:

$$M(\alpha, \beta, c) = \int_0^1 \frac{p^{\alpha-1}(1-p)^{\beta-1}e^{cp}}{B(\alpha, \beta)} dp$$

Derive the corresponding form for $\Pr(X = x|n)$ and evaluate $E[X|n]$.

6.32 The data of Table 6.17 were presented by Williams (1988b). He fitted the model:

$$\text{logit}(p_{ij}) = \alpha_i + \beta_1 n_{ij} + \beta_2 n_{ij}^2$$

using the quasi-likelihood approach, with the mean/variance specification as in equation (6.8), but now with i denoting treatment, and j denoting litter number. Here n_{ij} is the size of the jth litter given

Table 6.17 *Numbers of deaths per litter from a mouse teratology experiment*

Litter size s	Dosage in g/kg of body weight			
	0	0.75	1.5	3.0
3	—	—	3	—
5	—	—	4	—
7	—	0	4	—
8	—	—	—	—
9	1 2	2	2 2	2 2 5
10	—	0	0	1 6
11	0 1 1 5	0 1 2	—	2
12	0 0 1	0 1 1	0 1 1 2 3	0 0 1 1 2 2 2 5
13	0 0 1 3	0 1 1 2 5	0 0 1 2	1 1 1
14	0 2	0 1 2 4	1 2 3	2 4
15	0 1 2 2 2 5	0 0 0 1 5	0 2 2	3 3
16	0 2 2	—	2	2 9
17	3	—	—	—
19	—	5	—	—

the ith treatment, and p_{ij} is the corresponding probability of death. Discuss the resulting estimates, which were found to provide a satisfactory fit to the data:

$$\hat{\alpha}_1 - \hat{\alpha}_0 = -0.12\,(0.34), \quad \hat{\alpha}_2 - \hat{\alpha}_0 = 0.09\,(0.35)$$
$$\hat{\alpha}_3 - \hat{\alpha}_0 = 0.77\,(0.31), \quad \hat{\beta}_1 = -1.27\,(0.26)$$
$$\hat{\beta}_2 = 0.049\,(0.011), \quad \hat{\rho} = 0.059$$

Is there evidence that the compound administered increases incidence of deaths?

6.33 An urn contains r red and b black balls. Balls are removed singly and at random, and replaced with an extra ball of the same colour as the ball drawn. Show that the number of reds obtained after n balls have been drawn has a beta-binomial distribution.

6.34* Show that when the model of equation (6.22) is fitted to the data of Table 6.11, the maximum-likelihood estimates of the parameters are:

$$\hat{r} = 0.245\,(0.315), \quad \hat{b} = 0.793\,(0.422)$$

with an estimated asymptotic correlation of 0.943. Discuss the fit of the model to the data. Why does this model result in under-dispersion?

6.35 (Ridout and Morgan, 1991.) Close examination of the data of Table 6.12 suggests that responses may favour 6 months and 12 months. Such digit preference is indeed very strongly indicated in more extensive sets of fecundability data. Consider how you might model digit preference in this application.

6.36 Maruani and Schwartz (1983) consider the following mixture model for **prospective** fecundability data:

$$f(p) = (1 - \eta)\frac{p^{\alpha - 1}(1 - p)^{\beta - 1}}{B(\alpha, \beta)}, \quad \text{for } 0 \leqslant p \leqslant 1$$

with an additional mass of η at $p = 0$.

Why is this particular model inappropriate for **retrospective** data, as in fact have been presented in Table 6.12? For one set of data they estimate $\hat{\eta} = 0.00117\,(0.00023)$. Is it likely that the model requires $\eta > 0$ for these data? Discuss potential problems with fitting this model to data.

6.37 Discuss how the $\tilde{X}^2$ statistic of equation (6.24) has adjusted the Cochran/Armitage statistic. When will the historical data have the greatest impact on $\tilde{X}^2$?

6.38 (Hoel and Yanagawa, 1986.) The analysis of section 6.7.4 is based upon a logit model. Obtain the corresponding test statistic when an exponential model is used.

CHAPTER 7

Non-parametric and robust methods

7.1 Introduction

The history of the statistical analysis of quantal response data is documented by Finney (1971, pp 38–42). The maximum-likelihood fitting of the probit model stems from the papers of Fisher (1935) and Bliss (1938). However, non-parametric approaches to estimating the ED_{50} have a generally earlier origin, in papers such as Behrens (1929), Kärber (1931) and Spearman (1908). Computationally, the early non-parametric measures are far simpler than those which result from an iterative maximum-likelihood fit, and this was clearly a matter of great importance before the computer era.

As we saw in Chapter 4, the maximum-likelihood fitting of extended models can be difficult. However, today access to computers makes the fitting of simple models, such as the logit or probit, to 'well-behaved' data a formality. Nevertheless, modern developments in the areas of robustness and influence have resulted in a re-examination of the relative merits of parametric and non-parametric procedures.

The emphasis in this chapter is partly on measurement of the ED_{50} of substances presented for assay as part of a routine testing program. Such experiments may well be seen just as pilot studies, and doses may be set conservatively to cover a wide range, quite often resulting in a number of instances of 0% and 100% response. If too high doses are employed then there may be additional problems arising from high dose toxicity effects. Small numbers of individuals may be allotted to each dose level. In such an experimental context, data may well cease to be 'well behaved', and indeed standard parametric procedures may fail. Examples of this have already been noted in Exercise 2.27. If there is either no, or only one, dose which results in partial response (i.e. other than 0% or 100%) then maximum-likelihood fitting of standard models is either not possible (Silvapulle, 1981), or results in a fit with infinite slope parameter, β. The failure of standard procedures is normally an indication that further detailed investigation is necessary. In the context of routine testing, when only rough estimates are required, it can be a nuisance. In such a framework, more robust methods are needed. Robustness has already been encountered in Chapter 6, through the use of quasi-likelihood for describing over-dispersed data. In this chapter we start with a review of non-parametric procedures, with particular reference to estimation of the ED_{50}. We then proceed to consider robust parametric and non-parametric procedures, and describe their efficiency. The chapter concludes with a review of a number of alternative distribution-free procedures. As we shall see, some of the techniques which are proposed are very simple to operate, and of great practical importance. Others are of more academic interest, and unlikely to be used in a routine situation. The basis of much non-parametric work is a distribution-free estimate of the cumulative distribution function of the tolerance distribution, to be described in the next section.

7.2   The pool-adjacent-violators algorithm; ABERS estimate

Let $P(d)$ denote the probability of response to dose d, and let $P_i = P(d_i)$, for $1 \leqslant i \leqslant k$. If the only assumption made in the model is that $P(d)$ is a monotonic increasing function of d, then the maximum-likelihood estimate of $\mathbf{P} = (P_1, P_2, \ldots, P_k)$, corresponding to the log-likelihood,

$$l = \text{constant} + \sum_{i=1}^{k} r_i \log P_i + \sum_{i=1}^{k} (n_i - r_i) \log (1 - P_i)$$

results as an example of optimization under the order restriction:

$$P_1 \leqslant P_2 \leqslant \cdots \leqslant P_k$$

The necessary theory is to be found in Barlow *et al.* (1972, p 13). As it was originally propounded by Ayer *et al.* (1955), the resulting estimate is sometimes called the ABERS estimate. It is obtained by a straightforward algorithm called the pool-adjacent-violators algorithm, which proceeds as follows.

For any i we can form the maximum-likelihood estimator of P_i: $\hat{P}_i = r_i/n_i$. Naturally, in most cases we would like the $\hat{P}_i$ to be monotonically increasing as i increases. (An exception might arise if high-dose toxicity were anticipated.) If any two adjacent $\hat{P}_i$ are in the wrong order, i.e. $r_i/n_i > r_{i+1}/n_{i+1}$, then at the doses d_i and d_{i+1} we replace each of the responses $\hat{P}_i$ and $\hat{P}_{i+1}$ values by $(r_i + r_{i+1})/(n_i + n_{i+1})$. These two doses may now be considered as a **block**, and a number of such blocks may need to be formed. After the process of block formation is completed, it may be that adjacent block, or block and dose proportions may be found to be out of order. In such a case the relevant blocks and/or blocks and doses are pooled to give larger blocks, as appropriate, proceeding as in the original manner for proportions, until no further pooling is necessary. A simple illustration is given in Example 7.1.

Example 7.1 (Egger, 1979)

An illustration of the ABERS estimate, formed for a data set from Reed and Muench (1938). This is data set 3 of Exercise 1.3. The two non-monotonic proportions before adjustment are indicated by the arrows. Thus here there are just two blocks of doses, each of size two.

	Before adjustment			After adjustment		
Dose	No. of subjects	No. responding	Proportion	No. of subjects	No. responding	Proportion
1	6	0	0	6	0	0
2	6	0	0	6	0	0
4	6	1	↑0.1667	12	1	0.0833
8	6	0	↓0	12	1	0.0833
16	6	2	0.3333	6	2	0.3333
32	6	4	0.6667	6	4	0.6667
64	6	4	0.6667	6	4	0.6667
128	6	6	↑1.0	12	11	0.9167
256	6	5	↓0.8333	12	11	0.9167

In this case the adjustments are completed after just one stage. If additional adjustments are needed, then the order in which they are carried out is immaterial, as shown by Ayer *et al.* (1955) – see Exercise 7.1 Lexicographic confidence limits for the probabilities, illustrated for this data set, are given in Morris (1988).   □

The ABERS estimate is a natural preliminary to computing the Spearman–Kärber and related estimates of the ED_{50}. It has also been found to be useful when employed with the method of minimum logit chi-square (Hamilton, 1979; James *et al.*, 1984). As we can see from the illustration of Example 7.1, by itself the ABERS estimate is not particularly useful, and some additional smoothing is usually considered advantageous. A simple approach is to use linear interpolation. This is employed by Hamilton (1979) as a preliminary to constructing a trimmed Spearman–Kärber estimate of the ED_{50}, and it is used also in section 7.5.3 in constructing an influence curve. Other ways of proceeding have been suggested by Schmoyer (1984) and Glasbey (1987), and we consider their work in section 7.6. A FORTRAN program for the pool-adjacent violators algorithm is given in Cran (1980).

In mathematical terms, if $\tilde{P}_1, \ldots, \tilde{P}_k$ denote the distribution-free maximum-likelihood estimators of $P(d_1), \ldots, P(d_k)$ under the order-restriction, $P_1 \leqslant P_2 \leqslant \cdots \leqslant P_k$, then (Exercise 7.2),

$$\tilde{P}_i = \max_{1 < u < i} \min_{i < v < k} \left\{ \sum_{j=u}^{v} r_j \Big/ \sum_{j=u}^{v} n_j \right\}$$

In the special case when $n_j = n$ for all j, then

$$\tilde{P}_i = \max_{1 < u < i} \min_{i < v < k} \frac{\sum\limits_{j=u}^{v} r_j}{\{n(v - u + 1)\}}$$

7.3   The Spearman–Kärber estimate of the ED_{50}

If the tolerance distribution has cumulative distribution function $F(x)$, then its mean may be written as:

$$\mu = \int_{-\infty}^{\infty} x \, dF(x)$$

Let

$$\hat{P}_i = \frac{r_i}{n_i}, \quad 1 \leqslant i \leqslant k$$

and

$$\Delta_i = (d_{i+1} - d_i), \quad 1 \leqslant i \leqslant k - 1$$

then the Spearman–Kärber estimate for μ is given by:

$$\hat{\mu} = \hat{P}_1 \left(d_1 - \frac{\Delta_1}{2} \right) + \sum_{i=1}^{k-1} (\hat{P}_{i+1} - \hat{P}_i) \left(d_i + \frac{\Delta_i}{2} \right) + (1 - \hat{P}_k) \left(d_k + \frac{\Delta_{k-1}}{2} \right)$$

$$(7.1)$$

There are two features to note in connection with this expression. First, the effect of expression (7.1) is equivalent to introducing additional artificial doses, d_0 and d_{k+1}, where $d_0 = d_1 - \Delta_1$, and $d_{k+1} = d_k + \Delta_{k-1}$, with respectively, 0% and 100% responses. If in fact $\hat{P}_1 = 0$ and $\hat{P}_k = 1$, then, as can be seen from equation (7.1) this device is not used. The arbitrariness of this procedure if either $\hat{P}_1 \neq 0$ or $\hat{P}_k \neq 1$ has understandably led to criticism of the Spearman–Kärber approach. It has, additionally, been argued that efficient design of an experiment aimed at estimating the ED_{50} should in fact result in $\hat{P}_1 \neq 0$ and $\hat{P}_k \neq 1$. However, as we have seen in the discussion above, in the observational study of Table 1.4, and in Exercises 2.27 and 2.30, data encompassing both 0% and 100% response are frequently encountered, especially in the case of routine assessment of potentially toxic or beneficial substances, when a wide range of doses is typically adopted. In any case, this feature of the Spearman–Kärber estimate is avoided if suitable trimming is employed, as will be discussed in section 7.4.

Secondly, if $P_1 = 0$, $P_k = 1$ and $P_i = F(x_i)$, $l \leqslant i \leqslant k - 1$, $\hat{\mu}$ estimates μ indirectly, through estimating the mean, η, of the discrete random variable X, with distribution,

$$\Pr(X = d_i + \Delta_i/2) = P_{i+1} - P_i, \quad 1 \leqslant i \leqslant k - 1$$

The estimation of the probability $(P_{i+1} - P_i)$ by $(\hat{P}_{i+1} - \hat{P}_i)$ suggests that it may be useful to smooth the $\{\hat{P}_i\}$ before calculating the Spearman–Kärber estimate, to ensure that the $\{\hat{P}_i\}$ are monotonically increasing. The ABERS estimate may be used for this purpose if necessary. As we shall now see, this adjustment does not necessarily affect $\hat{\mu}$.

If we find that $\hat{P}_1 = 0$ and $\hat{P}_k = 1$, then equation (7.1) reduces to:

$$\hat{\mu} = \sum_{i=1}^{k-1} (\hat{P}_{i+1} - \hat{P}_i) \left(d_i + \frac{\Delta_i}{2} \right) \equiv \frac{1}{2} \sum_{i=1}^{k-1} (\hat{P}_{i+1} - \hat{P}_i)(d_i + d_{i+1})$$

which is a form in which the Spearman–Kärber estimate is often given. A simple rearrangement of terms gives the alternative expression:

$$\hat{\mu} = \frac{1}{2} \left\{ \sum_{j=2}^{k-1} \hat{P}_j (d_{j-1} - d_{j+1}) \right\} + (d_{k-1} + d_k) \qquad (7.2)$$

again making use of $\hat{P}_1 = 0$ and $\hat{P}_k = 1$. This form is convenient for evaluating the properties of $\hat{\mu}$. Further simplification results if the doses are equally spaced, i.e. $\Delta_i = \Delta$, for all i, leading to the expression:

$$\hat{\mu} = d_k + \frac{\Delta}{2} - \Delta \sum_{i=1}^{k} \hat{P}_i \qquad (7.3)$$

If additionally the same number, n, of individuals is employed at each dose, then if the ABERS estimate is formed this does not change $\hat{\mu}$, since

$$\sum_{i=1}^{k} \hat{P}_i = \frac{1}{n} \sum_{i=1}^{k} r_i, \quad \text{and} \quad \sum_{i=1}^{k} r_i$$

does not change under the ABERS averaging procedure. This is true also in the slightly more general case of an equal weight design (Exercise 7.4).

Working directly from the expression of equation (7.2), assuming $P_1 = 0$, $P_k = 1$, and without reference to any possible ABERS correction, we can immediately write down the variance of $\hat{\mu}$ as:

$$V(\hat{\mu}) = \frac{1}{4} \sum_{j=2}^{k-1} P_j (1 - P_j)(d_{j-1} - d_{j+1})^2 / n_j$$

which simplifies in the special case of equally spaced doses and equal numbers at each dose to give

$$V(\hat{\mu}) = \frac{\Delta^2}{n} \sum_{j=2}^{k-1} P_j (1 - P_j) \qquad (7.4)$$

In general, for the case when $P_1 = 0$ and $P_k = 1$, we obtain an unbiased estimate of $V(\hat{\mu})$ if we evaluate

$$\frac{1}{4} \sum_{j=2}^{k-1} \hat{P}_j (1 - \hat{P}_j)(d_{j-1} - d_{j+1})^2 / (n_j - 1)$$

as long as $n_j \geqslant 2$, for all j.

The explicit forms for $\hat{\mu}$ and its variance were clearly attractive when the iterations necessary for probit analysis, say, were onerous, prior to the arrival of computers. It is still attractive to have a simple expression for $V(\hat{\mu})$ as an aid to the design of experiments (Brown (1966), and Chapter 8). As the expressions for $\hat{\mu}$ are weighted sums of independent binomial random variables, confidence regions for μ may be based on the assumption of normality of the estimator $\hat{\mu}$. We note, however, that $E[\hat{\mu}] = \eta$ and in general, $\eta \neq \mu$, since η is the mean of the discrete random variable X with distribution specified in terms of the particular dose levels of an experiment. Thus $\hat{\mu}$ is a biased estimator of μ. In fact, as $\hat{\mu}$ is a function of the sufficient statistic, $(r_1, \ldots, r_k)$, which is complete, then $\hat{\mu}$ is the uniform minimum variance unbiased estimator of η. Additionally, $\hat{\mu}$ is the unrestricted non-parametric maximum likelihood estimator of η, as it is a linear combination of $P_1, \ldots, P_k$, and the unrestricted maximum likelihood estimator of $(P_1, \ldots, P_k)$ is $(\hat{P}_1, \ldots, \hat{P}_k)$. Several authors have demonstrated the close affinity between the logit model and the Spearman-Kärber estimate, which we shall encounter again in section 7.5. See, for example, Finney (1950, 1952), and Anscombe (1956). Cornfield and Mantel (1950) showed the asymptotic equivalence of the logistic maximum likelihood estimator of the ED_{50} and the Spearman-Kärber estimate.

Of course the Spearman-Kärber estimate estimates the **mean** of the tolerance distribution, and if we use $\hat{\mu}$ to estimate the ED_{50}, which is the **median** of the tolerance distribution, then we are additionally making the assumption of symmetry. The same is true of the Reed-Meunch and Dragstedt-Behrens estimators (Miller, 1973 – see Exercises 7.6 and 7.7).

The logic which results in the Spearman-Kärber estimator applies equally well to estimation of higher order moments of the tolerance distribution. Thus, for example, the Spearman-Kärber estimate of the jth non-central moment estimator is:

$$\hat{\mu}_j = \hat{P}_1\left(d_1 - \frac{\Delta_1}{2}\right)^j + \sum_{i=1}^{k-1} (\hat{P}_{i+1} - \hat{P}_i)\left(d_i + \frac{\Delta_i}{2}\right)^j$$

$$+ (1 - \hat{P}_k)\left(d_k + \frac{\Delta_{k-1}}{2}\right)^j \quad \text{for } j \geqslant 1$$

An interesting use is made of these estimators by Egger (1979). She replaced d by its Box-Cox transformed value, $(d^\lambda - 1)/\lambda$, and sought

Table 7.1 *Asymptotic efficiency values for the estimators of* μ *and* $\mu_{(2)}$, *resulting from the Spearman–Kärber approach (Chmiel, 1976) when a variety of distributional forms are assumed for the tolerance distribution. The angular distribution has cumulative distribution function:* $F(x) = \sin^2(x + \pi/4)$, *for* $-\pi/4 \leqslant x \leqslant \pi/4$

Distribution	$\hat{\mu}$	$\hat{\mu}_{(2)}$
Logistic	1.0000	1.0000
Normal	0.9814	0.9637
Laplace	0.9618	0.9928
Angular	0.8106	0.6677

λ as a root of the equation,

$$\hat{\mu}_{(3)} = 0$$

where now $\hat{\mu}_{(3)}$ estimates the third central moment, and is given by:

$$\hat{\mu}_{(3)} = \hat{\mu}_3 - 3\hat{\mu}_2\hat{\mu}_1 + 2\hat{\mu}_1^3$$

The resulting value for λ could then be used to symmetrize the data in a non-parametric way, in comparison with the approach of section 4.4, based on maximum-likelihood. This procedure was not in fact found to be straightforward in practice, and is discussed further in Exercise 7.8. The properties of $\hat{\mu}_{(2)}$ have been investigated in detail by Chmiel (1976) who presented the asymptotic efficiency values shown in Table 7.1. In the quantal response context, care is needed in how one takes asymptotic limits, and we return to these results in our discussion in section 7.5. Additionally, Chmiel (1976) indicates how to use $\hat{\mu}_{(2)}$ to estimate percentiles.

Example 7.2

As an illustration of computation of the Spearman–Kärber estimate of the ED_{50} we may use the data from Exercise 1.14 on the detonation of explosives. Here, $d_i = 4$ for all i, and $n_i = 16$ for all i. Thus formula (7.3) applies and we obtain $\hat{\mu} = 46.5$, with an estimated standard error of 0.85. This contrasts with the corresponding maximum likelihood estimated values under the logit model of 46.64 (0.95).

7.4   Trimming

7.4.1   Trimmed Spearman–Kärber

If one's aim is to analyse routinely and automatically many sets of quantal response data, and in each case obtain an estimate of the ED_{50} and of its error, then the Spearman–Kärber method may be made more robust by the addition of trimming, which is performed by analogy with the use of trimming to estimate the mean in direct samples from univariate distributions (Andrews *et al.* 1972, p. 31). The procedure for quantal response data was propounded by Hamilton *et al.* (1977) and has subsequently been evaluated by Hamilton (1979, 1980), Miller and Halpern (1980), James *et al.* (1984), Sanathanan *et al.* (1987) and Hoekstra (1989). We shall now state how the procedure operates, and present illustrations of the results, leaving efficiency considerations until section 7.5.

The first stage is to form the ABERS estimate, if necessary. What follows next is best illustrated first of all by example.

Example 7.3

Calculation of a 10% trimmed Spearman–Kärber estimate: we shall use the data of Example 7.1.

We start by forming the ABERS estimate $\{\tilde{P}_i\}$ from Example 7.1. The points on the graph of $\tilde{P}_i$ versus $x_i = \log_2(d_i)$ are joined by straight lines to give the polygon of Figure 7.1(a). The points marked $\otimes$ are then trimmed away, and the points marked $\bullet$ are added. The axis for proportions is then rescaled to give $\tilde{\tilde{P}}_i = (\tilde{P}_i - 0.1)/0.8$, with the result shown in Figure 7.1(b), and finally the Spearman–Kärber estimate is formed for the $\{\tilde{\tilde{P}}_i\}$. For the case of this example we obtain the 10% trimmed Spearman–Kärber estimate, which we denote by $\hat{\mu}_{10}$, as $\hat{\mu}_{10} = 27.7$. An estimated 95% confidence interval, obtained as shown in Figure 7.1, is (15.37, 49.93).

In mathematical terms, the $100\alpha\%$ trimmed Spearman–Kärber estimate is defined as follows (the notation is illustrated in Figure 7.1). For $0 \leqslant \alpha \leqslant 0.5$, let

$$L(\alpha) = \max\{i : \hat{P}_i \leqslant \alpha, \quad 1 \leqslant i \leqslant k\} \text{ and}$$
$$U(\alpha) = \min\{i : \hat{P}_i \geqslant (1 - \alpha), \quad 1 \leqslant i \leqslant k\}$$

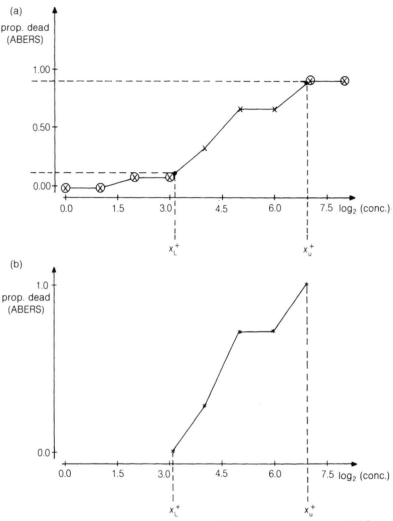

Figure 7.1 *An illustration of how the 10% trimmed Spearman–Kärber estimate is formed for the data of Example 7.1. In (a) the trimming is illustrated, while in (b) we see how the rescaling is done. Here* $L(10) = 4$, *and* $U(10) = 8$.

We can now specify the absicae of the added points by:

$$x_L^+ = x_{L(\alpha)} + (x_{L(\alpha)+1} - x_{L(\alpha)})(\alpha - \hat{P}_{L(\alpha)})/(\hat{P}_{L(\alpha)+1} - \hat{P}_{L(\alpha)})$$
$$x_U^+ = x_{U(\alpha)-1} + (x_{U(\alpha)} - x_{U(\alpha)-1})(1 - \alpha - \hat{P}_{U(\alpha)-1})/(\hat{P}_{U(\alpha)} - \hat{P}_{U(\alpha)-1})$$

Then the $100\,\alpha\%$ trimmed Spearman–Kärber estimator is defined as:

$$\hat{\mu}_\alpha = \left\{ (x_L^+ + x_{L(\alpha)+1})(\hat{P}_{L(\alpha)+1} - \alpha)/2 \right.$$
$$+ \sum_{i=L(\alpha)+1}^{U(\alpha)-2} (x_i + x_{i+1})(\hat{P}_{i+1} - \hat{P}_i)/2$$
$$\left. + (x_U^+ + x_{U(\alpha)-1})(1 - \alpha - \hat{P}_{U(\alpha)-1})/2 \right\} \bigg/ (1 - 2\alpha)$$

The estimator is only regarded as calculable if $\hat{P}_1 \leqslant \alpha$ and $\hat{P}_k \geqslant 1 - \alpha$.

The case of $\alpha = 0$ reduces to the original Spearman–Kärber estimator, and the case of $\alpha = 0.5$ corresponds to linear interpolation between the two central doses for which the $\hat{P}$s enclose 0.50. An illustration of this has already been provided in Figure 1.1.

The estimated variance of $\hat{\mu}_\alpha$, obtained by Hamilton (1980) using the delta-method (see Appendix A), is of the following form:

Case 1 $\alpha < 0.5$ and $U(\alpha) - L(\alpha) > 1$

$U(\alpha) - L(\alpha)$	$\hat{V}(\hat{\mu}_\alpha)$
2	$(v_1 + v_5 + v_6)/\{2(1 - 2\alpha)\}^2$
3	$(v_1 + v_2 + v_4 + v_5)/\{2(1 - 2\alpha)\}^2$
$\geqslant 4$	$(v_1 + v_2 + v_3 + v_4 + v_5)/\{2(1 - 2\alpha)\}^2$

where, suppressing the argument α of $L(\alpha)$ and $U(\alpha)$, we define,

$$v_1 = \{(x_{L+1} - x_L)(\hat{P}_{L+1} - \alpha)^2/(\hat{P}_{L+1} - \hat{P}_L)^2\}^2 \hat{P}_L(1 - \hat{P}_L)/n_L$$
$$v_2 = \{(x_L + x_{L+2}) + (x_{L+1} - x_L)(\alpha - \hat{P}_L)^2/(\hat{P}_{L+1} - \hat{P}_L)^2\}^2$$
$$\times \hat{P}_{L+1}(1 - \hat{P}_{L+1})/n_{L+1}$$
$$v_3 = \sum_{i=L+2}^{U-2} (x_{i-1} - x_{i+1})^2 \hat{P}_i(1 - \hat{P}_i)/n_i$$
$$v_4 = \{(x_{U-2} - x_U) + (x_U - x_{U-1})(\hat{P}_U - 1 + \alpha)^2/(\hat{P}_U - \hat{P}_{U-1})^2\}^2$$
$$\times \hat{P}_{U-1}(1 - \hat{P}_{U-1})/n_{U-1}$$
$$v_5 = \{(x_U - x_{U-1})(1 - \alpha - \hat{P}_{U-1})^2/(\hat{P}_U - \hat{P}_{U-1})^2\}^2 \hat{P}_U(1 - \hat{P}_U)/n_U$$
$$v_6 = [\{(x_U - x_{L+1})(1 - \alpha - \hat{P}_U)^2/(\hat{P}_U - \hat{P}_{L+1})^2\}$$
$$- \{(x_{L+1} - x_L)(\alpha - \hat{P}_L)^2/(\hat{P}_{L+1} - \hat{P}_L)^2\}$$
$$+ (x_L - x_U)]^2 \hat{P}_{L+1}(1 - \hat{P}_{L+1})/n_{L+1}$$

Case 2 $\alpha = 0.5$ or $U(\alpha) - L(\alpha) = 1$

$$\hat{V}(\hat{\mu}_{\alpha}) = (x_{U(\alpha)} - x_{L(\alpha)})^2 [\{(0.5 - \hat{P}_{U(\alpha)})^2 / (\hat{P}_{U(\alpha)} - \hat{P}_{L(\alpha)})^4\} \hat{P}_{L(\alpha)}(1 - \hat{P}_{L(\alpha)}) / n_{L(\alpha)}$$
$$+ \{(0.5 - \hat{P}_{L(\alpha)})^2 / (\hat{P}_{U(\alpha)} - \hat{P}_{L(\alpha)})^4\} \hat{P}_{U(\alpha)}(1 - \hat{P}_{U(\alpha)}) / n_{U(\alpha)}]$$

For two data sets we now present a number of $100\alpha\%$ trimmed Spearman–Kärber estimates of the ED_{50}, together with estimated 95% confidence intervals.

Example 7.4

In practice the extent of trimming may have less effect on precision than one might expect. Two examples of this are shown below, the data sets being taken from Exercise 1.3.

Data set	Level of trimming, α, given as a percentage	Estimate of ED_{50}	95% confidence interval
9	10	1.77	1.76, 1.79
	20	1.78	1.76, 1.79
	30	1.78	1.76, 1.80
	40	1.78	1.76, 1.80
22	20	0.70	0.64, 0.76
	30	0.71	0.65, 0.77
	40	0.72	0.68, 0.75

□

The level of trimming, α, may be a matter for personal choice, but it is likely also to be dictated by the features of each particular problem. As we shall see in section 7.5, a value of $\alpha = 0.05$ has been found to give good results in practice, when that is feasible. The need to take different values of α for different experiments may be thought to be undesirable in a system for routine analysis of several sets of data, and this was one motivating factor leading to the work of the next section.

7.4.2   Trimmed logit

The policy of trimming may, of course, also be used in conjunction with parametric model-fitting, to improve robustness. Furthermore, a parametric model fitted to trimmed data may be used to estimate ED_{100p} values for p other than 0.50.

A trimmed logit procedure is introduced by Sanathanan *et al.* (1987), and is shown to be more robust than the trimmed Spearman–Kärber method, even producing estimates of the ED_{50} from somewhat degenerate or limited data. Their procedure excludes doses for which the expected proportion of responses is outside the (0.001, 0.999) range. The logit model is fitted iteratively, with the aim of minimizing the heterogeneity factor, $X^2/(k-2)$, while simultaneously excluding doses as just explained. Care in initiating the iterative search is necessary lest too small a variance of the tolerance distribution is obtained, resulting in too much trimming. Error estimation based on the information matrix, without adjustment for trimming, was found to be satisfactory from a simulation study. The simulation study is the same as one designed by Hamilton (1979), and involved a logistic model and three contaminated logit models. As observed by Hoekstra (1989), the resulting 10 dose levels, while producing probabilities of response ranging from 0% to 100%, produce relatively few intermediate values for the probability of response. As a result, variance estimates can depend critically upon the precise location of the ED_{50}.

As we can see from the data of Table 7.2, the routine data encountered by Sanathan *et al.* (1987) are not very extensive, and typically arise as a pilot study.

The results from using the two trimming approaches are given in Table 7.3. The degree of trimming used for the Spearman–Kärber method varied between data sets, depending on the range of responses. For data sets D14 and D29, where the maximum response was 50%, it was not possible to obtain an estimate of the ED_{50} by the trimmed Spearman–Kärber approach. Note that, for data set D16, for the trimmed data sets the method of maximum likelihood, which was not used here, would have fitted the logit model with an infinite slope.

For data sets D8, D9, D12 and D16 it is suggested that the trimmed large responses at concentrations of 150 and 300 are likely to be due to toxic effects of high doses of the antibiotic. If indeed that is the case then in a pilot study it is sensible to trim them. (Cf. the approach of Simpson and Margolin, 1986.) There is no trimming of data set D14, while the data trimmed in an example such as D29 might alternatively have been described through natural mortality (section 3.2).

Table 7.2 Six sets of antibiotic assay data. In each case, groups of 10 mice were experimentally infected with a given strain of bacteria before being administered various dose levels of compounds. Recorded are number of deaths. The numbering of the data sets is that used by Sanathanan et al. (1987). Values which are trimmed are underlined

Concentration of antiobiotics (mg/kg)	Data set			Concentration	Data set		Concentration	Data set
	D8	D9	D12		D14	D16		D29
300	<u>3</u>	<u>8</u>	5	200	—	<u>6</u>	20	<u>0</u>
150	0	<u>0</u>	<u>10</u>	100	—	<u>0</u>	10	<u>0</u>
75	3	<u>0</u>	0	50	0	0	5	1
37.5	4	6	0	25	0	6	2.5	0
18.75	9	10	2	12.5	1	10	1.25	0
9.38	10	9	5	6.25	2	9	0.63	2
4.7	10	—	9	3.13	4	<u>10</u>	0.31	5
—	—	—	—	1.56	4	—	—	—
—	—	—	—	0.78	—	—	—	—

Table 7.3 ED_{50} estimates and 95% confidence intervals for the data of Table 7.2 (from Sanathanan et al. 1987)

Data set	Trimmed logit	Trimmed Spearman–Kärber
D8	36.8 (27.7, 48.9)	40.3 (25.9, 62.6)
D9	19.6 (16.8, 22.8)	22.1 (15.5, 31.6)
D12	10.0 (7.2, 13.9)	10.1 (6.6, 15.3)
D14	0.8 (0.3, 1.9)	–
D16	26.1 (22.5, 30.4)	29.5 (20.6, 42.2)
D29	0.3 (0.2, 0.5)	–

In the experimental framework described by Sanathanan *et al.* (1987), data sets such as these are used to indicate which dose levels to adopt for a further study. In such a context it is clearly important to be able to use a robust procedure which will return results, however imprecise these may be, rather than fail. For quantal response data with relatively few occurrences of 0% or 100% response, the suggested logit trimming procedure considered here has little or no effect. For example, a small amount of trimming occurs for the age of menarche data of Table 1.4 (Exercise 7.10). Unless there is clear interest in an ED_{100p} value for $p \neq 0.5$ then general use of the trimmed logit procedure, for an acceptance range other than $(0.001, 0.999)$, would seem unnecessary in view of the availability of the trimmed Spearman–Kärber method.

7.5 Robustness and efficiency

We expect to have to pay for robustness through a reduction in efficiency. The two approaches used to examine efficiency are through asymptotic results, or by simulation to evaluate small sample results. An illustration of asymptotic results has already been presented in section 7.3, and we start here with the asymptotic approach. Initially we shall consider the Spearman–Kärber estimator, which, as we shall see in section 7.5.2, may be regarded as both an L and an M estimator.

7.5.1 Asymptotic variance and the Spearman–Kärber estimate

We shall just consider the case of doses spaced equally, at a separation Δ, and equal numbers, n, of individuals at each dose.

In the standard dose-response experiment three different limits may be taken, viz. $n \to \infty$, for fixed k, $k \to \infty$, keeping n fixed, and both $n \to \infty$, $k \to \infty$. As $k \to \infty$, it is understood that also $\Delta \to 0$. It is standard to talk in terms of the **dose mesh**, and of the infinite mesh, which results in the limit as $k \to \infty$. Additionally, it is convenient to consider a dose mesh with dose levels given by $\{d_i, -k \leqslant i \leqslant k\}$.

From the work of section 2.2, if we just consider the case of a one-parameter tolerance distribution, $F(d - \mu)$, then the Fisher information for the maximum likelihood estimator $\hat{\mu}$ is given by:

$$I(\hat{\mu}) = \sum_{i=-k}^{k} \frac{n\left\{\dfrac{d}{d\mu} F(d_i - \mu)\right\}^2}{F(d_i - \mu)\{1 - F(d_i - \mu)\}}$$

Taking the limit, as $k \to \infty$ (Brown, 1961), results in:

$$I(\hat{\mu}) = \frac{n}{\Delta} \int_{-\infty}^{\infty} \frac{f^2(x)dx}{F(x)\{1 - F(x)\}} \tag{7.5}$$

where $f(x) = dF(x)/dx$, and we have retained $\Delta > 0$ for comparison with results for a range of other estimates. The value of equation (7.5) is the baseline against which we may compare other estimators, and is proportional to n/Δ, corresponding to the number of individuals per unit inverval in a fine dose mesh.

For the Spearman–Kärber estimator we have seen, from equation (7.4), that the variance of the estimator is given by

$$V(\hat{\mu}) = \frac{\Delta^2}{n} \sum_{i=-k}^{k} F(d_i - \mu)\{1 - F(d_i - \mu)\}$$

and as $k \to \infty$,

$$V(\hat{\mu}) \to \frac{\Delta}{n} \int_{-\infty}^{\infty} F(x)\{1 - F(x)\}dx \tag{7.6}$$

Note that in the logistic case,

$$f(x) = \frac{e^{-x}}{(1 + e^{-x})^2} = F(x)\{1 - F(x)\}$$

resulting in values of n/Δ and Δ/n respectively, in equation (7.5) and in (7.6). Thus the Spearman–Kärber estimate has maximum efficiency in this case. The logit model in fact provides the only distribution resulting in maximum efficiency.

It is now of much mathematical and statistical interest to evaluate and compare the integrals in equations (7.5) and (7.6) for a range of possible forms for $F(x)$. This work led to the results already presented in Table 7.1 and details are outlined in Exercises 7.11–7.13. In a number of cases numerical integration is required, and a full comparison will be presented in section 7.5.3. We shall now repeat this exercise for a range of robust estimators, which have been adapted from the case of direct sampling from a distribution.

7.5.2 L, M and R estimators

For the case of a direct random sample, $X_1, \ldots, X_m$ from a distribution with cumulative distribution function $F(x)$, and centre of symmetry θ, a range of robust estimators for θ is available, all of which may be based on the empirical cumulative distribution function. We shall denote the empirical cumulative distribution function by $\hat{F}(x)$, defined as: $\hat{F}(x) = $ (number of times $X_i \leqslant x$, $1 \leqslant i \leqslant m$)/$m$. For full discussion of robust estimators, see Huber (1977). We refer to this as the direct sampling situation, in contrast to the quantal assay application, in which one is not sampling directly from the underlying tolerance distribution. Robust estimators of the ED_{50} for quantal response data can be obtained by analogy with the direct sampling case, by defining a suitable empirical estimate, $\tilde{F}(x)$, of the threshold distribution cumulative distribution function. The form that is used is obtained from the ABERS estimate, followed by linear interpolation:

$$\tilde{F}(x) = \tilde{P}_i \quad \text{if } x = x_i, \ -k \leqslant i \leqslant k$$
$$= 0 \quad \text{if } x \leqslant x_{-k-1} = x_0 - (k+1)\Delta$$
$$= 1 \quad \text{if } x \geqslant x_{k+1} = x_0 + (k+1)\Delta$$

and $\tilde{F}(x)$ is linear and continuous in $[x_i, x_{i+1}]$, for all i. Efficiency comparisons may then be made as in the last section.

We start with a definition of L-estimators for the direct sampling case described above. (L-estimators are so called as they are obtained as linear combinations of order statistics.) Define $\{X_{(i)}\}$ as the ordered sample, and $J(u)$ as a non-negative symmetric function about $u = 0.5$, defined on the interval $[0, 1]$, with $\int_0^1 J(u)du = 1$. Then an L-estimator of θ is given by:

$$\hat{\theta} = \frac{1}{m} \sum_{i=1}^{m} J\left(\frac{i}{m+1}\right) X_{(i)} \tag{7.7}$$

This expression is similar to:

$$\hat{\theta} = \frac{1}{m} \sum_{i=1}^{m} J\left(\frac{i}{m}\right) X_{(i)}$$

which can be written in integral form as

$$\hat{\theta} = \int_{-\infty}^{\infty} x J\{\hat{F}(x)\} d\hat{F}(x) \qquad (7.8)$$

As $\int_{-\infty}^{\infty} x J\{F(x)\} dF(x) = x_0 + \int_{-\infty}^{\infty} (x - x_0) J\{F(x)\} dF(x)$, then equation (7.8) leads us to the L-estimator:

$$\hat{\theta} = d_0 + \Delta \sum_{i=-k}^{k+1} i J(\hat{P}_i)(\hat{P}_i - \hat{P}_{i-1})$$

where we define $\hat{P}_{-k-1} \equiv P_{-k-1} = 0$, and $\hat{P}_{k+1} \equiv P_{k+1} = 1$.

It is shown in Miller and Halpern (1980) that under mild regularity conditions, as $n \to \infty$, $\Delta \to 0$, $k \to \infty$ and $x_k \to \infty$, $\hat{\theta}$ is asymptotically normally distributed, with variance

$$\frac{\Delta}{n} \int_{-\infty}^{\infty} J^2\{F(x)\} F(x)\{1 - F(x)\} dx \qquad (7.9)$$

(Exercise 7.14).

We can see that if $J(u) \equiv 1$, then the resulting L-estimator is the Spearman–Kärber estimator, with asymptotic variance agreeing with equation (7.6). Alternatively, if $J(u) = (1 - 2\alpha)^{-1}$, for $\alpha \leqslant u \leqslant (1 - \alpha)$, and zero elsewhere, then the α-trimmed Spearman–Kärber method results, with asymptotic variance,

$$\frac{\Delta}{n(1 - 2\alpha)^2} \int_{F^{-1}(\alpha)}^{F^{-1}(1 - \alpha)} F(x)\{1 - F(x)\} dx \qquad (7.10)$$

The Spearman–Kärber estimator is also a particular example of an M-estimator for the ED_{50}. (M-estimators are so called because they are **maximum likelihood type** estimators. While they result from an optimization, not all M-estimators may be derived as maximum likelihood estimators.) In the direct sampling example, an M-estimator for θ is defined as a root of the equation,

$$\sum_{i=1}^{m} \Psi\left\{\frac{X_i - \theta}{s}\right\} = 0$$

for a suitable function Ψ, and scaling factor s. Trivially, this can be expressed as

$$\int_{-\infty}^{\infty} \Psi\left(\frac{x - \theta}{s}\right) d\hat{F}(x) = 0$$

which suggests an M-estimator for the quantal response case as a root of the equation,

$$\sum_{i=-k}^{k+1} \Psi\left(\frac{d_i - \theta}{s}\right)(\hat{P}_i - \hat{P}_{i-1}) = 0 \qquad (7.11)$$

where $\hat{P}_{-k-1} \equiv P_{-k-1} = 0$ and $\hat{P}_{k+1} \equiv P_{k+1} = 1$, by definition. Miller and Halpern (1980) establish that under suitable regularity conditions, the resulting estimate, $\hat{\theta}$, has an asymptotic normal distribution as $n \to \infty$, $\Delta \to 0$, $k \to \infty$, $x_k \to +\infty$. We may note that iteration is required for a solution to equation (7.11), and there is no guarantee that a unique solution may be obtained. If we use the Tukey biweight as the form for Ψ, then

$$\begin{aligned}\Psi(x) &= x(1 - x^2)^2 \quad \text{for } |x| \leqslant 1 \\ &= 0, \quad \text{for } |x| > 1\end{aligned} \qquad (7.12)$$

Miller and Halpern (1980) used a multiple of the mean absolute deviation (MAD) as the scale factor s, to be evaluated separately for each chosen form for $F(x)$. For comparison with the minimum variance resulting from the reciprocal of the Fisher information of equation (7.5) we have, in this case, the asymptotic variance,

$$\frac{\Delta}{n} \frac{\displaystyle\int_{-s}^{s} [\Psi'\{x - \theta)/s\}]^2 F(x)\{1 - F(x)\}\, dx}{\left[\displaystyle\int_{-s}^{s} \Psi'\{(x - \theta)/s\}\, dF(x)\right]^2}$$

where $s = k \times \text{MAD}$, and, directly from equation (7.12),

$$\Psi'(x) = (1 - x^2)(1 - 5x^2), \quad \text{for } |x| \leqslant 1$$

Two values of k, $k = 6$ and $k = 9$ were considered by Miller and Halpern (1980) and the results will be given in section 7.5.3.

R estimators involve ranks, and are based on non-parametric tests of location. In order to understand the rationale behind R-estimators, suppose that $X_1, \ldots, X_m$ and $Y_1, \ldots, Y_m$ are two independent random samples, from symmetric distributions with respective cumulative distribution functions, $F(x)$ and $F(x - \delta)$. A non-parametric test of the hypothesis that $\delta = 0$ can be based on the **score** function $S(u)$, $0 < u < 1$, by using the test-statistic,

$$U = \frac{1}{m} \sum_{i=1}^{m} S\left(\frac{R_i}{2m+1}\right)$$

where R_i is the rank of X_i in the combined sample. We also require $S(u)$ to be non-decreasing, and $S(1-u) = -S(u)$. This procedure generalizes both the sign test, for which $S(u) = -1$ for $u < 0.5$, and $S(u) = +1$ for $u > 0.5$, and the Wilcoxon test, for which $S(u) = (u - \frac{1}{2})$.

An estimator of the centre of symmetry, θ, of $F(x)$ can be obtained by choosing θ so that the samples, $(X_1 - \theta, \ldots, X_m - \theta)$ and $(-(X_1 - \theta), \ldots, -(X_m - \theta))$, when subjected to the above test, result in a test-statistic of zero. Thus we seek a root in θ to the equation,

$$0 = \frac{1}{m} \sum_{i=1}^{m} S\left(\frac{m\{F_m(X_i) + 1 - F_m(2\theta - X_i)\}}{2m+1}\right)$$

$$\approx \int_{-\infty}^{\infty} S\left(\frac{F_m(x) + 1 - F_m(2\theta - x)}{2}\right) d\hat{F}(x) \qquad (7.13)$$

Three robust estimators which result are the sample median, from using the sign test score function, the Hodges–Lehmann estimator, from using the Wilcoxon test score function, and the normal scores estimator, which results when $S(u) = \Phi^{-1}(u)$.

Once equation (7.13) is written in terms of an empirical distribution function then, once again, the way is open for use of the same ideas in the quantal response case. James *et al.* (1984) show that under certain regularity conditions the resulting estimator is asymptotically normally distributed, with variance,

$$\frac{d}{n} \frac{\displaystyle\int_{-\infty}^{\infty} (S'(F(x)))^2 f^2(x) F(x)\{1 - F(x)\} dx}{\left(\displaystyle\int_{-\infty}^{\infty} S'(F(x)) f^2(x) dx\right)^2} \qquad (7.14)$$

and consider in detail two cases, the Hodges–Lehmann case, and what they call the logistic scores estimator, when

$$S(u) = \log\{u/(1-u)\}$$

It is shown that the expression of equation (7.14) is minimized when $S(u)$ is of this form (Exercise 7.15), but unfortunately this $S(u)$ violates the required regularity conditions. However, simulation results indicate that it enjoys good properties (Exercise 7.16). Additionally, it is possible to approximate closely to $S(u)$ by functions which satisfy the regularity conditions. Computationally, it is necessary to solve the analogous analysis of equation (7.13) for the quantal assay case, and the numerical procedure for doing this is described by James et al. (1984).

7.5.3 Influence curve robustness

The robustness of estimates may also be discussed through consideration of their influence curves. In the case of a direct random sample, $\{X_1, \ldots, X_m\}$ from a distribution with cumulative distribution function $F(x)$, the influence curve of an estimator denoted by $T(F)$, where T is a functional on the space of distribution functions, is defined by Hampel (1974) as

$$\phi_{T,F}(x) = \frac{\lim_{\varepsilon \to 0} [T\{(1-\varepsilon)F + \varepsilon\delta_x\} - T(F)]}{\varepsilon} \tag{7.15}$$

where δ_x is the distribution of unit mass at ε, if the above limit exists for all real x.

Thus in equation (7.15) we consider the limit of the normalized effect on the estimator of an additional observation added at x. Subject to suitable regularity conditions, $\phi_{T,F}(x)$ is the limit, as $m \to \infty$, of Tukey's (1970) sensitivity curve defined by:

$$SC_{T,m}(x) = \frac{T\left(\frac{m}{m+1}\hat{F} + \frac{1}{(m+1)}\delta_x\right) - T(\hat{F})}{\left\{\frac{1}{(m+1)}\right\}} \tag{7.16}$$

(See also Andrews et al., 1972, p. 96.) For an illustration, see Exercise 7.18.

The influence curve for the quantal response case is defined by James and James (1983), by analogy with equation (7.16), but based on $\tilde{F}(x)$ defined in the last section. We shall assume a dose mesh D, with spacing Δ between doses, and n individuals treated at each dose.

An important difference between the work of this section and that of section 3.4 on influence, is that here we concentrate on the effect of adding a single binary response at any dose d. In section 3.4 we considered the effect of deleting both a dose level, and all of the information associated with that dose. In the quantal response case, we focus on the ED_{50}, and investigate the change in value of the estimator of the ED_{50} due to an additional response $y(y = 0$ or $y = 1)$ at dose level d. The definition of James and James (1983) is in two parts:

1. First of all, suppose the dose mesh, D, is fixed. Denote by $\hat{F}_{j1}(\hat{F}_{j0})$ the empirical distribution functions when we have one additional positive (negative) response at d_j. The influence curve of the estimator of the ED_{50} based on D, the function T and underlying tolerance distribution cumulative distribution function F is given by

$$IC_{T,F,D}(d_j, y) = \lim_{n \to \infty} \frac{n}{\Delta} \{T(\hat{F}_{jy}) - T(\hat{F})\}, \text{ almost surely,} \quad (7.17)$$

if the limit exists, for $y = 0, 1$, and $1 \leqslant j \leqslant k$. By analogy with the sensitivity curve for a direct random sample, $IC_{T,F,D}(d_j, y)$ is a function of the observed sample, and is a random variable for fixed y and d_j. However, it is only defined for a fixed dose mesh. The transition to a function defined for all values of d is accomplished by letting the dose mesh become dense, as $\Delta \to 0$, $d_k \to \infty$, and $d_1 \to -\infty$. We can then define:

2. The influence curve of T, under F, is given by:

$$IC_{T,F}(d, y) = \lim_{\Delta \to 0} IC_{T,F,D}(d, y)$$

if the limit exists, for $y = 0, 1$, and real d, with the convention that the limit is taken over those dose meshes D that contain d as a point.

As noted in section 7.5.1, the Fisher information in a fine dose mesh is proportional to n/Δ, which accounts for the presence of this term in equation (7.17).

In the direct sampling case, an estimator may be regarded as robust if its influence curve is bounded. In the quantal response application, James and James (1983) define T as influence-curve robust at F if both

$$\lim_{d \to \infty} IC_{T,F}(d, 0) = 0, \quad \text{and} \quad \lim_{d \to -\infty} IC_{T,F}(d, 1) = 0$$

For *L*-estimators, the influence curve is given by:

$$IC_{T,F}(d, y) = \{F(d) - y\} J(F(d))$$

where $J(u)$ is the weight function of section 7.5.2, subject to suitable regularity conditions (James and James, 1983). We can then demonstrate that the Spearman–Kärber estimator is not robust, but the α-trimmed version is (Exercise 7.19).

For *M*-estimators, the influence curve is:

$$IC_{T,F}(d, y) = (F(d) - y) \frac{\psi'(d - \theta)}{\displaystyle\int_{-\infty}^{\infty} \psi'(x - \theta) dF(x)}$$

where θ denotes the ED_{50}.

For *R*-estimators, the influence curve is:

$$IC_{T,F}(d, y) = \frac{\{F(d) - y\} J'(F(d)) f(d)}{\displaystyle\int_{-\infty}^{\infty} J'(F(x)) f(x) dF(x)}$$

where $f(x) = F'(x)$, and $J(u)$ is now the score function of section 7.5.2, subject to further regularity conditions.

It appears that the Tukey biweight and Hodges–Lehmann estimators (for standard threshold distributions) are influence-curve robust, but that the logistic scores estimator is generally not. James and James (1983) provide a modified form of the logistic scores estimator which is influence-curve robust. See Exercise 7.20 for further discussion. The maximum-likelihood estimator of θ under the logit model is also not robust (Exercise 7.21).

7.5.4 Efficiency comparisons

In Table 7.4 we present a range of asymptotic efficiencies, derived from the papers of Miller and Halpern (1980) and James *et al.* (1984). The contaminated distributions were 95%, 5% mixtures of the original distributions with distributions of the same form and mean, but a standard deviation 10 times larger. The slash distribution is the distribution of a unit normal random variable divided by an independent random variable uniformly distributed over the [0, 1] range. Like the Cauchy distribution, the slash is heavy tailed. By contrast, the angular distribution is short-tailed, with

Table 7.4 *Asymptotic efficiencies of seven estimators of the* ED_{50}, *for seven tolerance distributions. Presented are* $100 \times$ *ratio of the optimal variance of equation (7.5) to the variance of the estimator*

Tolerance	Spearman–Kärber (equivalent to assuming a logit model)	Trimmed Spearman–Kärber ($\alpha = 0.10$)	Trimmed Spearman–Kärber ($\alpha = 0.05$)	Tukey biweight ($k = 6$)	Tukey biweight ($k = 9$)	Hodges–Lehmann
Normal	98	75	86	72	89	80
Contaminated normal	75	85	95	76	88	87
Logistic	100	80	90	70	90	83
Contaminated logistic	75	88	96	73	88	89
Cauchy	0	84	74	61	70	88
Slash	0	88	80	65	74	94
Angular	81	58	68	63	74	67

cumulative distribution function,

$$F(x) = \sin^2(x + \pi/4), \quad \text{for } -\pi/4 \leqslant x \leqslant \pi/4$$

We can see that trimming the Spearman–Kärber approach avoids the poor performance for contaminated, slash and Cauchy distributions, at the cost of a certain loss in efficiency for the lighter tailed distributions. For the Tukey biweight approach the value $k = 9$ dominates the value $k = 6$. However, overall this is no reason to choose a Hodges–Lehmann or Tukey biweight estimator in favour of say a 5% trimmed procedure. As mentioned already, the logistic scores estimator is likely to dominate the others considered here, at a cost of computational complexity relative to the trimmed Spearman–Kärber procedure.

The asymptotic results of Table 7.4 can be misleading as a guide for small-sample behaviour. In Table 7.5 we have abstracted simulation results from James *et al.* (1984), for comparison. We can see that in this case the Spearman–Kärber estimate performs relatively well for the heavy tailed distributions. Additionally, the

Table 7.5 (James *et al.*, 1984) *Sample mean-square error (MSE) for the Spearman–Kärber estimator of the* ED_{50}, *and relative efficiencies for the other estimators, so that the table values are in that case the inverse ratios of the sample MSEs. Methods compared are the 10% trimmed (SK10) and 5% trimmed (SK5) Spearman–Kärber procedures, the Hodges–Lehmann estimator (HL) and the logistic scores estimator (LS), all of the* ED_{50}. *In all cases 1800 simulations were carried out, with* n = 10 *individuals per dose and 11 doses, the middle dose being the* ED_{50}. *Scale parameters were chosen so that 98% of the tolerance distributions fell between the third smallest and third largest doses. The contaminated distributions were contaminated with a 5% mixture of the same distributional form, but with a variance 100 times larger than the variance in the main component of the mixture*

	SK	SK10	SK5	HL	LS
Normal	0.0710	0.841	0.918	0.777	0.883
Cont. normal	0.0357	0.939	1.045	0.918	1.094
Logistic	0.0624	0.871	0.938	0.791	0.851
Cont. logistic	0.0368	0.954	1.057	0.923	1.069
Cauchy	0.0379	0.972	1.074	0.936	1.075
Slash	0.0379	0.972	1.074	0.936	1.075
Angular	0.0863	0.783	0.888	0.756	0.978

SK10 estimate MSE was consistently higher than that for the Spearman–Kärber measure without any trimming (Exercise 7.16).

Investigations by Finney (1950, 1953) showed how the precision of the ED_{50} can vary with regard to the location of doses relative to the ED_{50}. This point is emphasized by Hoekstra (1990), who also pointed out discrepancies between the simulation results of Hamilton (1979) and those of James *et al.* (1984). Hamilton (1979) simulated data from an experiment with 10 equally spaced doses, equal numbers, n, of individuals at each dose, for $n = 5, 10, 20$, and in all compared ten different methods of estimating the ED_{50}. In all cases the doses were symmetrically arranged about the known value for the ED_{50}; the doses used have since been employed by Sanathanan *et al.* (1987) and Kappenman (1987). They are primarily designed to mimic routine assay experiments with a wide dose range. Overall, Hamilton's main conclusion was that for the simulation experiments considered, moderately trimmed Spearman–Kärber estimators were to be recommended. A further feature which obscures comparisons in simulation experiments is that not all estimators are calculable for all sets of simulated data.

Extensive though these simulation results are, the conclusions are not generally applicable. For example, simulation was always from a symmetric model, with symmetrically placed doses, and the emphasis is very much on experiments with only a relatively small number of doses corresponding to intermediate probabilities of response. There is clearly a need for more research here.

7.6   Alternative distribution-free procedures

7.6.1   Sigmoidal constraint

Suppose we want to estimate an ED_{50}. On the one hand we may be prepared to make strong assumptions about the tolerance distribution, and use logit or probit analysis, though more flexible approaches have been described in Chapter 4. On the other hand we may only assume that the tolerance distribution exists, and estimate the ED_{50} by linear interpolation from the ABERS estimate. Intermediate methods between these two extremes have been described by Glasbey (1987).

Set $\Delta P_i = P_i - P_{i-1}, \quad 2 \leqslant i \leqslant k$

$$\Delta^2 P_i = P_i - \frac{(d_i - d_{i-2})}{(d_{i-1} - d_{i-2})} P_{i-1} + \frac{(d_i - d_{i-1})}{(d_{i-1} - d_{i-2})} P_{i-2}, \quad 3 \leqslant i \leqslant k$$

$$\Delta^3 P_i = P_i - \frac{(d_i - d_{i-2})(d_i - d_{i-3})}{(d_{i-1} - d_{i-2})(d_{i-1} - d_{i-3})} P_{i-1}$$

$$+ \frac{(d_i - d_{i-1})(d_i - d_{i-3})}{(d_{i-1} - d_{i-2})(d_{i-2} - d_{i-3})} P_{i-2}$$

$$- \frac{(d_i - d_{i-1})(d_i - d_{i-2})}{(d_{i-1} - d_{i-3})(d_{i-2} - d_{i-3})} P_{i-3}, \quad 4 \leqslant i \leqslant k$$

where the dose levels of an experiment are $\{d_i, 1 \leqslant i \leqslant k\}$. For the ABERS estimate the set of constraints on the $\{P_i\}$ is:

$$P_1 \geqslant 0$$
$$\Delta P_i \geqslant 0, \quad \text{for } 2 \leqslant i \leqslant k$$
$$P_k \leqslant 1$$

Now, except when we might anticipate non-homogeneity, it is natural to suppose that the tolerance distribution is unimodal. This imposes a stronger set of constraints on the $\{P_i\}$, and underlines how little is assumed when the ABERS estimate is formed. The unimodality assumption was considered by Schmoyer (1984). The cumulative distribution function is sigmoidal, so that there is one point of inflexion, to the left of which the function is convex, and to the right of which it is concave. Depending on the location of the point of inflexion, $(k-1)$ separate sets of constraints result, one for each value of s, lying in the range: $2 \leqslant s \leqslant k$. We have (Exercise 7.22):

$$P_1 \geqslant 0$$
$$\Delta P_2 \geqslant 0$$
$$\Delta^2 P_i \geqslant 0 \quad \text{for } 3 \leqslant i \leqslant s$$
$$\Delta^2 P_i \leqslant 0 \quad \text{for } s+1 \leqslant i \leqslant k$$
$$\Delta P_k \geqslant 0$$
$$P_k \leqslant 1$$

In this case we obtain the maximum likelihood under each of the $(k-1)$ sets of constraints, and select the $\{P_i\}$ which produce the overall maximum. Glasbey (1987) extended this idea, providing sets of constraints for symmetry, bell-shaped tolerance distributions, and for combinations of these assumptions (Exercise 7.23). Schmoyer (1984) found non-linear programming methods sensitive to starting values, but Glasbey (1987) simply used the NAG (1984) Fortran sub-routine, E04VDF, to perform the optimizations.

Example 7.5 (Glasbey, 1987)

Rather than obtain in $\widehat{ED}_{100p}$ value by linear interpolation from a fitted model, Glasbey (1987) defined an $\widehat{ED}_{100p}$ to be the range of doses at which the proportion of responding individuals can equal p, corresponding to the likelihood taking its maximum value, subject to the parameters being constrained. For the rotenone data set of Exercise 2.38, the results of Table 7.6 were obtained.

Table 7.6 ED_{50} *values (on a natural logarithmic scale) for the rotenone data of Exercise 2.38, for a range of assumptions about the tolerance distribution*

Assumptions	$E\widehat{D}_{50}$	95% confidence limits[a]	Width of confidence intervals
None	1.34–1.63	1.34–2.04	0.70
Symmetric	1.37–1.63	1.34–2.04	0.70
Unimodal	1.59–1.63	1.34–1.88	0.54
Unimodal and symmetric	1.57	1.45–1.68	0.23
Normal	1.58	1.48–1.67	0.19

[a]Bootstrap confidence intervals, based on 399 simulations. □

It is interesting to see from Example 7.5 how precision depends upon the assumptions made. Fitting extended models to the data set of this example did not affect the estimated precision of the ED_{50}. This was found to be true for all of the data sets in Exercise 1.3 (Morgan, 1985), but substantial differences can result for the ED_{99}, for example, when the discrepancies of Table 7.6 can also be expected to be greater. For further discussion, see Exercise 7.24.

7.6.2 Density estimation

The maximum-likelihood estimate of the probability, $P(d_j)$, of response to dose d_j where r_j individuals are observed to respond out of n_j treated is simply,

$$\hat{P}(d_j) = r_j/n_j$$

Less simply we can define $\hat{P}(d)$ for any real d in the form:

$$\hat{P}(d) = \frac{\sum_{i=1}^{k} r_i \delta(d - d_i)}{\sum_{i=1}^{k} n_i \delta(d - d_i)} \quad (7.18)$$

where $\delta(u) = 1$ if $u = 0$, and $\delta(u) = 0$, otherwise.

For most of the data sets presented in this book there is sufficient replication that the values $n_i \gg 1$. In some cases, however, it is necessary to pool over class intervals for d. Examples of this are illustrated in the data of Table 1.4 and Exercise 2.23, in each case for observational studies on the age of onset of menarche. The choice of class boundaries is arbitrary and any resulting plot can be sensitive to the particular class boundaries adopted. Copas (1983) suggested that a better method is to smooth using a kernel function $\psi(u)$, such as $\psi(u) = \exp(-u^2/2)$ to give,

$$\hat{P}(d) = \frac{\sum_{i=1}^{k} r_i \psi\{(d - d_i)/h\}}{\sum_{i=1}^{k} n_i \psi\{(d - d_i)/h\}} \quad (7.19)$$

where $h > 0$ is a smoothing parameter. The original intention was for equation (7.19) to be applied when most of the $\{n_i\}$ were unity, to enable a plot to be made of $\hat{P}(d)$ versus d, possibly with a view to assisting in model-selection. It was suggested that the plot should be made for a range of values of h. Copas (1983) also presents an estimate of $V(\hat{P}(d))$ and a method of bias correction. For a logistic regression application, see Kay and Little (1986).

This approach may also be applied when $n_i \gg 1$, for all i, and it has been developed for this application by Kappenman (1987), who presents a cross-validation procedure for estimating h. Based on the normal distribution kernel given above, the method reduces to solving the equation in h:

$$\sum_A r_j [p_{j1}(h) - p_{j1}(h)^2 - \{1 - p_{j1}(h)\}^2]$$

$$+ \sum_B (n_j - r_j)[1 - p_{j2}(h) - p_{j2}(h)^2 - \{1 - p_{j2}(h)\}^2] = 0 \quad (7.20)$$

where the first sum is over the set: $A = \{j : r_j > 0\}$, the second is over
the set: $B = \{j : r_j < n_j\}$, and

$$p_{j1}(h) = \{r_j - 1 + s_j(h)\}/t_j(h)$$

and

$$p_{j2}(h) = \{r_j + s_j(h)\}/t_j(h)$$

where

$$s_j(h) = \sum_{i \neq j} r_i e^{-(d_j - d_i)^2/2h^2}$$

$$t_j(h) = n_j - 1 + \sum_{i \neq j} e^{-(d_j - d_i)^2/2h^2}.$$

Unfortunately equation (7.20) sometimes has more than one
solution. An additional awkward feature is that the resulting
estimate of $F(x)$ is not always monotonic increasing. Ways around
these obstacles are provided by Kappenman (1987), who demon-
strates the good properties of the resulting method in a simulation
study. The method performs well, in terms of mean squared error
of the ED_{50}, in comparison with the 10% trimmed Spearman–Kärber
method, with the relative performance of the kernel estimator
improving as the sample size increases. However, the comments of
Hoekstra (1990) on the specific design of the simulation experiment
are germane here also. A further application of the kernel approach is
found in Staniswails and Cooper (1988).

7.7 Discussion

Computer power has opened up a rich range of new non-parametric
procedures for the analysis of quantal assay data. While some of
these methods are restricted to ED_{50} estimation, others, such as
those of section 7.6, are more widely applicable. As Glasbey (1987)
points out, the methods of his paper have application also to
estimating relative potency, and also in experimental design.
Schmoyer's (1984) sigmoidal constraint approach was originally
presented as potentially useful for low-dose extrapolation (section 4.6).
It is relevant here to note the discussion in Gaylor and Kodell (1980)
on the role of the sigmoid assumption in low-dose extrapolation.
For further related work, see Schmoyer (1986) and Schell and

Leysieffer (1989). An alternative method for fitting the standard logit model to quantal assay data has been provided by Cobb and Church (1983). Based on Spearman–Kärber type estimators, the method is shown to possess good small sample properties, subject to the equations for the parameter estimates having a solution.

Certain of the comparisons that have been made of the new methods mimic those made much earlier of old methods – see for example Finney (1950, 1953). For ED_{50} estimation, trimming the Spearman–Kärber estimate by about 5% seems to result in a substantial improvement, and a good general-purpose robust estimator. It is useful that suitable software is available on a floppy disk for evaluating the trimmed Spearman–Kärber estimate (Appendix E). For particular applications, methods such as those of sections 7.5.2 and 7.6, and the trimmed logit method will undoubtedly be of value, though their more general use may be hampered by computational complexity and the lack of off-the-peg computer software. However, the situation is bound to change in this regard.

The advantages of a parametric approach lie in the flexibility afforded, for instance, through the ready extension to wider models (as in Chapter 4), through the extension to the case of time-to-response data (as in Chapter 5), through the incorporation of natural mortality (as in Chapter 3), and through the addition of over-dispersion (as in Chapter 6). Particularly valuable, however, is the added perception, through the work reported in this chapter, of how non-robust parametric estimators may be, and, as shown in Table 7.6, the extent to which strong parametric assumptions can heighten estimates of precision.

7.8 Exercises and complements

The exercises vary in complexity. It is particularly desirable to attempt the five exercises marked with a †. Exercises marked with a * are generally more difficult or speculative.

7.1† Show that, when different orders of adjustment are possible to provide an ABERS estimate, they result in the same ABERS estimate.

7.2 Verify the mathematical expressions for $\{\tilde{P}_i\}$ given in section 7.2.

7.3 Provide a graphical illustration of the mechanics of the Spearman–Kärber estimate. Derive a formula for the variance of the tolerance distribution, using the same Spearman–Kärber approach, assuming a constant dose separation.

7.4[†] (Church and Cobb, 1973). In **equal weight** designs, $(d_{j+1} - d_{j-1})/n_j$ is constant, for $1 \leqslant j \leqslant k$. Show that under equal weight designs, if $\hat{P}_1 = 0$ and $\hat{P}_k = 1$, the Spearman–Kärber estimate is unaffected by the ABERS averaging procedure.

7.5 The Moving Average method is due to Thompson (1947), and proceeds as follows. Decide on a suitable interval (span) for taking moving averages, calculate these for the $\{P_i\}$ and the $\{d_i\}$, and then use linear interpolation between the averages d_j^* and d_{j+1}^* for which the P^*s enclose 0.50. Thus we set: $\hat{\mu} = d_j^* + a(d_{j+1}^* - d_j^*)$, where $a = (0.5 - P_j^*)/(P_{j+1}^* - P_j^*)$, and where, e.g., for a moving average of span s, $d_j^* = (d_j + d_{j+1} + \cdots + d_{j+s-1})/s$. How would you estimate $\text{Var}(\hat{\theta})$? Experiment with this method for a range of different spans and different data sets. Improvements have been suggested by Bennett (1952, 1963). The method is popular amongst ecotoxicologists (Stephan, 1977). It was outperformed by other methods in the simulation study of Engeman *et al.* (1986), but is useful for small or awkward data sets (Hoekstra, 1990). How does the method operate if (i) the span = 1, or (ii) the span increases?

7.6 (Miller, 1973). Consider quantal assay data with fixed spacing of doses, Δ, and n individuals per dose. If there is a value of s for which

$$\sum_{i=1}^{s} r_i = \sum_{i=s}^{k} (n - r_i)$$

then the Reed–Muench estimator of the ED_{50} is defined as

$$\hat{\theta}_R = d_k - \Delta(k - s).$$

Show that it is possible to write

$$\hat{\theta}_R = d_k + \Delta(1 - \hat{P}_s) - \Delta \sum_{i=1}^{k} \hat{P}_i$$

and discuss the similarity of this expression with the corresponding one for the Spearman–Kärber estimate. How would you proceed if there is no such value of s?

7.7 We use the same notation as for Exercise 7.6. Comment on the behaviour of the sequence

$$\tilde{P}_s = \frac{\sum\limits_{i=1}^{s} r_i}{\sum\limits_{i=1}^{s} r_i + \sum\limits_{i=s}^{k} (n - r_i)}, \quad 1 \leqslant s \leqslant k$$

as s increases.

If there is a value of s for which $\tilde{P}_s = 0.5$, then the Dragstedt–Behrens estimator of the ED_{50} is defined as,

$$\hat{\theta}_D = d_i - \Delta(k - s)$$

Show that if there is a value of s for which $\tilde{P}_s = 0.5$, then the Reed–Muench and Dragstedt–Behrens estimators coincide. How would you proceed if there is no such s?

7.8 Egger (1979) found multiple roots to be a problem when solving the equation for the Box–Cox parameter λ resulting from setting $\hat{\mu}_{(3)} = 0$, as in section 7.3. For data sets 2 and 3 of Exercise 1.3, single roots resulted, giving, respectively $\lambda = -0.97$, and $\lambda = -0.09$. In light of these results, discuss which of the many alternative models discussed in Chapter 4 you would expect to provide a good fit to the data.

7.9[†] Experiment with the trimmed Spearman–Kärber method, applying it to a range of the data sets of Exercise 1.3, using different levels of trimming.

7.10 Apply the trimmed-logit procedure to the data of Table 1.4.

7.11 Suppose the probability of response to dose d is given by $P(d) = F(d - \mu)$, for cumulative distribution function $F(x)$, of zero mean. Show that the asymptotic efficiency of the Spearman–Kärber estimator of the ED_{50} is:

$$e = 1 \bigg/ \left\{ \left(\int_{-\infty}^{\infty} F(x)\{1 - F(x)\} \, dx \right) \right.$$
$$\left. \times \left(\int_{-\infty}^{\infty} f^2(x)[F(x)\{1 - F(x)\}]^{-1} \, dx \right) \right\}.$$

Show that $e = 1$ if and only if $F(x)$ is logistic.

7.12 (Continuation.) Show that $e = 0$ if $F(x)$ is exponential, with $F(x) = 1 - \exp(-x/\mu)$, $x \geq 0$.

7.13[†] (Continuation.) Show that $e = 0.9814$ if $F(x)$ is normal.

7.14 Verify the expression of equation (7.9).

7.15* Prove that the variance expression of equation (7.14) is minimized when $S(u) = \log\{u/(1-u)\}$.

7.16 Discuss the following results, taken from James et al. (1984), based on 1800 simulations for each distribution, and make comparisons with Tables 7.4 and 7.5. There were 11 dose levels, and $n = 20$ individuals per dose. The middle dose is the ED_{50}. The scale parameters were chosen so that 98% of the tolerance distributions fell between the third smallest and third largest doses.

	Mean square error for Spearman–Kärber	Sample efficiencies, relative to Spearman–Kärber			
		SK10	SK5	HL	LS
Normal	0.0348	0.835	0.923	0.758	0.882
Cont. normal	0.0181	0.955	1.118	0.981	1.359
Logistic	0.0326	0.886	0.951	0.796	0.889
Cont. logistic	0.0188	0.978	1.126	0.985	1.295
Cauchy	0.0192	0.994	1.144	0.996	1.302
Slash	0.0192	0.993	1.143	0.995	1.299
Angular	0.0430	0.794	0.896	0.755	0.981

7.17 Hamilton (1979) carried out a simulation study of 10 estimators of the ED_{50}. His approach has been replicated in a number of subsequent studies. Five different models resulted in the probabilities of response shown below. Comment on the likely implications of this selection of models. Model I is logit. Compare the probabilities of response with those resulting from a logit model with $ED_{50} = 5.5$ and $\beta = 1$. This is one model used by Hoekstra

(1990) in a simulation study of the comparative performance of the method of Exercise 7.5. Cf. also Engeman *et al.* (1986).

Values of the Probability of a Response at Dose d, for each of five models

	Model				
Dose (d)	I	II	III	IV	V
1	0.00026	0.00157	0.00187	0.00188	0.00020
2	0.00160	0.00389	0.00440	0.00441	0.10128
3	0.01000	0.01000	0.01000	0.01000	0.10800
4	0.05969	0.03415	0.02232	0.02111	0.14775
5	0.28516	0.21945	0.12904	0.05016	0.32813
6	0.71484	0.78055	0.87096	0.94984	0.67187
7	0.94031	0.96585	0.97768	0.97889	0.85225
8	0.99000	0.99000	0.99000	0.99000	0.89200
9	0.99840	0.99611	0.99560	0.99559	0.89872
10	0.99974	0.99843	0.99813	0.99812	0.99980

7.18 Evaluate the influence curve of equation (7.15) and the sensitivity curve of equation (7.16) for the sample mean of a direct random sample of size *n*. Comment on the forms that result.

7.19 Show that for the Spearman–Kärber estimator the influence curve is:

$$IC_{T,F}(d, 1) = - \{1 - F(d)\};$$

$$IC_{T,F}(d, 0) = F(d)$$

Show further that the α-trimmed Spearman–Kärber estimator has the influence curve:

$$IC_{T,F}(d, y) = 0 \text{ if } d < F^{-1}(\alpha), \text{ or } d > F^{-1}(1 - \alpha),$$

$$= \frac{\{F(d) - y\}}{(1 - 2\alpha)}, \quad \text{if } F^{-1}(\alpha) < d < F^{-1}(1 - \alpha)$$

$$= \frac{\{F(d) - y\}}{2(1 - 2\alpha)} \quad \text{if } d = F^{-1}(\alpha), \text{ or } d = F^{-1}(1 - \alpha)$$

Draw conclusions from these findings.

7.20 Under suitable regularity conditions documented in James and James (1983), the influence curve for M-estimators is given by:

$$IC_{T,F}(d, y) = \frac{\{F(d) - y\}\, \psi'(d - \theta)}{\displaystyle\int_{-\infty}^{\infty} \psi'(x - \theta)\, dF(x)}$$

Verify that the Tukey biweight estimator is influence-curve robust.

7.21 Let $\hat{\theta}_{Ly}(d)$ denote the maximum likelihood estimator of the ED_{50} when one additional response of value y is observed at dose d under the logit model. Let the maximum likelihood estimator of the ED_{50}, under this model, without any additional response be $\hat{\theta}_L$. Discuss the influence curve adopted by James and James (1983):

$$IC_{L,F}(d, y) = \lim_{\Delta \to \infty} \lim_{n \to \infty} \frac{n}{\Delta}(\hat{\theta}_{Ly}(d) - \hat{\theta}_L), \quad y = 0, 1$$

These authors show further that

$$IC_{L,F}(d, 1) = -\frac{1}{1 + e^{(\alpha_0 + \beta_0 d)}}$$

$$IC_{L,F}(d, 0) = \frac{1}{1 + e^{-(\alpha_0 + \beta_0 d)}}$$

Where (α_0, β_0) are the parameters of the logistic distribution with the same two first moments as the threshold distribution, with cumulative distribution function $F(x)$. Deduce that $\hat{\theta}_L$ is not influence curve robust, and make a comparison with the conclusions to be drawn from Exercise 7.19.

7.22 Verify the validity of the sigmoidal constraints of section 7.6.1.

7.23* (Glasbey, 1987). Consider how you would constrain the $\{P_i\}$ to correspond to tolerance distributions which are (i) bell-shaped, (ii) symmetric, (iii) combinations of (i), (ii) and unimodal.

7.24[†] The table below is taken from Glasbey (1987), and provides, in the context of Example 7.5, an empirical investigation of the influence of a single additional positive response at different doses. Discuss the robustness implication of this table, and how the different assumptions regarding the form of the tolerance distribution result in differential use of the information available at the different doses.

1000 × change in the left and right limits of $\widehat{ED}_{50}$ *with the addition of a single positive response at log dose z, for a range of assumptions about the tolerance distribution*

Assumptions	$z = 0.8$	1.0	1.2	1.4	1.6	1.8	2.0	2.2	2.4
None	0, 0	0, 0	0, 0	0, −229	0, −29	0, 0	0, 0	0, 0	0, 0
Symmetric	0, 0	0, 0	0, 0	0, 0	29, −29	0, 0	0, 0	0, 0	0, 0
Unimodal	0, 0	−1, 0	−2, −28	−6, −40	−13, −46	0, 0	0, 0	0, 0	0, 0
Unimodal and symmetric	0	0	−15	−8	−4	−2	−1	0	0
Normal	−10	−9	−7	−5	−4	−2	−2	−1	−1

7.25 We have seen from Table 7.6 how strengthening assumptions regarding $F(x)$ can result in less robust and more precise estimators. How would a Bayesian analysis, as described in section 3.6, compare?

7.26 Show that if a sigmoid threshold distribution has differentiable cumulative distribution function $F(x)$, then the point of inflexion corresponds to the unique mode of the probability density function.

7.27 Schmoyer (1984) presented the example below, where the data result from an experiment performed to assess the lethality of diesel fuel aerosol smoke screens on rats.

		Maximum likelihood estimation	
Dose	*No. of rats*	*Proportion (no constraint)*	*Sigmoid constraint*
8	30	0.000	0.000
16	40	0.025	0.025
24	40	0.050	0.050
28	10	0.500	0.390
32	30	0.400	0.448
48	20	0.800	0.677
64	10	0.600	0.892
72	10	1.000	1.000

Provide the maximum likelihood estimates of the probabilities under the monotone constraint. Verify that the sigmoid constraint estimators do in fact satisfy the constraints of section 7.6.1, and plot the three sets of probabilities versus dose levels.

CHAPTER 8

Design and sequential methods

8.1 Introduction

Discussion of experimental design should start, rather than conclude, on a text on data analysis. It is more convenient, however, to have covered the material of earlier chapters first, since we make use of that material in what follows.

We saw in Chapter 7 that in routine testing of new and relatively unknown substances it is advisable to select a wide range of dose levels, to try to ensure that the full range of responses, from 0% to 100%, is obtained. If an experiment is conducted to estimate an ED_{50}, and prior knowledge suggests a likely interval for the ED_{50}, then more efficient use may be made of the experimental material. From this point of view, for estimation of the median age for onset of menarche, the age-distribution of girls questioned in the example of Exercise 2.23 is more efficient than that of Table 1.4. Considerations of this kind may be responsible for non-uniform features of data sets, such as the extra experimental effort devoted to the middle doses in Table 3.2, for example. It is clear also that design considerations will be very much determined by aims as well as by prior knowledge. Thus the experimental design for estimation of an ED_{50} would be different from the design for estimation of the location/scale pair of parameters, (α, β) in the logit model, as we shall see later.

In designing experiments we want to make the most efficient use of resources. A related feature, which has not been considered explicitly so far, is that experimental material may be expensive, and the cumulative cost of deaths of experimental animals may need to be taken into account. If the experimental facility is limited and/or costly, sequential methods may be needed. Additionally, as we shall see, some form of sequential procedure quite often develops naturally from a non-sequential approach. We shall start by considering approaches to optimal design in the non-sequential setting. Later these ideas will be carried over to sequential methods, and features such as differential costs will also be considered.

8.2 Optimal design

Suppose we have doses, $(d_1, d_2, \ldots, d_k)$, and a fixed number, m, of individuals to distribute over these doses.

We start by initially supposing **that the correct model is known**, with probability of response at dose d given by:

$$P(d) = F(\alpha + \beta d) = F(\beta(d - \theta)) \tag{8.1}$$

where θ is the ED_{50}. (Of course it is never the case that the model is known, for if it were experimentation would be unnecessary.)

How might we select the doses in an optimal way? There are various optimality criteria that have been considered, and Rosenberger and Kalish (1981) present D-, G-, E- and A-optimality results (Exercises 8.1–8.4). The book by Silvey (1980) gives a general introduction to the area. Finney (1971, p. 142) considered two- and three-point designs to minimize a Fiducial confidence-interval for the ED_{50}. There is further discussion of this approach in Ball (1990) and Jarrett *et al.* (1984). One criterion which has received particular attention is to select the doses to maximize the determinant of the expected Fisher information matrix for the parameters. This approach produces locally D-optimal designs (Minkin, 1987), sometimes simply described as D-optimal designs (Abdelbasit and Plackett, 1983). For non-linear models, such designs are given as functions of the **unknown** parameters, and so require good **prior** estimates of the parameters to be available. Alternatively, one might envisage a sequential approach, in which the results from one experiment may be used to determine the design for the next experiment, and we shall consider such a possibility later. Different optimality criteria will result in different designs. Thus a desire for precise ED_{50} estimation will result in a concentration of experimental effort which will produce less precise slope estimation than a design resulting from D-optimality. Seeking precise ED_{50} estimation is a particular instance of c-optimality, in which we try to minimize the variance of a single random variable. We shall now construct designs based on D-optimality.

It is shown by Minkin (1987) that for the standard logit model, parameterized in terms of (α, β) as in equation (8.1), the determinant of the Fisher information matrix, $I(\alpha, \beta)$, (equivalent in this case to the determinant of the expected Fisher information matrix) is given by:

$$\beta^2 |I(\alpha, \beta)| = \left(\sum_{i=1}^m w_i \sum_{j=1}^m w_j \eta_j^2 \right) - \left(\sum_{i=1}^m w_i \eta_i \right)^2 \qquad (8.2)$$

where $\eta_i = \alpha + \beta d_i$, $w_i = e^{\eta_i}(1 + e^{\eta_i})^{-2}$ and m is the total number of doses used (including repeated values) (Exercise 8.6).

The first term in equation (8.2) is maximized when

$$\eta_j = (e^{\eta_j} + 1)/(e^{\eta_j} - 1)$$

which is satisfied by $\eta_j = \pm 1.5434$.

Conveniently, when just one individual is treated at each dose level, then $I(\alpha, \beta)$ is clearly maximized, if $m = 2u$, say, by ensuring that u individuals are given the dose corresponding to $\eta_j = 1.5434$, and the other u are given the dose corresponding to $\eta_j = -1.5434$. Thus the m-point design (which may have included duplicate individuals at any dose) reduces to a 2-point optimal design given by:

$$\alpha + \beta d_1 = 1.5434,$$

so that

$$d_1 = (1.5434 - \alpha)/\beta,$$

and similarity

$$d_2 = -(1.5434 + \alpha)/\beta$$

clearly revealing the dependence of the optimal design on the parameters it is desired to estimate. These values correspond to $ED_{17.6}$ and $ED_{82.4}$ doses, and are derived directly for a symmetric, evenly divided 2-point design in Exercise 8.1. For probit analysis the corresponding optimal doses are $ED_{12.8}$ and $ED_{87.2}$.

This result generalizes work by Abdelbasit and Plackett (1983), who considered symmetric designs about the ED_{50}, θ, with equal numbers of individuals at each dose. Earlier derivations are given in the Ph.D. theses of White (1975) and Ford (1976). Conditional upon a two-point design being used, the optimal allocation of individuals was also determined by Kalish and Rosenberger (1978) (Exercise 8.1). Extensions to general forms for $F( )$ are considered by Khan and Yazdi (1988). When m is odd, the recommendation (Minkin, 1987; Khan and Yazdi, 1988) is again to use just the two doses corresponding to $\eta_j = \pm 1.5434$, ensuring that no fewer than $(m - 1)/2$ individuals are at each of the two doses, which improves

Table 8.1 (Abdelbasit and Plackett, 1983) *Robustness of the optimal two-point design for the logit model to poor initial estimates. The table gives 100 × ratio of the determinant of the Fisher information matrix at initial estimates* $(\theta_I \beta_I)$ *to the corresponding determinant at actual values* (θ_A, β_A)

	β_I/β_A				
$\beta_A(\theta_I - \theta_A)$	0.8	0.9	1	1.1	1.2
2.0	35.1	33.4	30.8	28.0	25.3
1.0	72.6	75.4	74.6	71.7	67.7
0.0	91.1	98.0	100.0	98.5	95.0

on an earlier suggestion of Abdelbasit and Plackett (1983) for the case of equal numbers of individuals at each of an odd number of doses, symmetrically placed about θ. Hoel and Jennrich (1979) show that the optimal design for a model with k parameters requires k dose levels.

It is interesting to investigate how the optimal strategy, based on poor initial estimates of (α, β), performs relative to using the correct parameter values. The results of Table 8.1 show how efficiency deteriorates as the initial estimates depart from the correct values. As a result of this work, Abdelbasit and Plackett (1983) recommend trying to avoid overestimation of β, e.g. by scaling down a best initial guess. For further discussion, see Exercise 8.5. A graphical illustration of this same feature is given by Minkin (1987).

The use of a pure optimal strategy is seen therefore as a high-risk approach, but it provides a useful baseline for comparisons. Furthermore, the risk may be reduced by adopting a sequential procedure, as we shall see below.

We may also consider how much efficiency is lost if multi-point, rather than 2-point designs are used. Rosenberger and Kalish (1981) derived best multi-point designs corresponding to equally spaced dose levels and equal numbers of individuals per dose. Thus, for example, the best (asymptotic) three-point design under D-optimality uses doses of $ED_{13.6}$, ED_{50} and $ED_{86.4}$ for the logit model, and requires 16% more observations than the optimal 2-point design for the same precision (an **efficiency ratio** of 1.16). For 5-, 10- and 75-point designs, the efficiency ratios are, respectively, 1.18, 1.19, and 1.19. The efficiency losses corresponding to these ratios may be regarded as a reasonable price to pay for being able to gauge the goodness-of-fit of the model, which of course is not possible when only two doses are employed (section 2.5).

We have seen that optimal design requires good prior knowledge of the model parameters. One way to proceed is by dividing the experiment into **stages**, later stages making use of the results available from the earlier stages.

The approach of Abdelbasit and Plackett (1983) for a **two-stage** design is as follows:

Under the logit model, we have seen that the optimal design points correspond to the two probabilities of response given by $P(d) = 0.176$, 0.824. Set $P = 0.176$.

1. Make an initial assessment of the parameter values. We shall denote this by: (θ_I, β_I).
2. Perform the first experiment with one quarter of the available individuals at each of the doses,

$$d_1 = \theta_I - \beta_I^{-1} \log \{P/(1-P)\}$$
$$d_2 = \theta_I + \beta_I^{-1} \log \{P/(1-P)\} \tag{8.3}$$

and form the maximum-likelihood estimates, $(\hat{\theta}, \hat{\beta})$, assuming that this is possible.
3. Now perform a second experiment, using the remaining half of the individuals distributed equally over doses,

$$d_3 = \hat{\theta} - \hat{\beta}^{-1} \log \{P/(1-P)\}$$
$$d_4 = \hat{\theta} + \hat{\beta}^{-1} \log \{P/(1-P)\} \tag{8.4}$$

4. Finally, estimate (θ, β) by maximum-likelihood using the responses to all of the doses, (d_1, d_2, d_3, d_4) (Exercise 8.7).

Table 8.2 presents the results of a simulation study of the efficiency of this two-stage procedure. As might be expected, the two-stage procedure is slightly less efficient than a single-stage procedure if the initial parameter estimates are correct. However, the efficiency of the two-stage approach relative to the single-stage approach increases as the accuracy of the initial parameter estimates deteriorates. The efficiency also increases as the total number of individuals increases. Initial overestimation of β, seen to be undesirable for a one-stage design from Table 8.1, corresponds to greater efficiency of the two-stage design.

Table 8.2 (Abdelbasit and Plackett, 1983) *Efficiency of 2-stage designs relative to a single-stage design for the logit model. Total number of subjects is m. Efficiency is the ratio of average determinants of Fisher information matrices, the averages being taken over the 1000 simulations that were used* (β_A, θ_A) *denote the parameter values used in the simulation*

$\beta_A(\theta_A - \theta_I)$	β_A/β_I						
	0.7	0.8	0.9	1.0	1.1	1.25	1.43
	m = 160						
1.5	1.853	1.640	1.492	1.405	1.358	1.348	1.424
1.0	1.355	1.230	1.158	1.120	1.108	1.139	1.239
0.5	1.137	1.052	1.013	0.999	1.007	1.051	1.167
0.0	1.074	1.004	0.974	0.967	0.975	1.029	1.142
	m = 360						
1.5	1.987	1.723	1.560	1.451	1.391	1.370	1.435
1.0	1.423	1.279	1.194	1.147	1.136	1.160	1.260
0.5	1.177	1.087	1.038	1.020	1.026	1.072	1.185
0.0	1.108	1.032	0.966	0.984	0.997	1.048	1.163

Clearly the two-stage procedure could be extended to many stages, and the case of three stages is also considered by Abdelbasit and Plackett (1983). This work is extended by Minkin (1987) in several ways. In particular, he observed that the allocation of individuals to doses d_3 and d_4 in equation (8.4) could be used to correct an unbalanced choice of d_1 and d_2 in equation (8.3), though at a slight cost in additional complexity. (Exercise 8.8.). Additionally, global D-optimal designs for the symmetric two-point case are evaluated numerically in Minkin (1987), and minimum values of m identified for local D-optimality to provide a good approximation to global D-optimality.

Other optimality criteria have also been employed, such as c-optimality, with particular reference to the ED_{50} of course. We shall consider such a criterion in a sequential context in section 8.5.

A refreshing approach is provided by Kalish (1990) who compares the relative efficiencies of D-optimal and c-optimal designs, in the latter case for both ED_{50} and the slope. An interesting feature of this work is the use of a second-order approximation for estimating variances, using the approach of Shenton and Bowman (1977) (cf.

the delta-method of Appendix A). It is shown that c-optimal designs for the ED_{50} for small m can be appreciably different from those for large m (Exercise 8.9). The recommendation of Kalish (1990) is to use a 3-dose design, dividing the available individuals uniformly between the (unknown but to be estimated) ED_{20}, ED_{50} and ED_{80} dose levels. For $m \leqslant 100$ this design is nearly D-optimal (cf. also the findings of Rosenberger and Kalish, 1981, mentioned above), and has efficiencies of at least 95% for estimating the ED_{50}, and at least 81% for estimating the slope (Exercise 8.10). With this design, therefore, a compromise is struck between the different optimality criteria. A sequential procedure based on this design is discussed in section 8.6.

Brown (1966) has presented an approach for designing a quantal assay experiment to obtain a particular precision on the ED_{50} when the experimenter has clear prior information regarding the likely ranges for the location/scale pair of parameters, (α, β). This work follows from an expression for the variance of the Spearman–Kärber estimator of the ED_{50}, given in equation (7.4). Similar guidelines for planning a comparative experiment are also provided.

The situation addressed by Müller and Schmitt (1990) is also one in which there is strong prior information which needs to be incorporated. In their case the ED_{50} is of prime interest, and is believed to be located roughly at the centre of a known dose interval. Suppose the threshold distribution has density function $f(x)$ and cumulative distribution function $F(x)$, and set

$$\Psi(x) = \frac{f^2(x)}{F(x)\{1 - F(x)\}}$$

We have seen, from equation (2.10), that when we parameterize the model in terms of the ED_{50}, then the asymptotic estimate of $V(ED_{50})$ is given by

$$\frac{m\sigma^2}{\sum_{i=1}^{m} \Psi(d_i)}$$

where m is the total number of individuals in the experiment, the ith tested at dose d_i, and σ^2 is the variance of the tolerance distribution. As usual, k will denote the number of distinct dose levels in the experiment. Suppose now that the doses $\{d_i\}$ are distributed between

0 and 1 through the equation

$$H(d_i) = \int_0^{d_i} h(y)\,dy = \left(\frac{i-1}{k-1}\right), \quad \text{for } k > 2$$

where $h(y)$ is a symmetric probability density function. Thus in particular,

$$d_1 = 0 \text{ and } d_k = 1 \tag{8.5}$$

For fixed m we need to choose k so that

$$\sum_{i=1}^{k} \frac{m}{k} \Psi\left[\left\{H^{-1}\left(\frac{i-1}{k-1}\right) - \theta\right\}\bigg/\sigma\right]$$

is a maximum, assuming that individuals are distributed evenly over doses. Müller and Schmitt (1990) show, for various forms for $H(d)$, that it is usually best to set $k = m$, i.e. have as many doses as possible, with just one individual at each dose. This finding is repeated also from small-sample simulation studies, one being based on the data set of Table 1.4. If one operates under the constraint of equation (8.5), then it is a good idea to use a large number of doses, because of the chance of violating Silvapulle's (1981) conditions with a small number of doses in this case (Exercise 2.27). This work does not investigate how conclusions may change if the initial dose interval is substantially asymmetric about the true ED_{50}.

Incorporation of prior information may also be accomplished through a formal Bayesian approach, and a detailed description is provided by Chaloner and Larntz (1989). Earlier work was given by Tsutakawa (1972, 1980), who considered the logit model, parameterized in terms of (η, θ_p), where θ_p denotes the ED_{100p} dose level, and $\eta = \beta^2$, and employed a prior distribution of the form

$$f(\theta_p, \eta) \propto \exp\left\{-(\theta_p - \mu)^2/(2\tau^2)\right\} \exp\left(-\omega v\eta/2\right)\eta^{1/2(\omega - 2)}$$

where (μ, τ, v, ω) are parameters. Numerical analysis was used to obtain optimal designs with equally spaced dose levels, and with equal numbers of observations at each dose. A less restricted approach was adopted by Chaloner and Larntz (1989). They approximated to the posterior distribution of the parameter, ξ, say, by a multivariate normal distribution, with mean the maximum likelihood estimate of the parameters, and dispersion matrix the inverse of the Fisher information matrix $I(\xi)$ (Berger, 1985, p. 224).

Two criteria for optimization were then investigated, viz.

$$\phi_1 = \mathrm{E}[\log\{\det I(\xi)\}]$$

and

$$\phi_2 = -\mathrm{E}[\mathrm{tr}\{\mathbf{B}I(\xi)^{-1}\}]$$

where in each case the expectation is taken w.r.t. the prior distribution of ξ. Under ϕ_2, the matrix $\mathbf{B}$ is to be chosen, and may, for example, be selected to focus attention on the ED_{50}.

Illustrations are provided for the logit model, with the two parameters (β, θ) given independent uniform prior distributions. It was found to be more computationally efficient to seek optima using the simplex method of Nelder and Mead (1965), rather than employ implementations of the specialist algorithms of Wynn (1972) and Fedorov (1972). Optimal designs were found to be more sensitive to the prior distribution for θ than to that for β. A general finding was that increasing uncertainty in the prior distributions had to be matched by an increased number of doses in the optimal design, which accords with common sense, and corresponding remarks made earlier in this chapter and in Chapter 7.

We have seen that pursuit of optimal designs leads to the incorporation of prior information or the use of sequential procedures. In the remainder of this chapter we shall now investigate a range of possible sequential approaches. A number of these are divorced from ideas of optimality, but in section 8.5 we shall again make use of optimality criteria.

8.3 The up-and-down experiment

The picture presented in Figure 8.1 is taken from Choi (1990), and results from an experiment conducted at the Medical College of Virginia Head Injury Centre. The experiment was designed to estimate the LD_{50} of mechanical head trauma, as measured by atmospheric pressure. Two factors are decided initially, viz. the starting dose, d_0, and the fixed spacing to be taken between subsequent doses, Δ. The first rat, given dose d_0, died, and as a result the dose was reduced to $d_1 = (d_0 - \Delta)$, and administered to the second rat, who also died. The dose was then reduced further to d_2, and subsequently again to d_3, at which point the treated rat survived. As a result, the next dose, for the fourth rat, was taken to be: $d_4 = d_3 + \Delta$, and so on. The experiment then continued for a

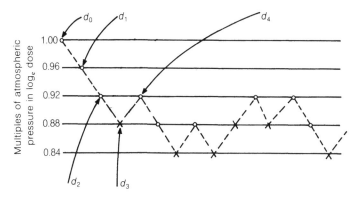

Figure 8.1 *(Choi, 1990)* Up-and-down experiment to estimate the LD_{50} of head trauma in rats. Doses are spaced by an amount $\Delta = 0.4$, so that, e.g., $d_1 = d_0 - \Delta$. etc. Death of a rat is indicated by o, while survival of a rat is shown by X. Reproduced with permission from the Biometric Society.

predetermined number of dose levels. Two things are clear from the outset:

1. the efficiency of the method rests on a sensible choice of Δ and d_0, and

2. once **turning points** (the peaks and troughs of the graph in Figure 8.1) are encountered, extreme dose levels are thereafter less likely to occur, and as a result experimental effort is concentrated in the neighbourhood of the LD_{50}.

The fixed dose procedure for evaluating toxicity (British Toxicology Society, 1984) is a form of up-and-down experiment, but the extent of movement between doses is restricted by a particular set of rules (Whitehead and Curnow, 1991).

For a fixed total number of experimental animals, one might well expect a sequential method such as the up-and-down experiment to be more efficient than a non-sequential method, which allocates animals to doses on a once-and-for-all basis. Indeed, Brownlee *et al.* (1953) found that two to three times as many observations were required in a particular non-sequential design to give the same MSE for the ED_{50} as a corresponding up-and-down approach, based on probit analysis. Davis (1971) reported similar findings, making a comparison with the Spearman–Kärber estimate in the non-sequential case.

However, there are other features to consider when sequential and non-sequential methods are compared. Sequential methods clearly require relatively rapid responses, and they could be well suited to experiments on large animals, say, for which a non-sequential experiment would simply take up too much laboratory space. Wu (1985) observed that sequential methods may also be well suited to certain engineering research problems, involving expensive equipment and short response times.

When an up-and-down method has terminated, how does one proceed? Rather than simply use a standard model, applied to the observed responses, much ingenuity has been devoted to the constuction of alternative, simple measures of the ED_{50}. Thus, for example, suppose there are $k + 1$ doses, being, in sequence, $\{d_0, d_1, \ldots, d_k\}$. If the dose spacing is Δ, then each dose d_i is of the form,

$$d_i = d_0 + i\Delta \quad 0 \leqslant i \leqslant k$$

We shall now use the notation of Kershaw (1987) to present a number of different estimates of the ED_{50}. One obvious measure to take is

$$E_M = \frac{\sum\limits_{i=0}^{k} d_i}{(k+1)} \tag{8.6}$$

Given d_k, we know what d_{k+1} is, although an individual has not been treated at that dose level. Brownlee et al. (1953) suggest replacing d_0 in equation (8.6) by d_{k+1}, since the latter level is dictated by the responses in the experiment, but the former level is not. This leads to the expression,

$$E_B = \sum\limits_{i=1}^{k+1} d_i / (k+1) \tag{8.7}$$

Measures such as E_M and E_B are susceptible to poor choice of d_0 and Δ. The E_{DB} measure of Brownlee et al. (1953), specified by:

$$E_{DB} = \frac{\sum\limits_{i=r}^{k+1} d_i}{\{k+1-(r-1)\}}$$

does not use the values: $\{d_0, \ldots, d_{r-1}\}$, where r indicates the dose level at which the first change of response, from that observed at d_0, occurs. In the illustration of Figure 8.1, $r = 3$.

Suppose the set Θ indexes the dose levels at which the responses at d_s and d_{s-1} are different, and suppose Θ has t members, then Wetherill *et al.* (1966) suggested the measure:

$$E_W = \sum_{s \in \Theta} (d_s + d_{s-1})/(2t)$$

This measure has been modified by Choi (1971), to omit the first turning point in the polygonal line joining successive doses, as drawn in Figure 8.1. A new method for estimating confidence intervals for this modified measure has been presented by Choi (1990), who found it preferable to the early method of Dixon and Mood (1948). The method utilizes a dispersion matrix from the theory of growth curves (Potthoff and Roy, 1964), which was discussed in section 5.2.

The four estimates of the ED_{50} presented here belong mainly to the pre-computer age, and to some extent reflect a need for simple measures which are easy to calculate. Once an experiment has run, one can also just apply maximum-likelihood, assuming a particular parametric model, to the observed responses. In contrast to maximum-likelihood applied to data obtained in a non-sequential manner, it is no longer a simple matter to deduce asymptotic properties, as successive observations are not independent. Results are given by Tsutakawa (1967a), who shows that the asymptotic dispersion matrix is similar to that which results from non-sequential experiments. However, proportions of individuals responding at various dose levels are replaced by equilibrium probabilities of being at those levels in an appropriate Markov Chain (Exercises 8.23 and 8.24).

For small-sample comparisons simulation is used but, as usual, conclusions depend on the particular configurations adopted, for the dose-spacing Δ for example. Kershaw (1987) reported his findings from comparing the above four estimators and the maximum likelihood approach for 24-, 48- and 96-step experiments, the maximum-likelihood approach being based on a logit model. As a general conclusion he recommended using the maximum-likelihood approach, with the proviso, especially for short experiments, that it may not always be calculable due to too few intermediate responses (Exercise 2.27). This conclusion is based on the assumption that the logit model is the correct one for the data (cf. here the robustness results of Chapter 7). The maximum-likelihood estimator had smaller bias than the competitors, and amongst these, the E_{DB} measure was preferred.

An attractive, practical feature of the standard up-and-down experiment is that it makes use of a limited number of doses, which may therefore be mainly made up in anticipation of the experiment. As we shall see shortly, other methods do not possess this property, and as a result may be of less practical utility, though this clearly depends on the experimental context. Much ingenuity has been devoted to devising variations on the up-and-down experiment, and we shall describe some of these below.

Using more than one individual per dose has been investigated by Tsutakawa (1967a) and Hsi (1969), in an attempt to combine the attractive features of both sequential and non-seqential experiments. Suppose the experiment runs for t trials, that the dose interval is Δ, and that n individuals are treated at each dose level, then the rule considered by Hsi (1969) is to define successive dose levels by the formula:

$$d_{i+1} = \begin{cases} d_i + \Delta & \text{if } 0 \leqslant r_i < n_1 \\ d_i & \text{if } n_1 \leqslant r_i < n_2 \\ d_i - \Delta & \text{if } n_2 \leqslant r_i \leqslant n \end{cases} \qquad (8.8)$$

where r_i denotes the number of individuals responding to the dose level d_i, and n_1 and n_2 are fixed integers with $0 \leqslant n_1 < n_2 \leqslant n$.

This approach is more wasteful than the standard up-and-down experiment if the step-length Δ and/or the initial starting dose are poorly chosen. One way of reducing this potential waste is to set $n = 1$ until the region for the ED_{50} is approximately located. Additionally, Hsi (1969) considered the effect of letting step-size vary in an adaptive manner, and also extended this approach for estimating an ED_{100p} value for $p \neq 0.5$. See also Exercise 8.11 and Example 8.1 below. Hsi (1969) computed the ED_{50} by means of the E_M measure of equation (8.6), and found greater efficiency, in mean-square error terms, for a sequential method with more than one individual per dose, as compared with the standard method of just one individual per dose level.

Example 8.1   Potato drop experiments

In experiments described by Kershaw and McRae (1985), potatoes were dropped from a range of heights, which were changed in the sequential manner described in Exercise 8.11. Due to time limitations, 144 potatoes, of each of 20 different varieties, were dropped in all, with changes of height after each 8 drops were made. The possible heights available for the experiment were (in mm): 50, 75, 110, 170, 255, 380, 570, 855, 1280 corresponding to equal separation on the log scale. Prior expectation dictated that all experiments started from 255 mm. It was noted that a mistake was made in applying the rule on only two occasions. Maximum-likelihood estimates of the EH_{10}, together with 95% Fieller confidence intervals are shown below. (Here, H refers to height.)

Variety	Lower limit	MLE of EH_{10}	Upper limit
Pentland Hawk	153	208	247
Pentland Squire	182	235	276
King Edward	63	88	105
Pentland Ivory	92	112	127
Record	110	189	239
Maris Piper	164	228	269
Pentland Crown	142	186	214
Pentland Dell	237	276	303
Désirée	133	167	190
Cara	94	109	120

For further discussion, see Exercise 8.12. □

An alternative approach for estimating the ED_{100p}, for $0.25 < p < 0.75$, was suggested by Wetherill (1963), resulting in a 'UDTR' rule. This rule applies to the sequential testing of single individuals at each dose. After each observation, for the current dose and those previous doses consecutive to it, i.e. back to the last change of response type, the proportion, p', of positive responses is evaluated. The change in dose levels is then given by:

$$d_{i+1} = \begin{cases} d_i + \Delta & \text{if } p' < p \\ d_i & \text{if } p' = p \\ d_i - \Delta & \text{if } p' > p \end{cases}$$

Wetherill and Glazebrook (1986, p. 194) suggest using the UDTR rule until the kth change of response type occurs, after which the sequence is restarted at an appropriate point suggested by the first run of trials, but thereafter using half the original step size. This approach could clearly be iterated further. Another related approach has been presented by Kershaw and McRae (1985), and is discussed in Exercise 8.11. A procedure incorporating variable step-sizes and using maximum-likelihood estimation of the parameters of an assumed logistic model is presented by Wharton and Srinivasan (1990).

8.4   The Robbins–Monro procedure

8.4.1   Introduction

The up-and-down method in its simplest form may be written as:

$$d_{i+1} = d_i - 2\Delta(r_i - 0.5)$$

where $r_i = 1$ if the individual exposed at dose d_i responds, and $r_i = 0$ otherwise. The sequential Robbins–Monro procedure for estimating an ED_{100p} value extends this rule, by allowing the step-length to decrease with increasing i, providing the sequence of doses:

$$d_{i+1} = d_i - \frac{c}{i}(r_i - p) \tag{8.9}$$

In the simplest form of this procedure, just one individual is treated at each dose, with $r_i = 1$ if the individual given dose d_i responds, and $r_i = 0$ otherwise. A number, t, of trials are made, as prescribed by equation (8.9), and then d_{t+1} is taken to estimate the ED_{100p} value. This 'stochastic approximation' formula derives from the work of Robbins and Monro (1951) and Chung (1954) and has applications more widely in the theory of sequential methods (Wetherill and Glazebrook, 1986, p. 162). More generally, r_i might be replaced by the proportion responding out of $n_i > 1$ individuals treated at dose d_i (Cochran and Davis, 1965). The approach has a clear intuitive appeal in adjusting dose levels by the discrepancy $(r_i - p)$, and doing so in a damped manner, so that the extent of the change is reduced as the experiment progresses. This procedure clearly assumes that there is no difficulty in administering the varied values of d_i that the formula will produce.

As with the up-and-down method, optimum efficiency requires good choice of starting dose and of a constant, in this case c. This has been investigated by Wetherill (1963) and Cochran and Davis (1965), who provided ranges for parameters which would produce reasonably small mean square errors. Wetherill (1963) concluded that quite large biases could result for $p \neq 0.5$, even for slight departures of p from 50%.

The basic procedure of equation (8.9) has seen many modifications; see for example, Cochran and Davis (1963), Davis (1971), Kesten (1958), and Exercises 8.13 and 8.14. As $t \to \infty$, d_{t+1} converges to the ED_{100p} in quadratic mean. The asymptotic variance of d_{t+1} is

$$\frac{p(1-p)c^2}{t(2c\gamma - 1)}$$

provided $2c\gamma > 1$, where $\gamma = F'(\theta_p)$, and θ_p denotes ED_{100p}. It is readily shown that the value of c to minimize this variance is $c = \gamma^{-1}$. Wu (1985) provides discussion and relevant references. Of course we do not know $\gamma = f(\theta_p) = F'(\theta_p)$. However, a crude approximation to $f(x)$ may be obtained from realizing that this measures the slope of the tolerance cumulative distribution function, $F(x)$, which is approximately linear over much of its range. Hence for the case when $r_i = 1$ or 0, we may estimate,

$$f(\theta_p) \approx \hat{\gamma}_i = \frac{\sum_{j=1}^{i} r_j(d_j - \bar{d}_i)}{\sum_{j=1}^{i} (d_j - \bar{d}_i)^2}$$

where

$$\bar{d}_i = \frac{1}{i} \sum_{j=1}^{i} d_j$$

This results in what is called the **adaptive** Robbins–Monro procedure:

$$d_{i+1} = d_i - \frac{1}{(i\hat{\gamma}_i)}(r_i - p) \qquad (8.10)$$

usually applied with a suitable truncation procedure for $\hat{\gamma}_i$ to ensure that wild fluctuations in $\hat{\gamma}_i$ are removed (Anbar; 1978; Lai and Robbins, 1979, 1981). Thus, for suitable constants, c and δ, with $c > \delta$, we set,

$$d_{i+1} = d_i - \frac{c_i}{i}(r_i - p)$$

where $c_i = \max[\delta, \min(c, \hat{\gamma}_i^{-1})]$.

Because of the similarity of equation (8.10) to the basic Newton–Raphson iteration (section 2.2), the adaptive Robbins–Monro procedure has been described as a stochastic Newton–Raphson method by Anbar (1978).

It might be supposed that better use could be made of all of the information accrued as the Robbins–Monro procedures progress towards termination. We shall now consider in detail the modification proposed by Wu (1985), which is one of several procedures for making use of all of the information available from the sequential experiment.

8.4.2 Wu's logit-MLE method

Wu (1985) proposed using a parametric model at each stage in a sequential design, in order to determine the next dose level to be used. For simplicity he adopted a logit model, but also discussed situations where it would be more appropriate to use extended models of the kind discussed in Chapter 4.

Suppose our interest focuses on the ED_{100p} value, θ_p. After the first i trials, we evaluate the ED_{100p} value, $\hat{\theta}_p^{(i)}$, say, using the logit model maximum-likelihood estimate, and all the information to date. We then obtain a constant κ_i by solving the equation:

$$\hat{\theta}_p^{(i)} = d_i - \frac{\kappa_i}{i}(r_i - p),$$

corresponding to setting $d_{i+1} = \hat{\theta}_p^{(i)}$ in equation (8.9).

Then the next dose level to be selected is given by

$$d_{i+1} = d_i - \frac{\kappa_i^*}{i}(r_i - p)$$

where $\kappa_i^* = \max[\delta, \min(\kappa_i, d)]$, and δ, d are chosen as truncation bounds, with $d > \delta \geqslant 0$. In fact, in a simulation study, Wu took $\delta = 0$. This rather elaborate procedure is introduced to dampen large changes in successive dose levels, which may arise early in the sequential procedure. The basis of Wu's approach is very simply to use sequentially fitted logit models, fitted by maximum likelihood, to estimate the tolerance distribution, thereby providing an estimate of θ_p, which is then used as the next dose level, subject to passing the above damping procedure. Extensions of this idea are to be found in McLeish and Tosh (1990) and Kalish (1990) and are described in the next two sections.

8.5 Sequential optimization

The optimization approach of section 8.2 has been given a fully sequential treatment by McLeish and Tosh (1990). Consider the general model of equation (8.1). In standard notation, the Fisher information matrix after k doses have been used, from Exercise 2.41, is given by:

$$\mathbf{J}_k(\theta, \beta) = \sum_{i=1}^{k} n_i \, \omega(d_i) \begin{bmatrix} \beta^2, & -\beta(d_i - \theta) \\ -\beta(d_i - \theta), & (d_i - \theta)^2 \end{bmatrix} \qquad (8.11)$$

where

$$\omega(d) = \frac{[F'\{\beta(d - \theta)\}]^2}{F\{\beta(d - \theta)\} \, [1 - F\{\beta(d - \theta)\}]}$$

and where

$$F'(z) = \frac{dF(z)}{dz}$$

for any z. The ED_{100p} values may be written as $ED_{100p} = g(\theta, \beta)$, estimated by $g(\hat{\theta}, \hat{\beta})$, where $(\hat{\theta}, \hat{\beta})$ is the maximum-likelihood estimate of (θ, β). Furthermore, the asymptotic estimate of the variance of $g(\hat{\theta}, \hat{\beta})$ is given, from the delta-method, by

$$V_k = \mathbf{c}' \mathbf{J}_k^{-1}(\theta, \beta) \mathbf{c} \qquad (8.12)$$

where

$$\mathbf{c}' = \left(\frac{\partial g}{\partial \theta}, \frac{\partial g}{\partial \beta} \right)$$

and we estimate (θ, β) by $(\hat{\theta}, \hat{\beta})$ (Appendix A).

Under the criterion of c-optimality, we want to minimize V_k, and McLeish and Tosh (1990) have adapted this approach to determining sequentially the dose levels to be employed. Suppose n individuals are to be used at each dose level. Once the kth dose has been determined, and the experiment at that dose concluded, the maximum-likelihood estimators of (θ, β) are, say, $(\hat{\theta}^{(k)}, \hat{\beta}^{(k)})$. They then suggest choosing the next dose, d_{k+1} to minimize $V_{k+1} = \mathbf{c}' \mathbf{J}_{k+1}^{-1}(\theta, \beta) \mathbf{c}$,

where

$$\mathbf{J}_{k+1} = \mathbf{J}_k(\theta^{(k)}, \beta^{(k)}) + n\,\omega(d_{k+1})$$

$$\times \begin{bmatrix} \beta^{(k)2} & -\beta^{(k)}(d_{k+1} - \theta^{(k)}) \\ -\beta^{(k)}(d_{k+1} - \theta^{(k)}), & (d_{k+1} - \theta^{(k)})^2 \end{bmatrix} \quad (8.13)$$

and throughout $(\theta^{(k)}, \beta^{(k)})$ is estimated by $(\hat{\theta}^{(k)}, \hat{\beta}^{(k)})$. We can see that equation (8.13) is a straightforward update of equation (8.11).

Example 8.2   (McLeish and Tosh, 1990)

Consider a logit model, for which $F(d) = (1 + e^{-\beta(\log d - \theta)})^{-1}$, and suppose we want to estimate the ED_{01} dose level, given by: $\theta_{0.01} = \theta - (\log_e(99))/\beta$. Suppose the number of individuals per dose level is given by $n = 1$. We start with preliminary estimates of $\hat{\theta} = 0$, $\hat{\beta} = 1$, and continue to use these until sufficient information has been gathered so that it is possible to obtain maximum-likelihood estimates of (θ, β) (Exercise 2.27). Thus we use $\hat{\theta}^{(1)}_{0.01} = -\log_e(99)$ until such a point is reached. If we start from $d_1 = 1$, then we choose d_2 to minimize equation (8.12), where

$$\mathbf{J}_2 = \omega(d_1) \begin{bmatrix} \hat{\beta}^{(1)2}, & -\hat{\beta}^{(1)}(\log d_1 - \hat{\theta}^{(1)}) \\ -\hat{\beta}^{(1)}(\log d_1 - \hat{\theta}^{(1)}), & (\log d_1 - \hat{\theta}^{(1)})^2 \end{bmatrix}$$

$$+ \omega(d_2) \begin{bmatrix} \hat{\beta}^{(1)2}, & -\hat{\beta}^{(1)}(\log d_2 - \hat{\theta}^{(1)}) \\ -\hat{\beta}^{(1)}(\log d_2 - \hat{\theta}^{(1)}), & (\log d_2 - \hat{\theta}^{(1)})^2 \end{bmatrix}$$

For the logit model, $\omega(d) = F(d)(1 - F(d))$ (section 2.2), so that

$$\mathbf{J}_2 = 0.25 \begin{bmatrix} 1, & 0 \\ 0, & 21.115 \end{bmatrix} + \frac{d_2^{-1}}{(1 + d_2^{-1})^2} \begin{bmatrix} 1, & -\log d_2 \\ -\log d_2, & (\log d_2)^2 \end{bmatrix}$$

and $\mathbf{c}' = [1, (\log_e 99)/\hat{\beta}_1^2] = [1, 4.5951]$. This is minimized at $\log_e d_2 = -2.60$. We then observe $\theta^{(2)}_{0.01}$, seek d_3, and continue in this way. If the first six responses to the selected doses are $= 0, 0, 0, 1, 0, 1$, then it is only after these six doses that maximum-likelihood estimates of (θ, β) can be obtained. Hence the first six log-dose levels will be calculated to be (Exercise 8.15): $0, -2.6, -2.75, 2.6, -2.35$ and -2.45. We then obtain $\hat{\theta}^{(6)} = 0.212$, and $\hat{\beta}^{(6)} = 0.537$, with $\hat{\theta}^{(6)}_{0.01} = -8.345$. This leads to $\log(d_7) = -4.30$, and the process continues until termination.   □

The approach of McLeish and Tosh (1990) is readily adapted to incorporate the **cost** of a positive response (resulting from the death of an expensive animal, say), and to more complex cases involving time-to-response, as discussed in Chapter 5, and natural morality (cf. section 3.2) (Exercises 8.16–8.18). Had the expression for J_k been omitted from equation (8.13), the approach then adopted would have clear similarities with the approach of Wu (1985), described in the last section, and in some cases is identical with it. The difference between using equation (8.13) with J_k, and without J_k is analogous to the difference between the approaches of Abdelbasit and Plackett (1983) and Minkin (1987), described in section 8.2 (Exercise 8.8).

Sequential design in a particular dilution series context (Exercise 2.37) is discussed by Ridout (1991).

8.6 Comparison of methods for ED_{100p} estimation

Because of the wealth of alternative sequential methods, and the great variety of alternative designs, there has correspondingly been a wide range of papers making comparisons between procedures and estimators. The early work (Davis, 1971; Wetherill *et al.*, 1966; Choi, 1971) concentrated on the simple measures, and has to some extent been superseded by the studies of Kershaw (1987), Wu (1985) and McLeish and Tosh (1990). Common to the more recent work is the finding that simple approaches, which do not use all of the information available in the way that the method of maximum-likelihood does, are frequently less efficient than approaches which assume a parametric form and which rest upon the sequential calculation of maximum-likelihood estimates. This is in spite of the potential difficulties which may arise due to the non-existence of maximum-likelihood estimates. Simulation studies may favour parametric approaches simply because data are simulated from assumed parametric forms, and both Wu (1985) and McLeish and Tosh (1990) discuss the use of extended models of the kind described in Chapter 4, to check on this feature. In particular, Wu (1985) simulated from underlying logit, probit, skewed logit (see Chapter 4), complementary log-log and cubic-logistic models, and also carried out robustness studies.

The main conclusion from Davis's (1971) comparison of various up-and-down and Robbins–Monro designs was the importance of incorporating an initial delay, corresponding to excluding initial trials when appropriate, and the good performance of delayed designs under fairly general circumstances. Choi (1971) found little to choose between Wetherill's estimators and a modified up-and-down method. Of the traditional designs, Kershaw (1987) found the Brownlee *et al.* (1953) delayed estimator to be best, but his overall recommendation was to base sequential methods on a parametric form and maximum-likelihood estimation of the parameters. Robbins–Monro (1951) designs were not included in Kershaw's comparison, but adaptive and non-adaptive Robbins–Monro designs were compared with the logit-MLE method and the up-and-down method by Wu (1985). The up-and-down approach was consistently outperformed. For standard designs, the logit-MLE method was found to be preferable to the Robbins–Monro procedures, though this finding must be balanced by the fact that the maximum likelihood estimators did not always exist. Wu (1985) also considered hybrid designs, starting from a Robbins–Monro approach and then switching to either a logit-MLE method or an adaptive Robbins–Monro method. In this case, relative performance was found to depend critically upon prior knowledge. McLeish and Tosh (1990) compared their approach with that of Wu (1985) in a simulation study. They found that for doses such as the ED_{50} and ED_{10} the two approaches were comparable, but for extreme dose levels, such as the ED_{01}, the approach of equation (8.13) was found to be preferable.

Kalish (1990) examined the performance of sequential adaptive versions of the two-point and three-point procedures mentioned in section 8.2 and discussed further in Exercise 8.10. In a simulation study, comparisons were also made with the Robbins–Monro and adaptive Robbins–Monro procedures, as well as with a modification of Wu's (1985) logit-MLE method. Truncation was employed when necessary. Data were simulated from logit, exponential and skewed-logistic models. In the sequential three-point procedure, for example, if $\hat{F}(x)$ denotes the current maximum-likelihood estimate of the assumed logistic tolerance distribution, then the next three doses are taken as $\hat{F}^{-1}(0.2)$, $\hat{F}^{-1}(0.5)$ and $\hat{F}^{-1}(0.8)$. See Kalish (1990) for discussion of alternative 'start-up' designs, and their relative performance, and also for the procedures adopted when problems arose with non-existence of the maximum-likelihood estimate, or if the estimate of the slope were negative. Except for certain poor initial designs, the occurrence of problems in obtaining a suitable maximum-likelihood estimate of $F(x)$ were found to be minimal (never $> 2\%$ of simulations, and usually far less frequent than that). The general conclusion is that the three-point sequential design performs best in terms of global description of the dose-response curve, and is at least comparable in perormance with other methods regarding ED_{50} estimation. It was found also that a standard Robbins–Monro procedure, with large step-size parameter c, can perform well if followed by maximum-likelihood estimation based on responses to **all** of the dose levels selected. It was suggested that increased performance might result from using a robust procedure, rather than one based simply on the logit model, possibly along the lines suggested by Sanathanan *et al.* (1987), described in Chapter 7.

8.7 Discussion

Work on sequential quantal analysis was initiated by Anderson *et al.* (1946). Since then research in the area has been extensive and continuous. A full discussion of design and sequential methods for quantal assay data would require far more space than we have devoted to the subject here. We have focused on designs for the estimation of single dose-response curves, or summaries of these. Design for comparisons is discussed in Brown (1966); design for mixtures is considered by Abdelbasit and Plackett (1982); design for the division of fixed resources between control and experimental animals is discussed in Thall and Simon (1990), with reference also to the weight to be placed on historical controls (Exercise 8.26). Response surface methodology is used by Carter *et al.* (1985) for describing responses to a mixture of three substances, and also for estimating toxic effects (Exercise 8.27).

A fundamental assumption of sequential methods of the Robbins–Monro kind is that it is possible to produce doses of the required strengths without difficulty. The investigations of Kalish (1990) and of Rosenberger and Kalish (1981) into gauging loss of efficiency resulting from sub-optimal assignments of doses is particularly valuable in the context of possible dose measuring errors (cf. section 3.9).

The poor performance of asymptotic criteria in small samples has been observed by Kalish (1990), and matches corresponding observations in Chapter 7. Simulation studies are most valuable for small-sample evaluations, but because of the many factors to be considered, it can be difficult to distil the important features from the results. For both the up-and-down and stochastic approximation sequential methods, the basic approaches have been found to be improved by the use of initial delays. Improvements may also result from the adoption of a simple parametric model, to be fitted to the data by maximum-likelihood, as the data become available.

There are difficult problems of inference from sequentially constructed designs, as it is difficult analytically to account for the sequential construction. This is discussed in Silvey (1980, section 7.4), and investigated in a particular simulation study by Ford and Silvey (1980). There is scope for more extensive investigations of this kind.

8.8 Exercises and complements

The exercises vary in complexity. It is particularly desirable to attempt the five exercises marked with a †. Exercises marked with a * are generally more difficult or speculative.

$8.1^{†}$ Note that Exercises 8.1–8.4 all derive from the paper by Rosenberger and Kalish (1981). In all cases we have a total of m individuals to allocate equally between two dose levels with respective probabilities of response $P(d_1)$, $P(d_2)$ where $P(d_1) = 1 - P(d_2)$. The assumed model is the logit model with $P(d) = (1 + e^{-(\alpha + \beta d)})^{-1}$. Verify that the asymptotic dispersion matrix for the maximum likelihood estimators $(\hat{\alpha}, \hat{\beta})$ is:

$$\mathbf{V} = \left[\frac{P(d_1)P(d_2)}{2} \begin{bmatrix} 2, & d_1 + d_2 \\ d_1 + d_2, & d_1^2 + d_2^2 \end{bmatrix} \right]^{-1}$$

Show that this result agrees with that of equation (8.2).

Show that the D-optimal design, resulting from minimizing $|\mathbf{V}|$ with respect to d_1 results from the solution to the equation:

$$P(d_1)P(d_2) \log \{P(d_1)/P(d_2)\}[1 + \{1 - 2P(d_1)\} \log \{P(d_1)/P(d_2)\}] = 0,$$

and verify that an approximate solution results when $P(d_1) = 0.824$.

8.2 (Continued.) Now evaluate the A-optimal design, resulting when d_1 is chosen to minimize the trace: $\operatorname{tr}(\mathbf{V})$. Verify that d_1 is the solution to:

$$P(d_1) - P(d_2) - \frac{(\alpha^2 + \beta^2)[2 - \log \{P(d_1)/P(d_2)\}\{P(d_1) - P(d_2)\}]}{[\log \{P(d_1)/P(d_2)\}]^3} = 0$$

8.3* (Continued.) The designs of Exercises 8.1 and 8.2 result from consideration of the product and sum, respectively, of the eigenvalues of **V**. Now derive the E-optimal design, in which d_1 is chosen to minimize the maximum eigenvalue of **V**.

8.4 (Concluded.) For a fitted model, let the predicted probability at dose d_0 be given by:

$$\hat{P}(d_0) = \frac{1}{1 + e^{-(\hat{\alpha} + \hat{\beta} d_0)}}$$

The G-optimal design results from choosing d_1 to minimize $\max_{d_0} [\mathbf{V}\{\hat{P}(d_0)\}]$.

Show that, for choice of $P(d_1)$, the resulting design is, like the D-optimal design of Exercise 8.1, and unlike the A- and E-optimal designs above, invariant with respect to choice of α and β, or scale of measurement. Comment on the desirability of this invariance. Discuss the difference between the D-optimal and G-optimal designs. Suggest a compromise design.

8.5 Discuss fully the implications of Table 8.1 for the choice of initial estimates (θ_I, β_I) to be used in deciding on a two-point design for the logit model.

8.6[†] Verify the expression for the determinant of the Fisher information matrix, given in equation (8.2). (Cf. Exercise 2.41.)

8.7* Write a computer program for estimating the parameters of the logit model, resulting from the two-stage sequential approach of Abdelbasit and Plackett (1983), and based on the doses given by equations (8.3) and (8.4).

8.8* (Minkin, 1987.) Give an example of how an unbalanced choice of d_1 and d_2 can arise from equations (8.3). Suggest an alternative approach to that of Abdelbasit and Plackett (1983) for selecting d_3 and d_4 in such a case.

8.9* Kalish (1990) considers a two-point design for estimating the ED_{50}, θ, with m individuals divided equally between the doses with probabilities of response, p and $q = (1 - p)$. A logit model is assumed, and the doses are placed symmetrically about θ. If $\lambda = \log(p/q)$, and β is the usual slope parameter, it is shown that

$$V(\hat{\theta}) = \frac{1}{\beta^2 pqm} + \frac{(3p^2 - pq + 3q^2)\lambda^2 - 6(p - q)\lambda + 3}{\beta^2(\lambda pq)^2 m^2} + o(m^{-2}).$$

Compare this with the expression given in equation (2.9).

Verify the following optimal values of p, working to order m^2:

m:	20	40	70	100	500	∞
p:	0.752	0.727	0.705	0.691	0.634	0.500

Comment on the speed of convergence of p to its value when $m = \infty$, and on the implications of this result for the use of asymptotic formulae.

8.10 Kalish (1990) presented the following results for equally weighted three-point designs, where the total number of individuals is m. The three design points correspond to p, $q = (1 - p)$ and θ, the ED_{50}, for each of three optimality criteria.

	Optimal values of p			Relative efficiencies (%)		
m	ED_{50} opt.,	D-opt.,	Slope-opt.	(i)	(ii)	(iii)
20	0.823	0.838	0.844	99.0	99.7	99.9
40	0.783	0.840	0.862	88.1	95.8	98.3
70	0.753	0.845	0.876	75.1	90.7	96.7
100	0.735	0.848	0.884	66.8	87.3	95.7
500	0.664	0.859	0.907	35.9	75.2	92.7
∞	0.500	0.864	0.917	0.0	64.7	91.5

Here relative efficiency (i) measures the efficiency of the ED_{50} optimal design for estimation of both of the parameters, while efficiencies (ii) and (iii) measure the efficiency of the D-optimal design for estimation of θ and the slope parameter β, respectively.

Compare and contrast the ED_{50} optimal designs with those in Exercise 8.9, and comment on the recommendation to use a three-point design which divides the individuals uniformly between the doses, ED_{20}, ED_{50} and ED_{80}.

8.11[†] Of interest to Kershaw and McRae (1985) was estimating the ED_{10}, in an experiment to determine the dropping height from which, on average, 10% of dropped potatoes were split. For a fixed value of Δ, in a sequential experiment successive heights were obtained from the formula:

$$h_{i+1} = \begin{cases} h_i + \Delta & \text{if } r_i \leqslant n_1, \\ h_i + \Delta z_i & \text{if } n_1 < r_i < n_2, \\ h_i - \Delta & \text{if } r_i \geqslant n_2, \end{cases}$$

where $z_i = 1(-1)$ if $h_{i-1} < (>)h_i$, $0 \leqslant n_1 < n_2 \leqslant n$ are all integers, observations being taken in groups of size n, and r_i potatoes split after n were dropped from height h_i.

Compare this rule with that of equation (8.8), and discuss the choice of n, n_1 and n_2 for estimating the ED_{10}.

8.12 (Continuation.) For the 10 species of potato in Example 8.1, the mean weights (in grams) are as follows: 98, 107, 167, 154, 150, 111, 137, 118, 113, 204. Consider how you might make use of this additional information. Jansen and Bowman (1988) investigated whether it was worth including size of bruising in a potato drop experiment.

8.13[†] The following **delayed** Robbins–Monro process was suggested by Cochran (Davis, 1971): replace equation (8.9) by

$$d_{i+1} = d_i - c(r_i - p)$$

until both responses and non-responses have been observed. Let t^* denote the number of the first trial for which this occurs. Then revert to the sequence:

$$d_{i+1} = d_i - \frac{c(r_i - p)}{(i - t^* + 2)}, \quad \text{for } i \geqslant t^*$$

Why might this be a sensible modification? Will this modification affect asymptotic properties?

8.14 Discuss the modification of the Robbins–Monro process given below, and suggested by Kesten (1958):
Replace c/i in equation (8.9) by $c/\psi(i)$, where

$$\psi(1) = 1, \ \psi(2) = 2,$$

and $\begin{cases} \psi(i) = \psi(i-1) \text{ if } (d_i - d_{i-1})(d_{i-1} - d_{i-2}) > 0, \\ \psi(i) = \psi(i-1) + 1, \text{ if } (d_i - d_{i-1})(d_{i-1} - d_{i-2}) \leqslant 0, \text{ for } i > 2. \end{cases}$

8.15[†] Verify that the first six log-dose levels in Example 8.2 are, in order, 0, -2.6, -2.75, 2.6, -2.35, -2.45.

8.16* (McLeish and Tosh, 1990). Suppose that D denotes the **extra** cost of a response, compared with non-response. Show that a total cost constraint for the expected cost can produce the constraint:

$$\sum_{i=1}^{k} n_i [1 + DF\{\beta(d_i - \theta)\}] = C$$

Discuss how this introduction of a differential cost affects the operation of the sequence given by equation (8.13).

8.17* Consider how to apply the approach of McLeish and Tosh (1990) to the case of observations being taken over time, as discussed in Chapter 5.

8.18* Consider how to apply the McLeish and Tosh (1990) approach when natural mortality is present, and described by Abbott's formula (section 3.2).

8.19* (Kalish, 1990). For the sequential three-point design discussed in section 8.6, consider how you would estimate $F(x)$ at any stage if either the maximum-likelihood estimate did not exist or if the slope estimate was negative.

8.20* Discuss factors which would need to be considered when start-up designs are devised for simulation experiments to compare alternative sequential methods.

8.21* Consider whether it might be feasible/advantageous to use the minimum logit chi-square rather than the method of maximum-likelihood in Wu's approach of section 8.4.2.

8.22   Compare and contrast the sequential methods of Kalish (1990) and of McLeish and Tosh (1990).

8.23   Suppose an up-and-down experiment starts from dose level d_0 and that the dose levels for the experiment can be written as:

$$d_i = d_0 + i\Delta$$

The progress through the dose levels can be modelled by means of a first-order Markov chain, with transition probabilities of $F(d_i)$, of moving from d_i to d_{i-1} and $1 - F(d_i)$ of moving from d_i to d_{i+1}, where $F(\ )$ denotes the tolerance distribution. Show that the Markov chain has an equilibrium distribution $\{\pi_i\}$, satisfying

$$\pi_i\{1 - F(d_i)\} = \pi_{i+1}F(d_{i+1})$$

8.24   (Continuation.) In order to obtain the asymptotic distribution of the measure E_W, given in section 8.3, it is first of all necessary to obtain the equilibrium distribution of a first-order Markov chain specified below. The required asymptotic distribution then follows from the application of theorems relating to functions of a Markov chain (Kershaw, 1985a; Chung, 1966).

There are four states, defined as follows:

$(d_i, 1)$:   observation of the sequence: d_i, d_{i-1}, d_i
$(d_i, 2)$:   observation of the sequence: d_{i-2}, d_{i-1}, d_i
$(d_i, 3)$:   observation of the sequence: d_i, d_{i+1}, d_i
$(d_i, 4)$:   observation of the sequence: d_{i+2}, d_{i+1}, d_i

Verify that this defines a four-state Markov chain, with equilibrium probabilities given by:

$$\pi_{i1} = F(d_i)\{1 - F(d_{i-1})\}\pi_i$$
$$\pi_{i2} = \{1 - F(d_{i-2})\}\{1 - F(d_{i-1})\}\pi_{i-2}$$
$$\pi_{i3} = \{1 - F(d_i)\}F(d_{i+1})\pi_i$$
$$\pi_{i4} = F(d_{i+2})F(d_{i+1})\pi_{i+2}$$

8.25*   Provide a Markov chain formulation of the alternation of heights in the experiment of Exercise 8.11.

8.26*   (Thall and Simon, 1990.) Consider an experiment involving a single treatment, and in which the problem is how to allocate a fixed number of individuals between treated and control groups. Obtain an expression for the variance of the treatment effect and hence derive an equation to be solved to provide the optimum allocation of individuals to treatment. Extend this approach to the case where historical control groups are also available and a decision has to be made concerning the relative weight to be placed on the historical information. (Cf. section 6.7.4.)

8.27   Exposed to the nerve agent *soman*, guinea pigs may die. Carter *et al.* (1985) exposed guinea pigs to a range of doses of *soman* (X_3), administered sub-cutaneously. One minute following dosing, animals were treated with various treatments, formed from mixing *atropine* (X_2) and *pralidioxime chloride* (X_1). The probability of animals surviving, p, was described by the logistic model:

$$p^{-1} = 1 + \exp\{ -(\beta_0 + \beta_1 X_1 + \beta_2 X_2 + \beta_3 X_3 + \beta_{11} X_1^2 + \beta_{22} X_2^2$$
$$+ \beta_{12} X_1 X_2 + \beta_{13} X_1 X_3 + \beta_{23} X_2 X_3 + \beta_{123} X_1 X_2 X_3)\}$$

After using maximum-likelihood, and fitting the model by the simplex method of Nelder and Mead (1965), the following parameter estimates were obtained:

Parameter	m.l.e	Estimated asymptotic standard error
β_0	3.0486	0.6084
β_1	0.0179	0.0097
β_2	0.0773	0.0114
β_3	-0.1095	0.0137
β_{11}	-0.0003	0.000032
β_{22}	-0.0001	0.000020
β_{12}	-0.0004	0.000122
β_{13}	0.0006	0.000170
β_{23}	-0.0005	0.000180
β_{123}	0.000006	0.0000021

Provide a full discussion of these results, with reference to the toxic, as well as the beneficial, effects of the two treatments.

8.28   (Robertson and Morgan, 1990). The signal-detection experiment has been described in Example 1.8 and Exercise 2.8. In the simple case of just two responses, Yes or No, a standard model is based on the probabilities: Pr(Respond Yes|signal presented) = $1 - \Phi(c - d)$; Pr(Respond Yes|noise presented) = $1 - \Phi(c)$. The results of the experiment may be described by notation analogous to that of Exercise 2.8. Show that the variance of the signal/noise distribution separation parameter d is given, for the normal model, by:

$$\mathbf{V}(\hat{d}) = \frac{\Phi(c)\{1 - \Phi(c)\}}{r_1.\{\phi(c)\}^2} + \frac{\Phi(c - d)\{1 - \Phi(c - d)\}}{r_2.\{\phi(c - d)\}^2}$$

Use this formula to predict values of r_1 and r_2. sufficient for a specified prior precision for $\hat{d}$, for a range of values for each of the probabilities of 'hit' and of 'false alarm'. Sequential versions of such experiments are considered by Wetherill and Levitt (1965) and Rose *et al.* (1970). A Bayesian approach is given by Watson and Pelli (1983). See also Ogilvie and Creelman (1968), McKee *et al.*, (1985), and Kershaw (1985b).

8.29*   Discuss the need for and effect of the truncation parameters (d, δ) in Wu's (1985) logit MLE method.

8.30*   (Khan, 1988.) Suppose the tolerance distribution has unit slope. Show that the Fisher information for the responses to a single dose d is a unimodal function of d when the underlying model is logit. Show that in general, if the underlying model has tolerance density function $f(d)$, which is unimodal and symmetric about 0, then the Fisher information function has a global maximum at $d = 0$, if and only if

$$\{F(d) - \tfrac{1}{2}\}^2 \leqslant \tfrac{1}{4}\{1 - f^2(d)/f^2(0)\}, \quad \text{for all } d > 0$$

where $F(d)$ denotes the tolerance distribution cumulative distribution function. Show that, corresponding to the probit model,

$$\Phi(d) \leqslant \tfrac{1}{2}[1 + \{1 - \exp(- d^2)\}^{1/2}], \quad \text{for all } d$$

and hence show that for the probit model the Fisher information function is unimodal.

8.31   (Abdelbasit and Plackett, 1983.) Assume an exponential model for a standard quantal assay experiment. Derive the optimum dose level to be used. Construct a sequential method and obtain a measure of efficiency of a two-stage design.

Approximation procedures

The delta-method

Based on a truncated Taylor series expansion, this procedure provides an approximation to the variance of $g(\zeta)$ when the dispersion matrix of ζ is known.

Suppose a model has p parameters, ζ. Under the usual regularity conditions for maximum-likelihood estimators, the asymptotic distribution of the maximum-likelihood estimator $\hat{\zeta}$ is multivariate normal, with mean ζ and dispersion matrix Σ, estimated by the inverse of Fisher's expected information matrix, evaluated at $\hat{\zeta}$.

If $g(\zeta)$ is a univariate function of ζ, with first-order partial derivatives, $\partial g/\partial \zeta_i$, which are continuous and not all zero at $\hat{\zeta}$, then $g(\hat{\zeta})$ is an estimator which is asymptotically normal, $N(g(\zeta), \sigma^2)$, where

$$\sigma^2 = \gamma' \Sigma \gamma$$

and

$$\gamma' = \left(\frac{\partial g}{\partial \zeta_1}, \frac{\partial g}{\partial \zeta_2}, \ldots, \frac{\partial g}{\partial \zeta_p} \right)$$

evaluated in practice at $\hat{\zeta}$. Resulting confidence intervals are invariant under continuously differentiable 1–1 transformations of the parameters, since both $g(\hat{\zeta})$ and σ^2 are invariant under such transformations.

Simple forms are given in Lindley (1965, pp. 134, 135), who extends the Taylor series expansion for $E[g(\hat{\zeta})]$. Extensions, proofs and examples are provided by Bishop *et al.* (1975, section 14.6). See also Cox (1984a) for further applications. GLIM macros are provided by Burn and Thompson (1981) and Vanderhoeft (1985).

One-step procedures

Suppose an estimator $\hat{\zeta}$ is based on n observations and is formed as the result of an iterative procedure. This procedure may require several iterations before the termination criterion for the iteration is satisfied. In a jack-knife context we need to form $\hat{\zeta}_{(i)}$, resulting from omitting the ith observation, $1 \leqslant i \leqslant n$. This too, of course, requires iteration. Chambers (1973) suggested that $\hat{\zeta}$ may be used as the starting value for the iteration for $\hat{\zeta}_{(i)}$ and that only a single step of the iterative procedure for $\hat{\zeta}_{(i)}$ might suffice, thereby greatly simplifying the numerical analysis. This approach has been taken up by Pregibon (1980, 1981), as described in sections 3.4 and 4.5.2, and was used also by Kleinman (1973) (Exercise 6.15). Detailed study, both asymptotic and by simulation, of the performance of one-step procedures for M-estimation (section 7.5.2) is provided in Bickel (1975). Pregibon (1982b) shows how score tests in GLIM can be accomplished by fitting a reduced model, followed by one-step in the fitting of the corresponding full model, starting from the maximum-likelihood estimate of the parameters in the reduced model. (Appendix D.) One-step principal component estimators are developed by Marx and Smith (1990) for use in generalized linear regression. A detailed analysis of one-step methods has been given by Jorgensen (1990), who also generalizes Pregibon's (1981) diagnostic measures.

GLMs and GLIM

In logistic regression, we write the probability of response of the ith individual, corresponding to a covariate vector x_i, as:

$$P(x_i) = (1 + e^{-x'_i\beta})^{-1}$$

as in the expression of equation (2.14). This is an example of a generalized linear model (GLM). The classical reference is McCullagh and Nelder (1989). An introduction to the subject is provided by Dobson (1990). Like all GLMs, the above example has three basic components:

1. a response distribution of a random variable, Y, say, of mean μ. The distribution is Bernoulli here (the **random** component);
2. a linear function of the covariates, $\eta_i = \beta'x_i$ (the **systematic** component);
3. a **link** function, linking the mean response μ_i of the ith individual to η_i: $\eta_i = h(\mu_i)$, where h is any monotonic differentiable function. In the logistic regression example, $\mu_i = P(x_i)$.

It may be simpler to think in terms of the inverse of h:

$$\mu_i = g(\eta_i)$$

and this is the natural approach for discussion of **composite** link functions. The important paper by Nelder and Wedderburn (1972) showed that a wide variety of statistical procedures could be regarded as GLMs, and fitted to data by means of the same iterative algorithm. Equally important was the first release, two years later, of GLIM, a computer package for fitting GLMs. GLIM stands for generalized linear interactive modelling. The success of this package derives from its wide applicability and interactive nature. Additionally it provides data-handling facilities and the fitting of user-defined models. This was originally through the OWN directive and its associated macros. However, this directive has been abolished in GLIM4, where the required macros are provided as further parameters of the $LINK and $ERROR directives (see below). In certain examples in the book the OWN directive is still in use. Similar features are now available in GENSTAT, and GLIM and GENSTAT Newsletters regularly publish macros for particular techniques. GLIM provides ingredients, rather than recipes. As a result it has proved remarkably flexible and of use in a wide range of areas. This is sometimes called GLIMNASTICS – for an illustration, see Aitkin and Clayton (1980), discussed in Appendix C.

The basic probability distribution (or probability function) for the random component of GLMs is of the form

$$f_Y(y) = \exp\left\{\frac{y\zeta - b(\zeta)}{a(\phi)} + c(y, \phi)\right\} \tag{A1}$$

for some functions a, b and c, and parameters ζ and ϕ. $E[Y] = \mu = b'(\zeta)$ and $V(Y) = a(\phi)b''(\zeta)$. Because of the way it enters the expression for $V(Y)$, ϕ is called the 'scale', or 'dispersion', parameter. For the normal distribution, $\phi = \sigma^2$, the normal variance, while for the Poisson and binomial distributions, $\phi = 1$. Over-dispersion for the binomial case may be simply accommodated by estimating ϕ – see the discussion of section 6.4, which, however, relates to the separate approach of quasi-likelihood. A similar procedure in the Poisson case can be used to approximate to a negative-binomial mean-variance relationship (Payne, 1986, p. 112). However, an exact procedure is given in section 6.5.2. For discussion, see Aitkin *et al.* (1989, pp. 214, 224). When ϕ is assumed known, $f_Y(y)$ belongs to the exponential family. Key examples are normal, Poisson, binomial, gamma and inverse Gaussian distributions.

The three main link functions, apart from the identity, are:

logit: $\eta = \log\{\mu/(1 - \mu)\}$,

probit: $\eta = \Phi^{-1}(\mu)$

complementary log-log; $\eta = \log\{-\log(1 - \mu)\}$.

The parameter ζ in (A1) is termed the **canonical** (or **natural**) parameter. For each distribution there is a **canonical link** function for which $\eta = \zeta$, i.e. $g(\mu) = \zeta$. For instance, for the normal, Poisson and binomial distributions, the respective canonical link functions are identity, log and logit. When canonical link functions are used there exists a sufficient statistic, equal in dimension to the dimension of $\boldsymbol{\beta}$ (McCullagh and Nelder, 1989, p. 32). Additionally, for canonical links the expected hessian of the log-likelihood equals the actual hessian, and so the iterative methods of Newton–Raphson and scoring coincide (McCullagh and Nelder, 1989, p. 43 – see Exercise 2.4 for an example of this). In GLIM, if a link is not specified the canonical link is taken as the default.

Instructions to the GLIM package are made through a series of **directives**, such as $ERROR NORMAL, $LINK LOG and $FIT, the last of which specifies the systematic part of the model through a structure formula. A simple illustration is shown in Example 2.3, and more complex procedures are provided in Exercises 4.6 and 6.6, and Example 6.2. These show also the specification of macros and the use of the OWN directive, which allows users to define their own link functions. Parameters may be fixed by use of the OFFSET directive. The original motivation for this was the dilution assay application of Exercise 2.37. Full detail is given in the GLIM manual (Payne, 1986). A very useful introduction to GLIM is given by Healy (1988) and more extensive coverage is found in Aitkin et al. (1989). Healy (1988), for example, describes the GLIM functions, system vectors, etc., that are used in some of the GLIM macros used in this book.

We can, for known ϕ, test the goodness of fit of a model by means of a likelihood ratio test, by comparison with a **saturated** model, with one parameter for each observation, and so fitting the observations exactly. Let $\hat{\zeta}$ and $\tilde{\zeta}$ denote the maximum-likelihood estimates of the canonical parameter under the null hypothesis and saturated models respectively:

$$\hat{\mu} = b'(\hat{\zeta}); \ y = b'(\tilde{\zeta})$$

Then the likelihood ratio statistic is:

$$-2\log\lambda = \frac{2}{\phi}\sum_{i=1}^{n} w_i\{y_i(\tilde{\zeta}_i - \hat{\zeta}_i) - b(\tilde{\zeta}_i) + b(\hat{\zeta}_i)\} \qquad (A2)$$

where $w_i = \phi/a_i(\phi)$. This is the **scaled deviance**. Without the divisor ϕ it is called the **deviance**. If the null-hypothesis is appropriate, the scaled deviance has, asymptotically, an appropriate chi-square distribution. In small examples the corresponding approximation for the difference of two scaled deviances, for comparing nested models, has been found to be more reliable than the approximation for the scaled deviance, used as a measure of fit of a single model (Payne, 1986, pp. 107, 111; note also McCullagh and Nelder, 1989, p. 36).

In GLIM the iterative procedure for maximum-likelihood model-fitting results from setting

$$\beta = (X'WX)^{-1}X'Wz \qquad (A3)$$

For a detailed derivation, see, for example, Dobson (1990, p. 41). Here X is the design matrix; W is a diagonal matrix of weights: $W = HV^{-1}H$, where V is a diagonal matrix with elements $\{v_{ii}\}$, where v_{ii} is the variance of the ith observation; H is a diagonal matrix of elements, $h_{ii} = d\mu_i/d\eta_i$; z is the working vector: $z = \eta + H^{-1}(y - \mu)$. The method is iterative since $\eta = X\beta$. Thus we see that equation (A3) generalizes the expressions of equations (2.2) and (2.19).

To start the iteration it suffices to estimate μ by the observation vector y. For the logit model, for example, this corresponds to using minimum logit chi-square estimates as starting values. When the OWN model facility is used in GLIM, then four macros have to be specified, providing the vectors, %FV, as $\mu_i = h(\eta_i)$, %DR, as $1/h_{ii}$, %VA as v_{ii}, from the values of the linear predictor %LP, which is η_i, and %DI, the individual deviance terms (used for deciding on termination). An illustration is provided in Exercise 4.7.

Several of the restrictions in the definition of GLMs can be relaxed, as discussed by Green (1984), Stirling (1984) and Cox (1984b). See also the comments of Nelder (1990). The extension to **composite** links is provided by Thompson and Baker (1981). Here the μ_i may be modelled by any function of the parameters, not just a linear function. In this case the same iteration of equation (A3) applies, but with H as a unit matrix, and X a matrix $\{x_{ij}\}$, where $x_{ij} = d\mu_i/d\beta_j$, which needs to be updated at each iteration. How to do this in GLIM is explained by Roger (1985). This is useful for example in modelling grouped data, as in section 3.5 in models incorporating natural response/immunity, as in section 3.2, and more generally in fitting models such as the Aranda-Ordaz asymmetric model of section 4.2. The approach of Exercise 4.1 will in general underestimate estimates of variances (see Appendix C). Ekholm and Palmgren (1989) present a direct and more general approach.

The adaptation of the iteration of equation (A3) to the case of quasi-likelihood has been described, for the binomial and Poisson cases, in section 6.4. Nyquist (1991) presents the theory for when the parameters in a generalized linear model have linear restrictions. A particular application is to quantal response comparisons when there is constant relative potency between substances tested.

Bordering Hessians

A model may have parameters: $(\alpha, \boldsymbol{\beta}') = (\alpha, \beta_1, \beta_2, \ldots, \beta_m)$. For fixed α it may be that likelihood maximization with respect to the elements of $\boldsymbol{\beta}$ is easily carried out, e.g. by using GLIM. The overall maximum-likelihood estimate, $(\hat{\alpha}, \hat{\boldsymbol{\beta}}')$ may then be obtained by maximizing the likelihood with respect to $\boldsymbol{\beta}$ for a range of values of α, and then plotting the resulting maxima versus α, to obtain a profile likelihood (cf. Figure 2.2). The maximum of this profile likelihood then provides the required $(\hat{\alpha}, \hat{\boldsymbol{\beta}}')$. However, error estimates for the elements of $\boldsymbol{\beta}$ will not reflect the variation in α, and it is necessary to account for this. The required adjustment is easily accomplished as follows.

Let l denote the log-likelihood, and set

$$\mathrm{E}\left[-\frac{\partial^2 l}{\partial \beta_j \partial \beta_k}\right] = a_{jk}$$

$$\mathrm{E}\left[-\frac{\partial^2 l}{\partial \alpha\, \partial \beta_j}\right] = b_j$$

$$\mathrm{E}\left[-\frac{\partial^2 l}{\partial \alpha^2}\right] = c$$

Let $\mathbf{A}$ denote the matrix $\{a_{jk}\}$, and $\boldsymbol{b}$ denote the column vector $\{b_j\}$. Assuming that $\mathbf{A}$ is non-singular, the asymptotic estimate of the dispersion matrix of the parameter estimators is then given bv:

$$\mathbf{V} = \begin{bmatrix} \mathbf{V}_{11} & \mathbf{V}_{12} \\ \mathbf{V}_{21} & \mathbf{V}_{22} \end{bmatrix} = \begin{bmatrix} \mathbf{A} & \mathbf{b} \\ \mathbf{b}' & c \end{bmatrix}^{-1}$$

Thus,

$$\mathbf{V}_{22} = 1/(c - \mathbf{b}'\mathbf{A}^{-1}\mathbf{b})$$
$$\mathbf{V}_{11} = \mathbf{A}^{-1} + \mathbf{A}^{-1}\mathbf{b}\mathbf{b}'\mathbf{A}^{-1}\mathbf{V}_{22}$$
$$\mathbf{V}_{12} = -\mathbf{A}^{-1}\mathbf{b}\mathbf{V}_{22}$$

An illustration of this technique in action is given in Aitkin and Clayton (1980). They show how GLIM may be used to analyse survival data, using a Poisson distribution and a log link. For the Weibull model the fit has to be iterated on the Weibull shape parameter α and the above adjustment is then needed to correct the estimate of V_{11} produced by GLIM, which is just A^{-1}, as well as to produce V_{22} and the components of V_{12}. It is not unusual to find that A^{-1} is incorrectly used for V_{11} – see for example Exercise 4.6 and the comments in Morgan (1983) and Roger (1985). The error involved in the application of Exercise 4.6 is investigated in Taylor (1988).

APPENDIX D

Asymptotically equivalent tests of hypotheses

Suppose inference concerns a parameter ζ, with p elements, a null hypothesis specifies, H_0: $\zeta = \zeta_0$, and we write the log-likelihood for a set of data as $l(\zeta)$, maximized at $\hat{\zeta}$.

The likelihood ratio test rejects H_0 at the $100\alpha\%$ level if

$$2\{l(\hat{\zeta}) - l(\zeta_0)\} > \chi^2_{p,\alpha}$$

where $\chi^2_{p,\alpha}$ denotes the $\alpha\%$ critical value for a random variable with a χ^2_p distribution. Examples are found throughout the text – see, for example, section 5.5.1.

We denote the vector of efficient scores by $U(\zeta)$, which has as its jth element,

$$\frac{\partial l(\zeta)}{\partial \zeta_j}.$$

Asymptotically, $\hat{\zeta}$ has the multivariate normal distribution given by $N(\zeta, \mathbf{I}^{-1}(\zeta))$, where $\mathbf{I}(\zeta)$ denotes the Fisher expected information matrix, with (j, k)th element,

$$-\mathrm{E}\left[\frac{\partial^2 l(\zeta)}{\partial \zeta_j \partial \zeta_k}\right].$$

Asymptotically, $U(\zeta)$ also has a multivariate normal distribution, but with dispersion matrix $\mathbf{I}(\zeta)$. The score test (sometimes called the Lagrange Multiplier test) rejects H_0 if

$$U(\zeta_0)'\mathbf{I}^{-1}(\zeta_0)U(\zeta_0) > \chi^2_{p,\alpha}.$$

The important point to note here is that the test statistic does not require $\hat{\zeta}$.

The Wald test rejects H_0 if

$$(\hat{\zeta} - \zeta_0)' I(\zeta_0)(\hat{\zeta} - \zeta_0) > \chi^2_{p,\alpha}.$$

An alternative version of the test replaces $I(\zeta_0)$ by $I\hat{\zeta})$. Examples are to be found in the solutions to Exercises 3.4, 3.5 and 3.6.

Asymptotically all three tests are equivalent. For certain hypotheses in logistic regression, the Wald test can behave in an aberrant manner, as shown by Hauck and Donner (1977).

If the parameter vector is $(\zeta, \phi)'$ and H_0 is unchanged, then we have nuisance parameters ϕ. We shall provide the required analysis for score tests. When $\zeta = \zeta_0$, let the maximum likelihood estimate of ϕ be $\hat{\phi}_0$.

Let us write $\mathbf{I}$, partitioned according to the partition of the parameter vector, as

$$\mathbf{I} = \begin{bmatrix} \mathbf{I}_{11} & \mathbf{I}_{12} \\ \mathbf{I}_{21} & \mathbf{I}_{22} \end{bmatrix}$$

and let $U(\zeta_0, \hat{\phi}_0)$ denote the vector of scores for ζ alone, evaluated when $\zeta = \zeta_0$ and $\phi = \hat{\phi}_0$. The score test statistic is now:

$$U(\zeta_0, \hat{\phi}_0)'(\mathbf{I}_{11} - \mathbf{I}_{12}\mathbf{I}_{22}^{-1}\mathbf{I}_{21})^{-1} U(\zeta_0, \hat{\phi}_0)$$

where $\mathbf{I}$ is calculated at $(\zeta_0, \hat{\phi}_0)'$. In particular, $\mathbf{I}_{22}$ is simply the Fisher expected information matrix for the parameters ϕ_0, when $\zeta = \zeta_0$. Reference is again to $\chi^2_{p,\alpha}$. Thus in this case the only optimization is done under H_0. A graphical illustration of the difference between these three tests is given by Buse (1982).

As an illustration, consider a score test of the logit model versus the Aranda-Ordaz asymmetric model of section 4.2 (this is Exercise 4.3). In this case, $\phi = (\alpha, \beta)'$, $\zeta = \lambda$, and $\zeta_0 = 1$. In order to perform the score test we just have to fit the logit model, corresponding to $\zeta_0 = 1$. Suppose the fitted probability of response to dose x_i is denoted by $\hat{p}_i$, then the additional terms needed for the evaluation of the score test statistic are:

$$\mathbf{I}_{11} = \sum_{i=1}^{k} \frac{n_i(1 - \hat{p}_i)}{\hat{p}_i} \{\hat{p}_i + \log(1 - \hat{p}_i)\}^2$$

$$I_{12} = \left(\sum_{i=1}^{k} n_i(1 - \hat{p}_i)\{\hat{p}_i + \log(1 - \hat{p}_i)\}, \right.$$

$$\left. \sum_{i=1}^{k} n_i x_i(1 - \hat{p}_i)\{\hat{p}_i + \log(1 - \hat{p}_i)\} \right)$$

$$\frac{\partial l}{\partial \lambda} = \sum_{i=1}^{k} \frac{(r_i - n_i\hat{p}_i)}{\hat{p}_i}\{\hat{p}_i + \log(1 - \hat{p}_i)\}$$

Examples of the result of this test, and comparisons with the likelihood-ratio, and goodness-of-link (section 4.5.2) tests are given in Table 4.7. As we test an hypothesis regarding the scalar parameter λ, we just require $\partial l/\partial \lambda$. This is a common example – for other illustrations, see Exercises 2.39, 3.38 and 6.10.

Asymptotic properties of score tests are reviewed by Tarone (1985). A detailed discussion is provided by Cox and Hinkley (1974, section 9.3). A particularly interesting article by Pregibon (1982b) shows how score tests may be performed in GLIM, amounting to a difference between chi-square goodness-of-fit statistics, one for the reduced model, and the other for the result of a one-step iteration when fitting the full model, starting from the maximum-likelihood parameter estimates from fitting the reduced model. This avoids the additional algebra, an illustration of which is given above, and does so by moving towards fitting the full model.

For applications in over-dispersion, see Breslow (1989). For further simplifications when likelihoods are based on exponential families of distributions, see Gart and Tarone (1983).

APPENDIX E

Computing

This appendix brings together various references to computing made throughout the text. It also describes relevant facilities in a number of computer packages. In this latter case we have concentrated on the main computer manuals, and make no reference to versions for personal computers. These can differ from the standard packages – e.g. SPSS/PC + does not contain the procedure PROBIT (see below). It should also be realized that computer packages regularly undergo changes.

As we shall see, it is now fairly standard for packages to give the option of using either a logit, probit or complementary log–log link, the option of fitting a model with natural mortality (and sometimes also immunity), and the possibility of testing for parallelism when results are available for more than one tested substance. The book by Afifi and Clark (1990) is most useful in discussing facilities available in different computer packages. Of particular relevance is Chapter 12 on logistic regression.

Function optimization

A library of FORTRAN routines such as NAG (see below) contains a wide range of programs for function optimization. In this text we have referenced the following NAG routines: Chapter 2, E04CCF (Nelder Mead simplex method); E04VCF (allows bounds to be placed on parameters; requires first order derivatives); Chapter 6, E04LAF (an easy-to-use modified Newton algorithm for bounded parameters); Chapter 7, E04VDF (requires specification of first order derivatives). Many statistical packages now provide the same facility. See for example the FITNONLINEAR directive in GENSTAT, and procedures AR and 3R in BMDP. Procedure NLIN in SAS fits non-linear regression models. As an illustration, procedure 3R in

BMDP uses a modified Gauss–Newton method, but in common with many procedures evaluates the required derivatives numerically, so that the user only has to supply the form of the function to be optimized.

The package MLP is discussed in Ross (1990) as well as in the MLP manual. In addition to providing a range of optimization tools, it has a module specifically designed for quantal response data, and which we shall now outline.

MLP

The FIT PROBIT module has 10 options. It is possible to fit probit, logit and complementary log–log links, to test for parallelism for several lines, fit models with natural mortality and/or immunity, fit a model to Wadley's problem (with an immunity option), fit a model with two covariates, fit a model with a mixture threshold distribution (cf. Exercise 4.28), and fit a dilution series model (cf. Exercise 2.37). Users are given the option of scaling standard errors using a heterogeneity factor. Details of MLP are available from NAG (see below).

SAS

We describe the facilities available in Version 6.03 and 6.04. Probit, logit and complementary log–log models may be fitted (with or without natural mortality) using the procedure PROBIT. There is automatic scaling of errors and covariances by the heterogeneity factor if the model fit is poor. Logistic regression is possible in procedure FUNCAT (Version 6.03) and procedure LOGIST (Version 6.04). In the latter case the link function may be either probit, logit or complementary log–log. There is mention of FUNCAT and LOGIST in Chapter 2, and PROBIT in Chapter 6. The CATMOD procedure can also be used for logistic regression.

SPSS-X

We describe the facilities available in the third edition of the manual, published in 1988. The key procedure here is PROBIT. By default a probit analysis will be carried out, but logit analysis, extending to logistic regression, is possible. It is possible to test for parallelism, and the NATRES subcommand allows for natural response.

BMDP

We describe the facilities available in the 1990 manual. This package provides a number of alternative procedures which may be used. The logit model may be fitted by procedure LE and as a non-linear regression by procedure 3R, which uses the Gauss–Newton method. It may also be fitted within LR which performs logistic regression with stepwise selection of variables. Case-control studies (section 1.7) may also be analysed by LR. Procedure AR is specifically designated for quantal bioassay, and provides measures of potency and relative potency. Procedure LE uses Newton–Raphson, and it is also possible to fit an ordered multinomial logistic model. Alternatively, for the latter one may use procedure PR (section 3.5).

Procedure 3R is referenced in Chapter 1; Procedure PLR, referced in Chapters 2 and 4, was an earlier version of LR. As one test of fit of the logit model it used a score test (Brown, 1982) within the Prentice (1976b) family of models – see Chapter 4 for discussion. An advocate for the use of BMDP is Cox (1987, 1990) who fitted, for example, the hockey-stick model and a logit model with natural response, using procedure 3R. The possibility of interfacing with FORTRAN programs greatly increases the flexibility and appeal of BMDP and other packages with the same facility.

GENSTAT 5

The possibilities within GENSTAT 5, as described by Payne *et al.* (1987), are greatly enhanced by the Procedure Library. Relevant procedures from the Release 2.2 are described below.

Fitting generalized linear models in GENSTAT 5 is done through the MODEL directive (Payne *et al.*, 1987, p. 350). For example:

    MODEL [DISTRIBUTION = Binomial; LINK = probit].

If the LINK option is not used then the default is the canonical link. Example 8.4.2 of Payne *et al.* (1987, p. 357) fits two probit lines; Example 8.4.1 of Payne *et al.* (1987, p. 354) fits a set of serial dilution data.

Details of the Procedure library can be obtained on-line from within GENSTAT by typing, e.g.:

    LIBINFORM [PRINT = contents, index, modules, errors]

For any procedure a description follows (illustrated for procedure FIELLER):

LIBINFORM [FIELLER],

and an example can be obtained by typing:

LIBHELP 'LIBEXAMPLE'; example = %Ex

##%Ex.

Of relevance are procedures: FIELLER, GLM, ORDINAL-LOGISTIC, PROBITANALYSIS and WADLEY.

In FIELLER either a logit, probit or complementary log–log link may be specified and ED_{100p} values together with Fieller confidence intervals can be obtained. Relative potencies can also be obtained. In ORDINALLOGISTIC the proportional odds model mentioned in section 3.5 is fitted to two-way contingency table data with ordered columns. (Just the logit link is available.) Procedure WADLEY provides the analysis for Wadley's problem, with the options of logit, probit, Cauchit or complementary log–log links. Distributional forms available are Poisson or negative-binomial, or a quasi-likelihood approach may be used in either a negative-binomial or a scaled Poisson form. A test for parallelism is included (Smith, 1991). Procedure PROBITANALYSIS also provides probit, logit and complementary log–log transformations. Natural mortality and immunity may be included in the model. Models for different substances tested may be fitted and compared. The models are fitted using the FITNONLINEAR directive.

In this book we have referenced GLIM macros when appropriate. However, the same analyses may be programmed in GENSTAT, making use of the GLM procedure, which allows specification of non-standard link functions and distributions. More complex applications (see for example Jansen, 1988, and Ridout, 1992) reveal the flexibility of the package.

GLIM

We have discussed GLIM in Appendix B. As we have seen throughout the book, it provides a flexible tool for the analysis of quantal response data. As well as being straightforward to use for the standard links and for making comparisons (Chapter 2), a macro exists (Baker, 1980) for conversational probit analysis. GLIM may be used to fit mixtures and to model misreporting (Chapter 3). Macros exist for fitting a range of extended models, and it is relatively easy to carry out score tests and goodness-of-link tests of model fit (Chapter 4). GLIM macros now exist for various aspects of Wadley's problem (Chapters 4 and 6), and for describing over-dispersion (Chapter 6). Diagnostics (Chapter 3; see Collett and Roger, 1988) and likelihood-ratio confidence intervals (Chapter 2) can be produced in GLIM. A general macro has been provided by Vanderhoeft (1985) for carrying out the delta-method (Appendix A). GLIM macros exist for aspects of survival analysis (Chapter 5) and case-control studies (Gilchrist, 1985; Whitehead, 1983, Chapter 1). Release 3.77 and later releases of GLIM contain the GLIM macro library, which is a compilation of a number of commonly used macros. To obtain the contents of the library, enter GLIM and type:

$ECHO $INPUT %PLC 80 INFO $ECHO

There is a range of macros for generalized linear models, which produce diagnostics such as leverage values, deviance residuals, Cook distance, etc. (Chapter 3). Particularly useful is the index of Reese (1989), which references not only articles appearing in the GLIM Newsletter, but also the GLIM macro library and GLIM conference proceedings, as well as articles involving GLIM which have appeared in the *Journals of the Royal Statistical Society, The Statistician* and *Biometrika*.

Other programs and packages

A range of dedicated programs and packages exist for the analysis of quantal response data. Two examples are POLO (Russell *et al.*, 1977) and PRODOS (Ihm *et al.*, 1987). Two examples of packages written specifically for use on IBM personal computers or compatibles, are PCPROBIT (Walsh, 1987) and QUAD (Morgan *et al.*, 1989). Special features of QUAD are that it fits symmetric and asymmetric extended models (Chapter 4) and presents a range of diagnostics (Chapter 3).

FORTRAN sub-routines exist for fitting the beta-binomial distribution (Smith, 1983) and for calculating the ABERS estimate (Cran, 1980).

A program for evaluating trimmed Spearman–Kärber estimates is available, for running on an IBM PC, from the Center for Water Quality Modeling, US Environmental Protection Agency, Environmental Research Laboratory, College Station Road, Athens, Georgia 30613, USA.

Useful addresses

Up-to-date information on the major packages can be obtained from the following addresses:

BMDP BMDP Statistical Software Inc.,
1440 Sepulreda Boulevard, Suite 316,
Los Angeles, CA 90025, USA.

BMDP Statistical Software,
Cork Technology Park,
Model Farm Road,
Cork, Ireland.

SPSS Marketing Department,
SPSS Inc.,
444 North Michigan Avenue,
Chicago, IL 60611, USA.

SPSS Europe B.V.,
P.O. Box 115,
4200 AC Gorinchem,
The Netherlands.

SAS SAS Institute Inc.,
Box 8000,
Cary, NC 27511-8000, USA.

GENSTAT, GLIM, MLP, NAG
Numerical Algorithms Group Ltd.,
Mayfield House,
256 Banbury Road,
Oxford, OX2 7DE, UK.

Numerical Algorithms Group Inc.,
1101 31st Street, Suite 100,
Downers Grove, IL 60515-1263, USA.

Ceanet Pty Ltd.,
4th Floor, 56 Berry Street,
North Sydney, 2060 NSW,
Australia.

Solutions and comments for selected exercises

Chapter 1

1.1

(a) Fryer and Pethybridge (1975) report the following data describing 220 children born in the English counties of Devon and Somerset in 1965.

Birth weight in ounces	No. of infants	No. of perinatal deaths in the group
34.5–38.5	18	8
38.5–42.5	24	18
42.5–46.5	21	13
46.5–50.5	32	19
50.5–54.5	39	23
54.5–58.5	43	15
58.5–62.5	46	12

(b) The data overleaf are taken from Ashford and Smith (1965). Here A and B correspond to independent assessments of two radiologists.

(c) Age versus decayed teeth. See also Exercise 1.14.

Note: examples (a) and (b) above are taken from Hubert (1980), which is a rich source of quantal response data sets.

No. of years spent working as a coal miner	No. of miners	No. of miners judged to be suffering from pneumoconiosis	
		A	B
2.25	43	0	0
7.0	29	4	3
12.0	27	6	6
17.0	50	22	24
22.0	24	17	16
27.0	23	14	16
32.0	12	5	5
37.0	7	5	5
42.8	6	6	6

1.2 We may write the Taylor series expansion as:

$$\psi(X) = \psi(\mu) + (X - \mu)\psi'(\mu) + 0\{(X - \mu)^2\}$$

For small σ, appeal to Chebychev's inequality may justify the approximation,

$$\psi(X) \approx \psi(\mu) + (X - \mu)\psi'(\mu), \text{ whence, } V\{\psi(X)\} \approx \{\psi'(\mu)\}^2\sigma^2.$$

Here,

$$\psi(R) = \log\left(\frac{R}{n - R}\right) = \log(R) - \log(n - R)$$

$$\psi'(R) = \frac{1}{R} + \frac{1}{(n - R)}; \quad \psi'(\mu) = \frac{1}{n}\left(\frac{1}{p} + \frac{1}{q}\right) = \frac{1}{npq}$$

$$\sigma^2 = V(R) = npq.$$

Hence, $V\left\{\log\left(\dfrac{R}{n - R}\right)\right\} \approx \dfrac{1}{npq}$, which we estimate by: $\dfrac{1}{r} + \dfrac{1}{(n - r)}$

1.3 Log transformations have taken place for data sets 4 (base 2), 7 (base 2), 9 (base 10), 10 (1 + logs base 2), 11 (base 2), 22 (base 10).

1.4 It is difficult here to distinguish between the two fits, to doses and to log doses. In fact fitting to doses is marginally better, while fitting to doses squared is better still (Figure 4.9).

1.6 For the control group, the proportion surviving varies from 0.7 to 1.0; mean = 0.89; median = 0.91. For the treated group, the proportion responding varies from 0.0 to 1.0; mean $= 0.76$; median $= 0.89$. Following an arc-sine square root transformation, a two-sample t-test resulted in a t-statistic of $t = 1.5$, resulting in $p = 0.15$ from applying the TWOSAMPLE test of MINITAB (Ryan *et al.*, 1985, p. 186). A 95% confidence-interval for the difference between the population means is: $(-0.067, 0.422)$. A Mann–Whitney test applied to the proportions results in a p-value of 0.25, and a 95% confidence interval of $(-0.02, 0.2)$ for the difference between proportions. A significant difference is obtained from the analysis presented in section 6.2.1, under test H5.

1.7 Increased concentration increases both death and speed of response. For the time-span of this experiment there is no suggestion that mortality levels off with time for any concentration. Crude analyses could be made at a number of intermediate times. Alternative procedures are described in Chapter 5.

1.9 For these data a log transformation does improve the fit. For further discussion, see Chapter 4. Note especially Exercises 4.35 and 4.36.

1.10 One possibility is to plot Receiver Operating Characteristic (ROC) curves, of Pr(Hit) versus Pr(False Alarm). For each subject two intermediate points on the ROC curves may be obtained, from counting 'not sure' responses as either 'yes' or 'no'.

1.11 For both Tables 1.10a and 1.10b there appear to be 'outlying' litters with a large number of deaths. More generally, as is readily verified, variances appear to be larger than expected under a binomial model.

1.12 One possibility considered by Anderson (1985a,b) was the use of sequential methods discussed in Chapter 8. Some experiments were of the 'forced-choice' kind – presented with prickly and non-prickly fabrics, subjects had to specify which they thought was prickly. The use of different size patches was investigated. There was a large amount of variation between subjects.

1.13 This data set is frequently presented to illustrate techniques of analysis, but without any discussion. As given, the data have been condensed. If the full data were available a logistic-regression model could be fitted to the binary responses for each girl (see section 2.8). If only endpoints are available, probabilities of response could be obtained by integrating the age threshold distribution over the age-classes (e.g. section 5.5.1). The last age-class clearly includes all girls greater than a certain age. Cf. also the first age-class. Cf. the way the data are presented in Exercise 2.23.

1.14   Data are of standard quantal response form.

1.15   Suppose weights have p.d.f. $f(w)$. Let $\tilde{P}(x) = \text{Pr (response}|\text{dosage}$ x per unit weight). Let $P(d) = \text{Pr (individual responds to dose } d)$, then

$$\tilde{P}(x) = \int_0^\infty P(wx)f(w)dw$$

1.16   High dose toxicity can result in the kind of responses shown here. Non-monotonic response can also result from mixtures of substances (Finney, 1971, p. 265). See also Exercise 8.27.

1.17
$$P(d; h, \zeta) = \sum_{j=h}^\infty \frac{(\zeta d)^j e^{-\zeta d}}{j!}$$
$$= \int_0^d \frac{\zeta^h x^{h-1} e^{\zeta x}}{\Gamma(h)} dx$$

as may be verified from integration by parts.

The above reverses the usual summation representation of the incomplete gamma function. It is clearly equivalent to assuming a gamma tolerance distribution as $P(d; h, \zeta)$ is obtained by integrating a gamma p.d.f.

1.18   Solution follows from writing down the Kolmogorov forward equations, and proceeding in the usual way (Bailey, 1964, p. 94). We assume n is sufficiently large that $N(t)$ is not subject to large downturns once $N(t) \geqslant n$. Then

$$\Pr(T \leqslant t) = \Pr\{N(t) \geqslant n | N(\infty) \neq 0\}$$
$$\approx \Pr(N(t) \geqslant n)/\Pr(N(\infty) \neq 0)$$
$$= (1 - \alpha)\beta^{n-1}(1 - \mu/\lambda)^{-1}$$

1.19   For small d, $\log(P(d) - P(0)) \approx \zeta_0 + \log\zeta_1 + \log d$. In low dose extrapolation it is of interest to estimate $P(d)$ for small d. Under this model, with $\zeta_1 > 0$, this would correspond to linear extrapolation. However, data may indicate slopes > 1 (Armitage, 1982).

1.21   Otto used groups of brown painted beer cans, placed in group sizes 1, 2, 4 and 8. Groups were placed at random about transect lines, but none at a distance greater than 20 metres. Nine observers walked each transect (Otto and Pollock, 1990). Exponential power models were fitted to the data.

1.22   Base a model on the Poisson distribution (Exercise 3.8).

1.23   Differing group sizes may simply reflect intentional different importance given to different dose levels (cf. the design work of Chapter 5). However, they may conceal features of response, as discussed in section 6.6.5.

1.24   Suppose we have two thresholds, h_1 and h_2, and a **bivariate** c.d.f., $F(h_1, h_2)$, with marginal c.d.f.s $F_{H_i}(x)$, $i = 1, 2$. The four probabilities required are then given by:

$$F(d, d); F_{H_1}(d) - F(d, d); F_{H_2}(d) - F(d, d);$$
$$1 - F_{H_1}(d) - F_{H_2}(d) + F(d, d)$$

See section 3.8 for the results from fitting this model when a bivariate normal distribution is assumed. Individuals with severe problems might not be at work due to sickness. It is therefore not possible to extrapolate to the larger population of miners.

1.25   A parallel lines model provides a good fit, when logits are plotted versus age. See Healy (1988, p. 86) and Example 2.5.

1.26   Fluconazole does not perform well *in vitro*, but the situation is reversed *in vivo*.

1.27   See Chapter 8 for discussion of design. Cf. Exercise 1.23 concerning group sizes.

Chapter 2

2.2   Cf. Appendix E.

2.4   For the logit model,

$$l(\alpha, \beta) = \text{constant} - \sum_{i=1}^{k} r_i \log_e \{1 + e^{-(\alpha + \beta d_i)}\}$$

$$- \sum_{i=1}^{k} (n_i - r_i) \log_e \{1 + e^{(\alpha + \beta d_i)}\}$$

$$= \text{constant} - \sum_{i=1}^{k} n_i \log \{1 + e^{(\alpha + \beta d_i)}\} + \sum_{i=1}^{k} r_i(\alpha + \beta d_i)$$

Hence the sufficient statistics are $\left(\sum\limits_{i=1}^{k} r_i \right)$ and $\left\{ \sum\limits_{i=1}^{k} (r_i d_i) \right\}$. As shown in section 2.2, $\dfrac{\partial^2 L}{\partial \alpha^2}$, etc., are not functions of $\{r_i\}$. This follows directly from:

$$\frac{\partial l}{\partial \alpha} = \sum_i n_i(p_i - P_i); \quad \frac{\partial l}{\partial \beta} = \sum_i n_i d_i(p_i - P_i)$$

For the extension to logistic regression, see McCullagh and Nelder (1989, pp. 32 and 115). The result does not extend to the complementary log–log and probit links as the links are not the canonical links for the binomial distribution. Corresponding to the complementary log–log link, the threshold density function is of the extreme value form:

$$f(x) = \exp(y - e^y)$$

2.6

(a) For the Newton–Raphson method, the Hessian is given by:

$$
\begin{bmatrix}
\dfrac{\partial^2 L}{\partial \alpha^2} & \dfrac{\partial^2 L}{\partial \alpha \partial \beta} \\[2mm]
\dfrac{\partial^2 L}{\partial \alpha \partial \beta} & \dfrac{\partial^2 L}{\partial \beta^2}
\end{bmatrix} =
$$

$$
2 \begin{bmatrix}
\sum\limits_{i=1}^{n} \dfrac{(x_i - \alpha)^2 - \beta^2}{\{\beta^2 + (x_i - \alpha)^2\}^2}, & -2\beta \sum\limits_{i=1}^{n} \dfrac{(x_i - \alpha)}{\{\beta^2 + (x_i - \alpha)^2\}^2} \\[4mm]
-2\beta \sum\limits_{i=1}^{n} \dfrac{(x_i - \alpha)}{\{\beta^2 + (x_i - \alpha)^2\}^2}, & -\dfrac{n}{2\beta^2} + \sum\limits_{i=1}^{n} \dfrac{\beta^2 - (x_i - \alpha)^2}{\{\beta^2 + (x_i - \alpha)^2\}^2}
\end{bmatrix},
$$

corresponding to a random sample: $(x_1, x_2, \ldots, x_n)$. For the method of scoring, the expected Hessian is given by:

$$-\frac{n}{2\beta^2}\begin{bmatrix} 1, & 0 \\ 0, & 1 \end{bmatrix}$$

For further details and examples, see Morgan (1978).

(b) For a GLIM program, we may make use of the fact that if U is uniformly distributed over $(0, 1)$ then $X = \tan(\pi U)$ has a Cauchy distribution, with parameters $\alpha = 0$, $\beta = 1$ (Morgan, 1984, p. 285). Hence use the OWN command (or its replacement in GLIM4) to specify as link function, $h(\mu) = \tan(\pi\mu)$.

2.8   The number of degrees of freedom = the number of parameters = 4, and so the model is saturated. The cell occupancy probabilities are estimated by maximum likelihood as $r_{11}/r_{1.}$, etc., and the parameters $(z_1, z_2, \alpha, \beta)$ are obtained as a 1–1 transformation of the cell occupancy probabilities.

2.10   Set $a = 0.5$.

2.13   Let θ denote the ED_{50}, then

$$0.5 = \{1 + \lambda e^{(\alpha + \beta\theta)}\}^{-1/\lambda}$$
$$2^\lambda = 1 + \lambda e^{(\alpha + \beta\theta)}$$
$$\theta = \frac{1}{\beta}\left[\left\{\log_e\left(\frac{2^\lambda - 1}{\lambda}\right)\right\} - \alpha\right]$$

2.14   The ED_{50} is given by $x = \mu$, since when $x = \mu$, $P(x) = 0.5$.

2.15   See also Ross (1990, p. 67).

2.17   Little (1968) showed that the slope estimated by maximum likelihood was greater than the slope estimated by minimum logit chi-square.

2.19   Gauss–Newton is equivalent to the method of scoring used for maximum likelihood estimation of the parameters of the probit and logit models, for example.

2.20 The required interval is given by:

$$c_1 \tan[\hat{\phi} \pm \lambda I_\phi^{-1/2}] + c_2$$

where

$$c_1 = (V_{11}V_{22} - V_{12}^2)^{1/2}/V_{22}$$
$$c_2 = -V_{12}/V_{22}$$

and λ is the value of a standard normal variate chosen to give a confidence interval of the desired level.

2.22 The log-likelihood function is:

$$l(\boldsymbol{\beta}) = -\frac{1}{2\sigma^2}(y - \mathbf{X}\boldsymbol{\beta})'(y - \mathbf{X}\boldsymbol{\beta}) - \frac{k}{2}\log(2\pi\sigma^2)$$

$$\frac{\partial l}{\partial \boldsymbol{\beta}} = \frac{1}{\sigma^2}\mathbf{X}'(y - \mathbf{X}\boldsymbol{\beta}) \quad \text{(Morrison, 1976, p. 71)}$$

Hence if $r = (y - \mathbf{X}\hat{\boldsymbol{\beta}})$, then the m.l.e. $\hat{\boldsymbol{\beta}}$ is the solution of: $\mathbf{X}'r = \mathbf{0}$. Assuming $\mathbf{X}'\mathbf{X}$ is non-singular, it is then usual to write:

$$\hat{\boldsymbol{\beta}} = (\mathbf{X}'\mathbf{X})^{-1}\mathbf{X}'y$$

2.25

$$\frac{\partial Q}{\partial \beta_i} = -2\sum_{i=1}^{k} w_i\{y_i - g(x_i;\boldsymbol{\beta})\}\frac{\partial g}{\partial \beta_i}$$

Set $\dfrac{\partial Q}{\partial \beta_i} = 0$, and compare with (from section 2.2)

$$\frac{\partial l}{\partial \zeta} = \sum_{i=1}^{k}\frac{\partial P_i}{\partial \zeta}\frac{n_i(p_i - P_i)}{P_i(1 - P_i)} = 0$$

These equations are the same, if we make the equivalences:

$$P_i = g(x_i;\boldsymbol{\beta})$$
$$p_i = y_i$$
$$n_i/\{P_i(1 - P_i)\} = w_i$$
$$\zeta = \beta_i$$

Hence an algorithm for model-fitting by weighted least squares, such as the Gauss–Newton algorithm of Exercise 2.19, can be used to fit models to quantal response data by maximum-likelihood. Note the difference between $Q(\boldsymbol{\beta})$ and $\tilde{S}(\alpha, \beta)$ of section 2.6. In $Q(\boldsymbol{\beta})$ the weight is the reciprocal of an exact variance, but the numerator terms involve r_i, rather than $\log\{r_i/(n_i - r_i)\}$ and $g(x_i; \boldsymbol{\beta})$ is non-linear in $\boldsymbol{\beta}$. See Charnes *et al.* (1976) for a generalization to the exponential family case.

2.26   The estimated variance-covariance matrix for $\hat{\boldsymbol{\beta}}$ is the inverse of the matrix $\{b_{ij}\}$, where

$$b_{ij} = \sum_{i=1}^{k} w_i \frac{\partial g}{\partial \beta_i} \frac{\partial g}{\partial \beta_j}, \quad 1 \leqslant i, j \leqslant k$$

With the equivalences of the solution to Exercise 2.25, this is equivalent to using the Fisher information matrix (section 2.3).

2.27   When there is only one partial response (i.e. not 0% or 100%) a perfect maximum likelihood fit results from setting $\beta = \infty$. An interesting test of computer packages results from trying them with such data. While $\hat{\beta} = \infty$, we can still obtain the m.l.e. of the ED_{50}. However, that is not true if there are no partial responses. (Try drawing diagrams to illustrate these two cases.) See also Williams (1986b) and Pate (1989).

2.30   See the discussion above for Exercise 2.27. The one additional response for Example B suffices to ensure that $\hat{\beta} \neq \infty$ (draw a graph to see why). See Ashton (1972, p. 39) for discussion of the effect of high and low responses on precision.

2.32   An example of ordered categorical data (cf. Exercise 2.8 and section 3.5 for discussion).

```
2.35   $UNITS 8 $DATA COUNT $READ
       12  3
       30 13
       18 27
        4 10
       $CALC SEX=%GL(2,4) : VOT=%GL(2,2) : ATT=%GL(2,1) $
       $FACT SEX 2 : VOT 2 : ATT 2 $
       $LOOK SEX VOT ATT COUNT $
       $ERROR POISSON $YVARIABLE COUNT $
       $FIT VOT*SEX+SEX*ATT+VOT*ATT $DISP E $
       $FIT -VOT.ATT $DISP E $
       $FIT VOT*ATT $DISP E $
       $STOP                                        (Cont.)
```

```
$UNITS 4 $DATA APP DISS $READ
12    3
30   13
18   27
 4   10
$CALCULATE SEX=%GL(2,2) : VOT = %GL(2,1) $
$FACTOR SEX 2 : VOT 2 $
$CALCULATE N=APP + DISS $
$LOOK SEX VOT APP DISS  N $
$ERROR BIN N $YVAR APP $
$FIT VOT + SEX $DISP E $
$FIT -VOT $DISP E $
$STOP
```

2.36   Set $f(x) = \dfrac{dP(x)}{dx}$ and integrate.

2.37   Let π_i denote the probability that the organism is present in a subsample at the ith dilution. Randomness implies the Poisson process, which gives:

$$\pi_i = 1 - \exp(-\lambda v_i)$$

Hence X_i has the binomial distribution, Bin($n_i\pi_i$). Ridout (1990) gives the GLIM commands, using the OFFSET facility (see Appendix B):

```
$CAL   LGV=%LOG(V) $
$YVAR  Y $
$ERR   B N $
$LINK  C $
$OFF   LGV $
$FIT $
$DIS E R $
```

2.39   The log likelihood is:

$$l = \text{constant} + \sum_i r_i \log P(d_i) + \sum_i (n_i - r_i) \log \{1 - P(d_i)\}$$

$$\frac{\partial l}{\partial \beta} = \left[\sum_i \frac{r_i d_i}{P(d_i)} - \sum_i \frac{(n_i - r_i)d_i}{\{1 - P(d_i)\}} \right] P'$$

We now set $\beta = 0$ and $\alpha = \hat{\alpha}_0$, the maximum likelihood estimate of α when $\beta = 0$. $P(d_i)$ no longer involves i, and so we can write

$$\frac{\partial l}{\partial \beta} = \frac{P'}{P(1-P)} \left(\sum_i r_i d_i - P \sum_i n_i d_i \right)$$

The elements of the expected Fisher information matrix are, from section 2.2, and using the notation of Appendix D,

$$I_{11} = \sum_i \frac{n_i \left(\frac{\partial P_i}{\partial \beta}\right)^2}{P_i(1 - P_i)}$$

$$I_{21} = \sum_i \frac{n_i \frac{\partial P_i}{\partial \alpha} \frac{\partial P_i}{\partial \beta}}{P_i(1 - P_i)}, \quad I_{22} = \sum_i \frac{n_i \left(\frac{\partial P_i}{\partial \alpha}\right)^2}{P_i(1 - P_i)}$$

and with $\beta = 0$, $\alpha = \hat{\alpha}_0$, these become:

$$I_{11} = \frac{(P')^2 \sum_i n_i d_i^2}{P(1 - P)}; \quad I_{21} = I_{12} = \frac{(P')^2 \sum_i n_i d_i}{P(1 - P)};$$

$$I_{22} = \frac{(P')^2 \sum_i n_i}{P(1 - P)}$$

Hence the score test statistic is:

$$\frac{\left(\sum_i r_i d_i - P \sum_i n_i d_i\right)^2}{\left\{\sum_i n_i d_i^2 - \frac{\left(\sum_i n_i d_i\right)^2}{\sum_i n_i}\right\} P(1 - P)}$$

Conveniently the P' term cancels. Also we see that we must have $P = \dfrac{\sum r_i}{\sum n_i}$, since this is the maximum likelihood estimate of P when $\beta = 0$. The above is the form for X^2 in equation (2.21).

Chapter 3

3.1 λ may be estimated by the proportion that die in the control group, say $\hat{\lambda} = r_0/n_0$. At any dose d we observe a proportion r/n responding. Using equation (3.1), this proportion may then be adjusted to give:

$$\left(\frac{r}{n} - \frac{r_0}{n_0}\right) \bigg/ \left(\frac{n_0 - r_0}{n_0}\right)$$

An adjusted number responding would then be $(n_0 r - n r_0)/(n_0 - r_0)$. For the data of Table 3.1, $\hat{\lambda} = 0.036$, and adjusted numbers responding are thus: 0.2, 6.8, 9.6, 58.9. See also Barlow and Feigl (1985).

3.2   For the model of equation (3.2) a simple correction is not possible.

3.3 $P(d) = \lambda + (1 - \lambda - \zeta)\tilde{P}(d).$

Here ζ is the probability of immunity. See also Finney (1971, Ch. 7) for further discussion.

3.4   Results suggest a significant effect of dose but not of sex. An exponential model would not seem appropriate.

3.5   There appears to be a site difference. The distribution of subjects over doses varies between sites, and this is probably responsible for a spurious dose effect if the data are pooled over sites.

3.6   Let $F(t|x)$ denote the Weibull c.d.f. corresponding to covariate x. Suppose the censoring time is τ days in general (we have $\tau = 10$). Suppose the observed life-times are: $\{t_i, 1 \leqslant i \leqslant m\}$, and n fish survive the experiment.
  The probability of surviving is:

$$1 - P(x) + P(x)\{1 - F(\tau|x)\}$$

The likelihood is:

$$L(\zeta|x) = \prod_{i=1}^{m} \{P(x_i)f(t_i|x_i)\} \prod_{j=1}^{n} [1 - P(x_j) + P(x_j)\{1 - F(\tau|x_j)\}].$$

Both acclimatization and concentration appear to be useful covariates in both $P(x)$ and $F(t|x)$. Here and in the solutions to Exercises 3.4 and 3.5 we are performing Wald tests (Appendix D), but in the presence of nuisance parameters. For goodness-of-fit, Farewell (1982b) compared estimates of $\Pr(T > t_i)$ with corresponding observed proportions. We see that $\hat{\delta} > 1$ (the value for an exponential model), but no standard error is given.

3.8   Measures like an ED_{50} relate to a known number of treated individuals, which is not the case in this example.

3.9 A standard example of a compound distribution. Let numbers of organisms be X, with probability generating function $G_X(z)$. Let numbers of live organisms be Y, with probability generating function $G_Y(z)$. Then $G_Y(z) = G_X(G(z))$, where

$$G(x) = \{p + (1 - p)z\}$$

But $G_X(z) = e^{\zeta(z-1)}$. Hence $G_Y(z) = \exp\{\zeta(1 - p)(z - 1)\}$, so that Y is Poisson, of mean $\zeta(1 - p)$.

3.10 Two possibilities are to have the number of encounters of a host with a wasp to be either Poisson or binomial. Then an egg is supposed to be laid at first encounter with probability 1, but with probability $\delta < 1$ at all subsequent encounters. A review of the field is given by Daley and Maindonald (1989).

3.11 A simple model for polyspermy is based on a Poisson process of rate λ until the first fertilization, and then with reduced rate μ thereafter. A number of added complexities (such as the rate dropping to zero after a fertilization membrane is formed) are considered by Morgan (1982). These models result in *under*-dispersion (cf. Chapter 6). For a further example, on sex proportions, see Brooks *et al.* (1991).

3.14 $\sum_i h_{ii} = \text{tr}(\mathbf{H}) = \text{tr}\{\mathbf{X}(\mathbf{X}'\mathbf{X})^{-1}\mathbf{X}'\} = \text{tr}\{\mathbf{X}(\mathbf{X}'\mathbf{X})^{-1}\}\text{tr}(\mathbf{X}')$

$= \text{tr}(\mathbf{X}'\mathbf{X})(\mathbf{X}'\mathbf{X})^{-1} = \text{tr}(\mathbf{I}) = 2$, as here the dimension of $\mathbf{X}'\mathbf{X} = 2$, the number of parameters in the model.

3.22 The change, compared with Table 3.9, is striking. Overall the fit is much better and no one case dominates the set of Cook distances, for example.

3.29 $f(\theta|r)$ is obtained by integrating the joint density for the pair (α, β) with respect to β. The Jacobian of the transformation from (α, β) to (θ, β) is β. In equation (3.23), β is constrained to be non-negative. However, β is not so constrained in Figure 3.8.

3.31 One possibility is to match first- and second-order moments.

3.32 The key part of the exponent of a multivariate distribution of mean $\boldsymbol{\mu}$ and dispersion matrix $\boldsymbol{\Sigma}$ is: $x'\boldsymbol{\Sigma}^{-1}x - 2x'\boldsymbol{\Sigma}^{-1}\boldsymbol{\mu}$. For the posterior distribution, we have, correspondingly:

$$x'(\Sigma_0^{-1} + \Sigma^{-1})x - 2x'(\Sigma_0^{-1}\mu_0 + \Sigma^{-1}\mu)$$

which gives a multivariate normal form, of dispersion matrix

$$\Sigma^* = (\Sigma_0^{-1} + \Sigma^{-1})^{-1} \text{ and mean } \Sigma^* (\Sigma_0^{-1}\mu_0 + \Sigma^{-1}\mu)$$

to match the form given above.

3.33   Let $y = (l-1)\log x + (m-l-1)\log(1-x)$

$$\frac{dy}{dx} = \frac{l-1}{x} - \frac{(m-l-1)}{(1-x)}$$

Set

$$\frac{dy}{dx} = 0, \text{ to give } x = (l-1)/(m-2)$$

3.34   The Jacobian of the transformation is:

$$J = \frac{\partial(\pi_1, \pi_2)}{\partial(\alpha, \beta)}$$

i.e. $$J = \begin{vmatrix} \phi(\alpha + \beta d_1), & d_1\phi(\alpha + \beta d_1) \\ \phi(\alpha + \beta d_1), & d_2\phi(\alpha + \beta d_2) \end{vmatrix}$$

i.e. $$J = (d_2 - d_1)\phi(\alpha + \beta d_1)\phi(\alpha + \beta d_2)$$

3.36   The text shows how to equate the models of equations (3.20) and (2.21). Equation (3.20) is the particular logistic form of the signal-detection model of Exercise 2.8.

3.37 $(I - H)Y = (I - H)(X\beta + \phi z + \varepsilon)$
                $= \phi(I - H)z + (I - H)\varepsilon_2 + (I - H)X\beta.$

but $(I - H)X\beta = (I - x(X'X)^{-1}X')X\beta = 0.$

3.38   Evidence for $\alpha = 0$ is not due to responses at just one or two dose levels, but a feature of the entire data set.

3.39 $$\sigma\sqrt{2\pi} I = \frac{1}{\sqrt{2\pi}} \int_{-\infty}^{\infty} \int_{-\infty}^{\alpha' + \beta' u} e^{-\frac{x^2}{2}} e^{\frac{1}{2}(\frac{z-u}{\sigma})^2} dx\, du$$

and

$$2\pi I = \sigma \int_{-\infty}^{\infty} \int_{-\infty}^{\alpha' + \beta' z - \beta \sigma y} e^{-\frac{1}{2}(x^2 + y^2)} dx \, dy$$

Hence, by the circular symmetry,

$$I = \frac{1}{\sqrt{2\pi}} \int_{-\infty}^{\kappa} e^{-\frac{1}{2}x^2} dx = \Phi(\kappa)$$

By equating two expressions for triangle area, we get

$$\frac{(\alpha' + \beta' z)^2}{\sigma \beta'} = \kappa(\alpha' + \beta' z) \left(1 + \frac{1}{(\sigma \beta')^2}\right)^{1/2}$$

$$\kappa = \frac{(\alpha' + \beta' z)}{(1 + \sigma^2 (\beta')^2)^{1/2}}$$

and the result follows.

A simple proof is as follows.

Consider a nominal dose level, z, say. The actual dose received is: $d = z + e$, where $e \sim N(0, \sigma^2)$. We are assuming that the tolerance distribution has the form: $F(d) = \Phi(\alpha' + \beta' d)$, i.e. if an individual's tolerance is T

$$\Pr(T \leqslant d) = \Phi(\alpha' + \beta' d)$$

But

$$d = z + e, \quad \text{and hence}$$
$$\Pr(T \leqslant d) = \Pr(T \leqslant z + e)$$
$$= \Pr(T - e \leqslant z)$$

Note that $(T - e)$ is Normal, with the same mean (and hence median) as T (which is why the LD_{50} does not change), but with increased variance:

$$T, e \quad \text{are independent; } \text{Var}(T - e) = \sigma^2 + \text{Var}(T)$$
$$= \sigma^2 + (\beta')^{-2}$$

Hence,

$$\Pr(T - e \leqslant z) = \Pr\left\{\frac{T - e + \alpha'/\beta'}{(\sigma^2 + (\beta')^{-2})^{1/2}} \leqslant \frac{\beta' z + \alpha'}{(1 + \sigma^2 (\beta')^2)^{1/2}}\right\}$$

as required.

3.40 Under the logit model,

$$F(h) = 1/(1 + e^{-(\alpha + \beta h)}).$$

Let D denote diagnosis, and let W, NW denote, respectively, wheeze and no wheeze. Then we have,

$$\Pr(D = W|x) = F(x)(1 - \varepsilon_1) + (1 - F(x))\varepsilon_0 \quad \text{and}$$
$$\Pr(D = NW|x) = (1 - F(x))(1 - \varepsilon_0) + F(x)\varepsilon_1$$

Suppose at age x we have $n_w(x)$ that wheeze, and $n_{nw}(x)$ that do not. Then the likelihood of just the wheeze data is:

$$L(\alpha, \beta, \varepsilon_0, \varepsilon_1) \propto \prod_x [F(x)(1 - \varepsilon_1) + \{1 - F(x)\}\varepsilon_0]^{n_w(x)}$$

$$\times \prod_x [\{1 - F(x)\}(1 - \varepsilon_0) + F(x)\varepsilon_1]^{n_{nw}(x)}$$

Comparing the fits of the two models, we obtain the deviance difference:

$$(8.20 - 3.73) = 4.47, \quad \text{with reference to } \chi_2^2.$$

The result is not significant at the 10% level.

We observe that the (α, β) pair of parameters are similar in each case, but the change in $\hat{\beta}$ is $+0.012$, which is quite substantial relative to its initial estimated standard error of 0.002. The ED_{50} values are: before adding mis-diagnosis (65.54), after adding mis-diagnosis (54.55), which is a substantial change. (Previously extrapolation was involved.) This is because of $\hat{\varepsilon}_1 = 0.322$ (0.13), suggesting that a large fraction of wheeze cases are not reported as such. Note that $0.322 > 2 \times 0.13$. Note the large increase in standard errors of $\hat{\alpha}$ and $\hat{\beta}$ when $(\varepsilon_0, \varepsilon_1)$ are added to the model. There is no evidence to suggest $\varepsilon_0 \neq 0$. It would be interesting to fit the model with $\varepsilon_0 = 0$. Burn (1983) did this, and obtained the results:

$$\hat{\varepsilon}_1 = 0.321 \ (0.060)$$
$$\hat{\alpha} = -4.189 \ (0.068)$$
$$\hat{\beta} = 0.0762 \ (0.004)$$

with a deviance of 3.718 on 6 d.f.

3.41 Let $P(d) = 1 - e^{-\zeta(d + d_0)}$. This is the approach which equates natural mortality to a baseline dose of level d_0. Then

$$P(d) = 1 - e^{-\zeta d}e^{-\zeta d_0}$$
$$= (1 - e^{-\zeta d})e^{-\zeta d_0} + 1 - e^{-\zeta d_0}$$
$$= \lambda + (1 - \lambda)\tilde{P}(d), \text{ as in equation (3.1), where}$$
$$\lambda = (1 - e^{-\zeta d_0}) \text{ and } \tilde{P}(d) = (1 - e^{-\zeta d})$$

Chapter 4

4.4 The data may require fitting by an extended symmetric distribution, such as those in the quantit family. The Weibull is skewed, except for $\alpha \approx 3.6$ – see Looney (1983). As is to be expected, the ED_{50} is barely affected by model choice, but extreme estimated doses can change appreciably. The widest intervals are given by the 3-parameter Prentice model. Is a 3-parameter model required here? (Example 4.2.) All of the $\hat{ED}_{01}$ values and two of the $\hat{ED}_{99}$ values involve extrapolation.

4.5 We could consider expressions for kurtosis. Aranda-Ordaz (1981) considered two alternative criteria: C1 chose λ to give equal slopes at the ED_{50}; C2 chose λ to guarantee equal quartiles, e.g. 80% quartiles. Resulting values of λ were: C1: 0.3955; C2: 0.3800.

4.7 Symmetry about zero follows from the form of the power series expansion of $h(\mu)$ as a function of $|2p - 1|$.

4.11 (b) Expand the integrand in a power series and then integrate.

4.13 Make use of the following results:
$$\int x^n \log x \, dx = \frac{1}{(n+1)}\left\{x^{n+1}\left(\log x - \frac{1}{(n+1)}\right)\right\}$$
For $p > 0$ and $q > 0$,
$$\int_0^1 \frac{x^{p-1}}{1 - x^q} \log x \, dx = -\frac{1}{q^2}\psi'\left(\frac{p}{q}\right)$$
where $\psi(x)$ is Euler's psi function, and
$$\psi'(x) = \sum_{k=0}^{\infty} (x + k)^{-2}$$

4.14 For data set 7 the cubic logistic model provides a fitted curve with a slowly changing second-order derivative. However, for data set 18 this is not the case and the curve passes through most of the data points. In this last example the curve is overfitting the data. The same is true for data set 10.

4.15 It would be convenient to use the same shape parameter for different data sets, if possible.

4.16 In the cubic-logistic case, when $x - \theta = \varepsilon$ we can write the probability of response as $P(\theta + \varepsilon) = 1/(1 + e^{-\eta})$. When $x - \theta = -\varepsilon$, we have $P(\theta - \varepsilon) = 1/(1 + e^{\eta})$. Hence $P(\theta + \varepsilon) + P(\theta - \varepsilon) = 1$ and so the underlying threshold p.d.f. is symmetric. That is not the case if we replace $(x - \theta)^3$ by $(x - \theta)^2$ in the exponents.

4.26 The GLIM program below is reproduced from Baker *et al.* (1980), with permission from the GLIM Newsletter.

```
$SUBFILE WADLEY
$C EXAMPLE ON WADLEY'S PROBLEM USING
$C     (1) THE IDENTITY LINK FOR CONTROLS
$C     (2) A PARAMETRIC LINK FOR OTHERS
$UNITS 25 $DATA S $READ
219 228 202 237 228
167 158 158 175 167
105 123 105 105 105
 88  88  61  61  88
 61  44  35  35  44
$DATA X $READ
  0  0  0  0  0
  1  1  1  1  1
  5  5  5  5  5
 10 10 10 10 10
 50 50 50 50 50
$CAL I2=1-I1=%EQ(X,0) : A=%NE(X,0) : X=%LOG(X) $
$YVAR S $OWN M1 M2 M3 M4 $SCALE 1
$MAC M1 !
$CAL %S=%NE(%PL,0)+1 $SWITCH %S INIT MEXT !
$CAL TA=I1+I2*(1-%NP(LP)) !
  :  TB=%X*%EXP(-0.5*LP*LP)*%N
  :  M=I1+I2*(-TA/TB)
  :  %FV=%N*TA
  :  %LP=LP+%N*M
$ENDMAC
$MAC M2 $CAL %DR=I1+I2*(-1/TB) $ENDMAC
$MAC M3 $CAL %VA=%FV $ENDMAC
$MAC M4 $CAL %DI=2*(%YV*%LOG(%YV/%FV)-%YV+%FV) $ENDMAC
$MAC INIT $SORT TA %YV !
$CAL %N=TA(%NU)*1.1 : LP=I2*%ND(1-(%YV+1)/%N) !
$CAL %X=1/%SQRT(2*%PI) $ENDMAC
$MAC MEXT $EXTRACT %PE $CAL %N=%PE(%P) : LP=%LP-%N*M $ENDMAC
$CAL %LP=M=0 : %P=1
$FIT M+A+X-%GM $DIS EC $
$RETURN
$FINISH
```

4.27   See the paper by Taylor (1988) for relevant discussion. Highly revealing graphs are provided by van Montfort and Otten (1976).

4.28   The MLP package (Appendix E) fits such a mixture model as one of its options.

4.30

(i)
$$f(x) = \int_0^\infty f(x|\beta)g(\beta)d\beta$$

$$= \int_0^\infty \alpha\beta x^{\alpha-1} e^{-\beta x^2} \frac{\delta^\nu \beta^{\nu-1} e^{-\delta\beta}}{\Gamma(\nu)} d\beta$$

$$= \frac{\alpha x^{\alpha-1}\delta^\nu}{\Gamma(\nu)} \int_0^\infty \beta^\nu e^{-\beta(\delta+x^2)} d\beta \qquad (*)$$

We know that $\int_0^\infty g(\beta)d\beta = 1$

i.e.
$$\frac{\delta^\nu}{\Gamma(\nu)} \int_0^\infty \beta^{\nu-1} e^{-\delta\beta} d\beta = 1$$

$$\int_0^\infty \beta^\kappa e^{-\theta\beta} d\beta = \Gamma(\kappa+1)\theta^{-(\kappa+1)}$$

Using this in (*) gives us:
$$f(x) = \frac{\alpha x^{\alpha-1}\delta^\nu}{\Gamma(\nu)} \Gamma(\nu+1)(\delta+x^\alpha)^{-(\nu+1)}$$

i.e.
$$f(x) = \frac{\nu\alpha\delta^\nu x^{\alpha-1}}{(\delta+x^\alpha)^{\nu+1}}, \quad \text{for } x \geq 0$$

as required.
    The c.d.f. is given by:
$$F(x) = \nu\alpha\delta^\nu \int_0^\infty \frac{x^{\alpha-1}}{(\delta+x^\alpha)^{\nu+1}} dx$$

$$= \delta^\nu \left[\frac{-1}{(\delta+x^\alpha)^\nu} \right]_0^x$$

$$= 1 - (1+x^\alpha/\delta)^{-\nu}$$

(ii) Let
$$Z = \underset{i}{\text{Min}} \{Y_i\}$$

$$\Pr(Z > z) = \prod_{i=1}^{n} \Pr(Y_i > z) = \left(\frac{1}{1 + e^z}\right)^n$$

$$\therefore \qquad \Pr(Z \leqslant z) = 1 - (1 + e^z)^{-n} \qquad (\dagger)$$

Finally, suppose $X = e^z$

$$\Pr(X \leqslant x) = \Pr(e^z \leqslant x) = \Pr(Z \leqslant \log x)$$
$$= 1 - (1 + x)^{-n}$$

We see, from above, that this is a particular example of a Burr distribution, with $\alpha = \delta = 1$, $v = n$.

(iii) Follows from ($\dagger$).

4.34 The high absolute correlations between the pair $(\hat{\alpha}, \hat{\beta})$ and $\hat{\lambda}$ are to be expected, as x enters into the model through the form: $(\alpha + \beta x^\lambda)$. An interesting contrast is provided by the correlation between $(\hat{\alpha}, \hat{\beta})$ and $\hat{v}$ in Quantit analysis (Table 4.2). Imprecision in $\hat{\lambda}$ (typically high for small data sets) will be translated into imprecision in $(\hat{\alpha}, \hat{\beta})$ (Bickel and Doksum, 1981). Because of this we may view the choice of λ as selecting a suitable measurement scale for x (Box and Cox, 1982). However, because the ED_{50} is a stable parameter, we would not expect estimation of λ greatly to reduce its precision. It is likely to have low correlation with other parameters. See the values given below, taken from Goedhart (1985).

Data set	$E\hat{D}_{50}$		$E\hat{D}_{95}$	
	Logit	*Logit + Box–Cox*	*Logit*	*Logit + Box–Cox*
3	3.371 (0.302)	3.238 (0.385)	5.603 (0.660)	6.234 (2.158)
6	4.080 (0.012)	4.104 (0.016)	4.277 (0.025)	4.239 (0.018)

Egger (1979) and Guerrero and Johnson (1982) do drop a degree of freedom for estimating λ. There should be a cost for using the data.

4.35 For either the logit or probit models there is a significant improvement in fit from introducing the Box–Cox transformation. The fit to $\log(z/(1 - z))$ is attractive as it compares well with the fit to $\log z$ and $\log(1 - z)$ and involves one fewer parameter. Certain comparisons may not be made by likelihood-ratio tests as hypotheses are not nested, but in terms of X^2 for example, the $\log(z/(1-z))$ fit is better than the logit fit to the Box–Cox transformed x, and one fewer parameter is involved. Motivation for the log-based transformations is given by Kay and Little (1987).

4.37 The generalized gamma distribution is given by:

$$f(x) = \left\{ \frac{b}{\Gamma(k)} \right\} a^{-bk} x^{bk-1} \exp\left\{ -\left(\frac{x}{a}\right)^b \right\},$$

for $x > 0$,   and for $a, b, k > 0$

Special cases include gamma ($b = 1$), Weibull ($k = 1$) and log normal ($k \to \infty$).

If $\log x = \log a + w/b$, then

$$f(x) = \Gamma(k)^{-1} e^{kw} e^{-e^w}$$

Thus the log of a r.v. with a generalized gamma distribution is of location-scale form, with error term having a log gamma distribution. Rather than let $k \to \infty$, we set $k = \lambda^{-2}$ and let $\lambda \to 0$. Finally, we set $x = \lambda^{-1}(w + 2\log\lambda)$, which has the form given in the question (extended to the case of $\lambda < 0$).

4.38   In the multi-hit model, we have from Exercise 1.17:

$$P(d; h, \zeta) = \int_0^d \frac{\zeta^h x^{h-1} e^{-\zeta x}}{\Gamma(h)} dx$$

This is an incomplete gamma integral, which we may write as:

$$P(d; h, \zeta) = \Gamma(h)^{-1} \sum_{j=0}^\infty (-)^j \frac{\{\zeta d\}^{j+h}}{(h+j)j!}$$

For small d,

$$P(d; h, \zeta) \approx \Gamma(h+1)\zeta^h d^h$$

From Exercise 4.4, in the simple case of $\gamma = 0$, under the Weibull model,

$$P(d) = 1 - \exp\left[-(d/\beta)^\alpha \right]$$

and so for small d,

$$P(d) \approx \frac{d^\alpha}{\beta^\alpha}$$

revealing the similarity with the multi-hit model – set $\alpha = h$.

Also, under the Weibull model,

$$\log P(d) = \log(e^{(d/\beta)^\tau} - 1)$$

$$\approx \alpha \log\left(\frac{d}{\beta}\right), \quad \text{for small } d$$

$$= \alpha \log d - \alpha \log \beta$$

which is equivalent to the logit model with a log dose transformation. The slope parameter, α, can be used to approximate the number of hits in the multi-hit model, as $\alpha = h$ is required to match multi-hit and Weibull models.

4.39 In both comparisons there is a good agreement between the three alternative test statistics. When the logit model is improved by the Aranda-Ordaz model (data sets 6, 10, 23), the latter does not improve on the complementary log–log model, suggesting that the simpler complementary log–log model would suffice. In these cases the data just require an asymmetric model, or a suitable dose transformation. Note here also the discussion of Exercise 6.26 and its solution.

Chapter 5

5.1 The striking feature is the low value at time between 0.1 and 0.5 hours. This requires checking, with other times in the interval also being used. Less molluscicide is needed for short treatment time, if the figure represents a true effect. The relatively constant value of the product of concentration and time for large times is reassuring, suggesting that dilution of molluscicide, which might occur in a river over time will be balanced by time of exposure, to maintain mortality.

5.3 In equation (5.1) there is an order relationship involving the $\{r_{ij}\}$, as these are cumulative sums:

$$r_{i1} \leqslant r_{i2} \leqslant \ldots \leqslant r_{im}$$

This does not accord with the multivariate normal distribution for $\{\varepsilon_{ij}\}$, particularly when applied to data of the dimensions of Table 5.2 ($m = 3$, $n_i = 10$ for all i). Additionally, as pointed out by Petkau and Sitter (1989), the dispersion matrix cannot be regarded as constant.

5.4 Hewlett's LD_{50} estimates are:

> Males: 0.2576
> Females: 0.3664

giving a 'susceptibility ratio' of $3664/2576 = 1.42$. (Males are more susceptible as they have a lower LD_{50}.) The logit analysis results are similar:

> Males: 0.2564
> Females: 0.3708

giving a ratio of 1.45.

5.5 From equation (5.2)

$$2 = e^{\beta\theta}\left(1 + \delta\frac{\theta}{\mu}\right)^{1 - \beta\mu/\delta}$$

$$\log_e 2 = \beta\theta + \left(1 - \frac{\beta}{\rho}\right)\log_e(1 + \rho\theta)$$

where

$$\rho = \frac{\delta}{\mu}.$$

5.7 We can describe μ by a simple monotonic decreasing function of dose–logistic could be used, at least halving the number of parameters for μ. For γ the situation is less clear.

5.8 Consider, for example, the case of a 7-day experiment. Consider the proportion of eggs dying up to day 7, say s_7/n, and, for the eggs that die, the average number of days dead up to day 8 (i.e. we count 1 for those that die on day 7, 2 for those that die on day 6, etc). This is

$$\sum_{i=1}^{7}(8 - i)d_i/s_7 = S/s_7$$

say. In this notation, the MICE index is a function of S/n.

5.9   Let m denote the number of times and r denote the total number of individuals. As $m, r \to \infty$, then consistent estimators of μ and γ result from solving equations (5.10). The first of these equations is:

$$\hat{p}_m = \hat{\alpha}R(\hat{\alpha})$$

and from a first-order Taylor series expansion we have, for suitably large m and r

$$\hat{p}_m \approx \alpha R(\alpha) + (\hat{\alpha} - \alpha)\frac{\partial p_m}{\partial \alpha}$$

and so

$$V(\hat{\alpha}) = V(\hat{p}_m)\left(\frac{\partial p_m}{\partial \alpha}\right)^{-2}$$

(i.e. the result of the delta-method – see Appendix A.)
    After some algebra, we have

$$\frac{\partial p_m}{\partial \alpha} = R(\alpha)(1 + \alpha^2) - \alpha$$

but

$$V(\hat{p}_m) = \hat{p}_m(1 - \hat{p}_m)/r$$

and so

$$V(\hat{\alpha}) \approx \frac{\hat{p}_m(1 - \hat{p}_m)}{r\{R(\hat{\alpha})(1 + \hat{\alpha}^2) - \hat{\alpha}\}^2}$$

Following similar first-order Taylor series expansions and further algebra, we obtain, for suitably large r and m,

$$V(\hat{\mu}) = \left\{\sum_{i=1}^{m-1} c_i^2 \hat{p}_i - \left(\sum_{i=1}^{m-1} c_i \hat{p}_i\right)^2 + \hat{p}_m(1 - \hat{p}_m)(\nu\kappa)^2 + 2\hat{p}_n\nu\kappa \sum_{i=1}^{m-1} c_i\hat{p}_i\right\}\frac{1}{r\omega^2}$$

$$\mathrm{Cov}(\hat{\alpha}, \hat{\mu}) \approx \left\{\frac{1}{r}\sum_{i=1}^{m-1} c_i^2 \hat{p}_i - \frac{1}{r}\left(\sum_{i=1}^{m-1} c_i\hat{p}_i\right)^2 - V(\hat{\alpha})\nu^2 - V(\hat{\mu})\omega^2\right\}\frac{1}{2\nu\omega}$$

where

$$\kappa^{-1} = R(\hat{\alpha})(1 + \hat{\alpha}^2) - \hat{\alpha}$$

$$v = \frac{1}{\hat{\mu}} \left\{ \frac{(1 + \hat{\alpha}^2)}{\kappa} + 2\hat{\alpha}(\hat{p}_m - 1) \right\}$$

$$\omega = -\frac{1}{\hat{\mu}^2} \left\{ \hat{\alpha}R(\hat{\alpha})(1 + \hat{\alpha}^2) - \hat{\alpha}^2 \right\}$$

the $\{c_i\}$ are given by equation (5.9), and

$$\hat{p}_i = 1 - \sum_{j=1}^{i} r_j/r, \quad 1 \leqslant i \leqslant m$$

5.10   The recursion formula for the incomplete gamma function is

$$I_{k,\theta}(t) = I_{k-1,\theta}(t) - W_{k-1} \quad \text{for } k \geqslant 1$$
$$I_{0,\theta}(t) = 1$$

where

$$W_k = \frac{\theta t}{k} W_{k-1}, \quad \text{for } k \geqslant 1$$

Suppose $S(t) = S_1(t) + S_2(t)$, with

$$S_1(t) = \sum_{k=0}^{\infty} \frac{1}{k!} A_k \left(\frac{\beta\mu}{\delta} - 1 \right) \left(\frac{\delta d}{\mu + \delta d} \right)^k e^{-\beta d} I_{k,\mu + \delta d}(t)$$

and

$$S_2(t) = \sum_{k=0}^{\infty} \frac{1}{k!} A_k \left(\frac{\beta\mu}{\delta} \right) \left(\frac{\delta}{\beta} \right)^k I_{k+1,\beta}(d)t^k \exp\left[-(\mu + \delta d)t \right]$$

Then

$$S_1(t) = \sum_{k=0}^{\infty} F_k I_{m,\mu + \delta z_d}(t) e^{-\beta z_d}$$

where the $\{F_k\}$ satisfy the recursion

$$F_0 = 1,$$

$$F_k = \frac{1}{k} \left(\frac{\beta\mu}{\delta} + k - 2 \right) \left(\frac{\delta d}{\mu + \delta d} \right) F_{k-1}, \quad \text{for } k \geqslant 1$$

and

$$S_2(t) = \sum_{k=0}^{\infty} G_k I_{k+1,\beta}(d) e^{-(\mu + \delta d)t}$$

where the $\{G_k\}$ satisfy the recursion

$$G_0 = 1,$$

$$G_k = \frac{G_{k-1}}{k\beta} \left(\frac{\beta\mu}{\delta} + k - 1 \right) t\delta \quad \text{for } k \geqslant 1$$

For both $S_1(t)$ and $S_2(t)$, the summation was terminated if an incomplete gamma integral was found to be less than 10^{-15}, equivalent to zero for the degree of accuracy of the computer used. Apart from this feature, summations proceeded for a minimum of u terms, subsequent terms being added until a new term was found to be less than ε, when the series was truncated. A useful combination is: $u = 20$ and $\varepsilon = 10^{-10}$.

5.11 First we describe how to fit the multi-hit model without observations recorded over time.

If there is no discussion of how long a study was run, then the duration of the study is subsumed under the rate parameter ζ, and we write

$$\text{Pr(response to dose } d_i) = P_i(h, \zeta) = \int_0^{\zeta} \frac{x^{h-1} e^{-x} dx}{(h-1)!} \equiv \sum_{j=h}^{\infty} \frac{(\zeta d_i)^j e^{-\zeta d_i}}{j!}$$

The usual format for quantal response data has k groups of individuals exposed to doses $\{d_i\}$, with r_i responding out of the n_i exposed to dose d_i, for $1 \leqslant i \leqslant k$. Standard maximum-likelihood estimation of (h, ζ) is then as follows: assumptions of independence produce the product binomial form for the likelihood:

$$L(h, \zeta) = \prod_{i=1}^{k} \binom{n_i}{r_i} P_i(h, \zeta)^{r_i} (1 - P_i(h, \zeta))^{n_i - r_i}$$

For simplicity we write $P_i = P_i(h, \zeta)$ and $p_i = r_i/n_i$. The log-likelihood has the form:

$$l = l(h, \zeta) = \text{constant} + \sum_{i=1}^{k} r_i \log P_i + \sum_{i=1}^{k} (n_i - r_i) \log(1 - P_i)$$

Iterative maximum-likelihood by the method-of-scoring is then standard, and requires the following terms: let η denote either h or ζ:

$$\frac{\partial l}{\partial \eta} = \sum_{i=1}^{k} \frac{\partial P_i}{\partial \eta} \frac{n_i(p_i - P_i)}{P_i(1 - P_i)}$$

$$E\left[\frac{\partial^2 l}{\partial \eta^2}\right] = -\sum_{i=1}^{k} n_i \frac{\left(\frac{\partial P_i}{\partial \eta}\right)^2}{P_i(1 - P_i)}$$

$$E\left[\frac{\partial^2 l}{\partial h \partial \zeta}\right] = -\sum_{i=1}^{k} \frac{n_i\left(\frac{\partial P_i}{\partial h} \frac{\partial P_i}{\partial \zeta}\right)}{P_i(1 - P_i)}$$

$$\frac{\partial P_i}{\partial \zeta} = \frac{d_i(\zeta d_i)^{h-1} e^{-\zeta d_i}}{(h-1)!}$$

$$\frac{\partial P_i}{\partial \theta_1} = \int_0^{\zeta d_i}\left[\frac{t^{h-1}e^{-t}\log t}{(h-1)!} - \frac{e^{-t}\psi(h)t^{h-1}}{(h-1)!}\right]dt$$

$$= \int_0^{\zeta} \frac{t^{h-1}e^{-t}\log t}{(h-1)!} - \psi(h)P_i$$

where $\psi(\ )$ denotes the di-gamma function.

In this work we could relax the restriction of h to positive integer values. However, analysis is clearly simplified with h restricted to positive integer values.

Because of the nature of the model we might expect the likelihood surface to exhibit a ridge for $h \propto \zeta$, and the estimators $\hat{h}$ and $\hat{\zeta}$ correspondingly to be highly positively correlated.

The same approach is adopted for data recorded over time, using the expressions given in section 5.4.

5.12 As the proportion of individuals surviving decreases (from 0.96 to 0.21), then the selected s value also decreases. As s decreases, more emphasis is placed on $\hat{p}_m$ in $\hat{\psi}(s)$, and so as $\hat{p}_m$ becomes more 'informative', the method responds by reducing the selected s value.

5.14 The local maxima of curves are particularly intriguing and require explanation. If we take the curve, $h = 8$ as an illustration then simulated responses out of 10 000 are shown below for $\zeta = 1$ and $\zeta = 1.6$.

Dose	$\zeta = 1$ Time			$\zeta = 1.6$ Time		
	0.25	1	2.0	0.25	1	2.0
1	0	0	17	0	5	160
2	0	11	515	0	188	3224
3	0	119	2668	0	1127	7323
4	0	499	5456	0	3129	9371
5	0	1337	7819	16	5466	9887

In both cases the first sampling time provides little information, but what is critical is the spread of responses for the endpoint. When $\zeta = 1.6$, this spread covers a wider range than when $\zeta = 1$, emphasizing the substantial amount of information present at the endpoint when $\zeta = 1.6$. We would expect this feature to be most pronounced for estimation of the scale parameter ζ, as indeed is seen to be the case.

5.15

(i) The likelihood is given by:

$$L \propto \prod_{i=1}^{k} \left[\left\{ \sum_{j=1}^{m} (F(t_j; d_i) - F(t_{j-1}; d_i))^{r_{ij}} \right\} \{1 - F(t_m; d_i)\}^{n_i - r_{i.}} \right]$$

A log-logistic survival distribution would suffice for the females, but not the males.

(ii) Deviance differences due to fitting $\beta_i = \beta$ are: $(40.85 - 37.75) = 3.10$ for males, and $(49.92 - 41.94) = 7.98$ for females, both on 3 d.f. For males we can take $\beta_i = $ constant. For females the result is just significant at the 5% level, but Pack and Morgan (1990a) argue in favour of setting $\beta_i = $ constant there too. Fitting model (b) with a common value for α_2 provides a test of parallelism, of the logit lines. The resulting test-statistic is 1.25, referred to χ_1^2. We conclude that the data are consistent with a common value of α_2, but that α_1 and the survival distributions are different for the two sexes.

(iii) The curves of Figure 5.1 possess two asymptotes. This is a consequence of the model being a mixture model, and of the time distribution being taken independent of dose. For example, a very low dose, say d_l, cannot produce an expected proportion responding greater than $a(d_l)$, however long the experiment runs. Also, whatever the dose, the time to response follows a fixed distribution. If t is fixed, increasing the dose will not, ultimately, result in a probability p of response, for any p. Figure 5.7 illustrates a clear difference between mixture and non-mixture models.

5.16 Model (d) does improve the fit to the endpoint mortalities, but the overall improvement in the model is not significant.

5.18 The model appears to be possibly compatible with a Weibull model ($\lambda = 0$). A mixture model may be appropriate, but we would need longer experiments at the low doses before this can be discussed further. Ostwald's equation does not hold.

5.19 If either d_i or t_j is fixed then in an odds ratio the component involving the fixed term cancels. For example,

$$\{p_{ij_1}/(1 - p_{ij_1})\}/\{p_{ij_2}/(1 - p_{ij_2})\} = (t_{j_1}/t_{j_2})^\lambda$$

irrespective of the value of d_i.

5.21 Cumulative mortalities appear to level out with respect to t and then increase again. Larvae which escape initial infection may be infected by viruses which are released by the explosion of dead larvae that die after initial infection.

Chapter 6

$$
\begin{aligned}
6.1 \quad \Pr(X_j = x_j | n_j) &= \binom{n_j}{x_j} \frac{B(\alpha + x_j, n_j + \beta - x_j)}{B(\alpha, \beta)} \\
&= \binom{n_j}{x_j} \frac{\Gamma(\alpha + x_j)\Gamma(n_j + \beta - x_j)}{\Gamma(n_j + \beta + \alpha)} \frac{\Gamma(\alpha + \beta)}{\Gamma(\alpha)\Gamma(\beta)} \\
&= \binom{n_j}{x_j} \frac{\displaystyle\prod_{i=1}^{x_j} (\alpha + x_j - i) \prod_{i=1}^{n_j - x_j} (n_j + \beta - x_j - i)}{\displaystyle\prod_{i=1}^{n_j} (n_j + \beta + \alpha - i)}
\end{aligned}
$$

(If $x_j = 0$ or $x_j = n_j$, the product term $\prod\limits_{i=1}^{0}$ is taken as unity.)

Divide numerator and denominator by $(\alpha + \beta)^{n_j}$ to give:

$$\binom{n_j}{x_j} \frac{\prod\limits_{i=1}^{x_j} \left\{ \left(\frac{\alpha}{\alpha+\beta} \right) + \left(\frac{x_j - i}{\alpha + \beta} \right) \right\} \prod\limits_{i=1}^{n_j - x_j} \left\{ \left(\frac{\beta}{\alpha+\beta} \right) + \left(\frac{n_j - x_j - i}{\alpha + \beta} \right) \right\}}{\prod\limits_{i=1}^{n_j} \left\{ 1 + \frac{(n_j - i)}{(\alpha + \beta)} \right\}}$$

$$= \binom{n_j}{x_j} \frac{\prod\limits_{r=0}^{x_j - 1} (\mu + r\theta) \prod\limits_{r=0}^{n_j - x_j - 1} (1 - \mu + r\theta)}{\prod\limits_{r=0}^{n_j - 1} (1 + r\theta)}$$

6.2   Use induction!

6.3   Let $Z_{jk} = 1$ if the kth foetus responds
          $= 0$ if it does not, $k = 1, 2, \ldots, n_j$.

Under the beta-binomial model

$$\Pr(Z_{jk} = 1) = \mu$$
$$\Pr(Z_{jk} = 0) = 1 - \mu$$

Hence

$$E(Z_{jk}) = \mu$$
$$\operatorname{Var}(Z_{jk} | n_j) = \mu(1 - \mu)$$

But

$$\operatorname{Var}(X_j | n_j) = n_j \mu(1 - \mu)\{1 + \zeta(n_j - 1)\}$$

and because $X_j \equiv \sum\limits_{k=1}^{n_j} Z_{jk}$, we have

$$\operatorname{Var}(X_j | n_j) = \sum\limits_{k=1}^{n_j} \operatorname{Var}(Z_{jk}) + \sum\limits_{k=1}^{n_j} \sum\limits_{\substack{k'=1 \\ k' \neq k}}^{n_j} \operatorname{Cov}(Z_{jk}, Z_{jk'} | n_j)$$

The first term is just $n_j \mu(1 - \mu)$, while the second (assuming the X_{jk} are equicorrelated) is $n_j(n_j - 1)\operatorname{Cov}(Z_{jk}, Z_{jk'} | n_j)$. Equating the two expressions for the variance we obtain

$$\operatorname{Cov}(Z_{jk}, Z_{jk'} | n_j) = \mu(1 - \mu)\rho$$

so that finally we have

$$\mathrm{Corr}(Z_{jk}, Z_{jk'}) = \rho$$

Knowledge that $Z_{jk} = 1$ provides information on p: we now know that $p > 0$, with clear implications for the distribution of $Z_{jk'}$.

6.4   See Exercise 6.15 and its solution.

6.5   We can solve directly for $\hat{\mu}$. Iteration is then needed for $\hat{\theta}$.

6.6   Helpful comments are given in Brooks (1984).

6.7   The model cannot account for the small number of litters with high mortality, but the qualitative fit is generally good. Using a Monte Carlo approach, Pack (1986a) found a binomial mixture provided a better fit to the Haseman and Soares data. Cf. Exercises 6.25 and 6.27. For the Aeschbacher *et al.* data set the beta-binomial and mixture models provide similar fits.

6.8   Williams (1988b) suggests using a permutation test, for example to test for a dose response, or for a monotonic regression. For further details, see Williams (1988d).

6.9   The data are too dispersed. Apart from the clear outlying litters with high mortality, the zero column is frequently under-estimated.

6.10   The beta-binomial log-likelihood is given by equation (6.4). For a score test of $\theta = 0$ we require (Appendix D), $\partial L/\partial\theta$ and the terms of the expected Hessian evaluated at $\theta = 0$, $\mu = \hat{\mu}$ (the maximum-likeihood estimate of μ when $\theta = 0$). The required derivatives are given in section 6.1.2. When $\theta = 0$ we obtain $\hat{\mu} = x./n.$, the ordinary binomial estimate. With $\theta = 0$ and $\mu = \hat{\mu}$ we obtain:

$$\frac{\partial l}{\partial \theta} = \frac{1}{2} \sum_{j=1}^{m} \left\{ \frac{x_j(x_j - 1)}{\hat{\mu}} + \frac{(n_j - x_j)(n_j - x_j - 1)}{(1 - \hat{\mu})} - n_j(n_j - 1) \right\}$$

When $\theta = 0$, we can show that $\mathrm{E}\left[\dfrac{\partial^2 l}{\partial\mu\partial\theta}\right] = 0$, and

$$\mathrm{E}\left[\frac{\partial^2 l}{\partial\theta^2}\right] = -\frac{1}{2} \sum_{j=1}^{m} n_j(n_j - 1)$$

The score statistic therefore simplifies to:

$$T = -\left(\frac{\partial l}{\partial \theta}\right)^2 \bigg/ E\left[\frac{\partial^2 l}{\partial \theta^2}\right]$$

Simplification of $\dfrac{\partial l}{\partial \theta}$ gives

$$\frac{\partial l}{\partial \theta} = \frac{1}{\hat{\mu}(1-\hat{\mu})}\left\{\sum_{j=1}^{m} x_j^2(1-\hat{\mu}) + \hat{\mu}\sum_{j=1}^{m}(n_j - x_j)^2\right\}$$

$$- \sum_{j=1}^{m} n_j(n_j - 1) - 2\sum_{j=1}^{m} n_j$$

resulting in

$$T^{1/2} = \frac{\displaystyle\sum_{j=1}^{m} \frac{(x_j - n_j\hat{\mu})^2}{\hat{\mu}(1-\hat{\mu})} - \sum_{j=1}^{m} n_j}{\left\{2\displaystyle\sum_{j=1}^{m} n_j(n_j - 1)\right\}^{1/2}}$$

which is the statistic Z of section 6.1.3.

6.12 $\theta < \mu \Leftrightarrow \alpha > 1$; $\theta < (1-\mu) \Leftrightarrow \beta > 1$.
Bell-shaped $\Leftrightarrow \alpha, \beta > 1$.

6.15 $E[\tilde{p}] = \mu$, hence set $\hat{\mu} = \tilde{p}$. Set

$$\left(\frac{x_i}{n_i} - \tilde{p}\right)^2 = \left(\frac{x_i}{n_i} - \mu + \mu - \tilde{p}\right)^2$$

$$= \left(\frac{x_i}{n_i} - \mu\right)^2 + (\mu - \tilde{p})^2 + 2\left(\frac{x_i}{n_i} - \mu\right)(\mu - \tilde{p})$$

Hence

$$S = \sum_i w_i\left(\frac{x_i}{n_i} - \mu\right)^2 + w.(\mu - \tilde{p})^2 + 2\sum_i w_i\left(\frac{x_i}{n_i} - \mu\right)(\mu - \tilde{p})$$

$$E[S] = \sum_i w_i \, \text{Var}\left(\frac{X}{n_i}\right) - w.E[(\mu - \tilde{p})^2]$$

Set $S = E[S]$ to give:

$$S = \sum_i \frac{w_i \mu(1-\mu)\{1 + \rho(n_i - 1)\}}{n_i} - w_. E[(\mu - \tilde{p})^2]$$

$$S = \mu(1-\mu)\sum_i \frac{w_i}{n_i} + \rho\mu(1-\mu)\sum_i w_i\left(1 - \frac{1}{n_i}\right)$$

$$- w_.^{-1}\sum_i \frac{w_i^2}{n_i}\mu(1-\mu)\{1 + \rho(n_i - 1)\}$$

leading directly to the desired expression.

$$\text{Var}(x_i/n_i) = \mu(1-\mu)\{1 + \rho(n-1)\}/n_i$$

from equation 6.8. Since $\mu(1-\mu)$ is just a multiplicative constant, the result follows.

Williams's iterative procedure works as follows.

Set $\hat{\rho} = 0$ and fit the binomial model. All $\{w_i\}$ are unity. Next, if X^2 is appreciably $> (n - p)$, from the first of equations 6.10, set

$$\hat{\rho} = \{X^2 - (n - p)\}/\sum_{i=1}^{k} \{(n_i - 1)(1 - v_i q_i)\}$$

Recalculate weights: $w_i = \{1 + \hat{\rho}(n_i - 1)\}^{-1}$, estimate $\boldsymbol{\beta}$ iteratively, and recalculate X^2. If $X^2 \approx (n - p)$ accept $\hat{\rho}$. If not, re-estimate $\hat{\rho}$ from the second of equations (6.10) – this explains why we do not cancel w_i and ρ in this equation. Estimate $\boldsymbol{\beta}$, and continue until $X^2 \approx (n - p)$.

For the link with Kleinman's procedure, see Pack (1986a).

6.18   From Appendix A, $V(\psi(p_i)) \approx \{\psi'(\mu)\}^2\sigma^2$, where $\mu = E[p_i]$ and $\sigma^2 = V(p_i)$. Here,

$$\psi(p_i) = \log p_i - \log(1 - p_i); \; p_i \sim B(\mu_i, \theta_i)$$
$$\psi'(p_i) = 1/\{p_i(1 - p_i)\}; \; E[p_i] = \mu_i$$
$$V(p_i) = \mu_i(1 - \mu_i)\rho_i$$

Hence

$$V(\text{logit}(p_i)) \approx \frac{\mu_i(1 - \mu_i)\rho_i}{\{\mu_i(1 - \mu_i)\}^2}$$

If $\rho_i = \rho^*\{\mu_i(1 - \mu_i)\}^{\zeta - 1}$, it is clear that when $\zeta = 2$, $V(\log(p_i)) \approx$ constant.

6.19   Fixing ζ greatly reduces the error of ρ^*. The corresponding estimates have high correlation. However, neither appears to be related to $(\hat{\alpha}, \hat{\beta})$ – see also Moore (1986). Similar comments are found in Chapter 4 – see also Appendix C.

6.21   (Morgan and Smith, 1992) The X^2 statistic consists of two components, one due to variation about the mean for a particular dose level, and the other due to discrepancy between the observed means at each dose level and their estimated values from the fitted model. Using both components to estimate the heterogeneity factor is misleading, as it overestimates this factor, and leaves no scope for judging the goodness-of-fit of the model.

6.22   Different models are not nested, but we may suppose all of the models to be embedded in some extended parametric family (cf. the work of Chapter 4), with a fitted deviance < 65.50. Such an argument would lead us to reject the Cauchit and complementary-log–log models.

6.23   A bi-modal form is possible, because of the quadratic function of x_j in the expression for $\Pr(X_j = x_j \mid n_j)$ for the correlated-binomial model in section 6.6.1. An example is given by Pack (1986a).

6.24   In the correlated–binomial model, the correlation between littermate responses is $\rho = \phi/\{p(1 - p)\}$, where ϕ is the covariance between the binary responses of any two foetuses. Hence there is a constraint: $\phi \leqslant p(1 - p)$. There are also data-dependent bounds on p.

$$E[X \mid n_i] = n_i\mu$$

$$V(X \mid n_i) = n_i\mu(1 - \mu)\left\{1 + (n_i - 1)\left(\rho + \frac{\theta}{1 + \theta}\right)\right\}$$

where (μ, θ) are the parameters of the beta distribution:

$$\mu = 1/(\alpha + \beta); \quad \theta = \alpha/(\alpha + \beta)$$

6.25   When we mix a binomial with a beta-binomial, the binomial may play the role of an outlier detector/description. This occurs for HS1. The beta-binomial parameters change little, but there is a substantial increase in log-likelihood $(-777.70 \rightarrow -771.16)$. Similar comments apply to HS3. For HS2 the fit is not of the outlier description type, though an alternative fit to the HS2 data was. When a mixture model incorporates an outlier detector element, the mixing parameter may be matched to the actual number of outlying litters observed. There are interesting comparisons to be drawn with the mixtures of binomials models. We can legitimately compare the mixtures of this exercise with the two-component binomial mixtures. When this is done, in all cases there is a clear significant improvement from fitting the model of this exercise. One approach to testing to see whether the beta-binomial fits are significantly improved by mixing with a binomial distribution is to examine profile log-likelihoods for the mixing parameter.

6.26   In this case we know that an extended model containing the BB and CB models exists. Its maximum-likelihood will be at least the larger maximum-likelihood from fitting the BB and CB models. If twice the modulus of the difference of the two maximum-likelihood values exceeds the appropriate percentile of the χ_1^2 distribution then we can infer that the model with the smaller maximum is inappropriate. Had we fitted the BCB model and performed a likelihood ratio test then we would reject the sub-model with the smaller maximum at a lower percentage level. For example, for the HSI data set, we can reject the CB model without having to fit the BCB model. We also see that the BB model is not significantly improved by the BCB model. The binomial model is clearly not appropriate. In fact, apart from the Aesbacher *et al.* example, the CB model is always rejected. In two cases (HS2 and HS3) the BCB model improves on the BB model. Cf. the conclusions in Exercise 4.39.

6.27   For HS1, from the beta-binomial simulations, the beta-binomial model appears satisfactory for the data. In view of the rank of the real data difference when simulating from the mixture model, we reject the mixture model in favour of the beta-binomial model for these data. For HS2, the situation is reversed. Cf. the conclusions of Exercise 6.25.

6.28   The model is under-dispersed if $\theta > 1$, and over-dispersed if $\theta < 1$.

$$\mathrm{E}[X] = np(p + q\zeta)^{n-1} f\left(\frac{p}{p+q\zeta}, \zeta, n-1\right) \Big/ f(p, \zeta, n)$$

where $p + q = 1$ and

$$E[X(X - 1)] = n(n - 1)p^2(p + q\zeta^2)^{n-2} f\left(\frac{p}{p + q\zeta^2}, \zeta, n - 2\right)\bigg/ f(p, \zeta, n)$$

6.29 $E[X] = np$

            $V(X) = np(1 - p)\{n - (n - 1)c\}$

where

$$c = \frac{1}{(1 + n\gamma)} \sum_{i=0}^{n-2} \frac{(n - 2)!}{(n - 2 - i)!}\left(\frac{\gamma}{1 + n\gamma}\right)^i$$

6.30 $E[X] = np$

            $V(X) = np(1 - p)\{1 + \rho(n - 1)2^{n-2}\}$

6.31 $\Pr(X = x \mid n) = \binom{n}{x}\dfrac{B(x + \alpha, n - x + \beta)M(x + \alpha, n - x + \beta, c)}{B(\alpha, \beta)M(\alpha, \beta, c)}$

        $E[X \mid n] = \left(\dfrac{\alpha}{\alpha + \beta}\right)\dfrac{M(\alpha + 1, \beta, c)}{M(\alpha, \beta, c)}$

6.32   Williams (1988b) used a Wald-test argument in favour of a clear effect of the compound: $(0.77 > 2 \times 0.31)$. However, this does not take account of the multiple-comparison feature of the tests carried out. See Williams (1988d) for relevant discussion.

6.33   The urn model suggested is a particular example of one in which s extra balls are added after each drawing. In this case the probability of k reds in n drawings is given by the Polya–Eggenberger distribution:

$$\binom{n}{k}\frac{r(r + s)\dots\{r + (k - 1)s\}b(b + s)\dots\{b + (n - k - 1)s\}}{(r + b)(r + b + s)\dots\{r + b + (n - 1)s\}}$$

The derivation is obvious, as is the reduction to the beta-binomial distribution when $s = 1$ – see equation (6.1), when (α_i, β_i) are integers – simply set $B(\alpha, \beta) = \Gamma(\alpha)\Gamma(\beta)/\Gamma(\alpha + \beta)$ etc.

6.35   The basic model of Ridout and Morgan (1991) assumed that the probability of misreporting, say $X = k$ for $k \neq 6$, would decrease as $|k - 6|$ increases. One approach is to set:

$$p_{k,6} = \begin{cases} \alpha^{|k-6|} & \text{for} \quad 1 \leqslant k \leqslant 11 \\ 0 & \text{otherwise} \end{cases}$$

$$p_{k,12} = \begin{cases} \beta^{|12-k|} & \text{for} \quad 7 \leqslant k \leqslant 17 \\ 0 & \text{otherwise} \end{cases}$$

for additional parameters, $0 < \alpha, \beta < 1$. Here p_{ij} is the probability of an interval i (resulting under the beta-geometric model) being misreported as j. How to fit this model in GENSTAT is described in Ridout (1992).

6.36 Retrospective data start from a pregnancy (cf. case-control studies), and so do not require the infertility feature of an added η. For the example, a Wald test would suggest we do need $\eta > 0$. If $E[p]$ is small, it can be difficult to distinguish $\eta \neq 0$, as the flexibility of the beta distribution will extend to accounting for sterile couples. In two examples considered by Maruani and Schwartz the mean fecundabilities of non-sterile individuals were 0.16 in each case.

6.37 The $\tilde{X}^2$ statistic of equation (6.24) is the Cochran/Armitage statistic, but with an adjustment to the control data. If the historical rates vary greatly then the historical data will make relatively little difference ($\hat{\alpha} + \hat{\beta}$ will be small). However, if historical rates vary little, ($\hat{\alpha} + \hat{\beta}$) will be large and the historical data may have an appreciable effect. The effect of historical data is greatest when the current control group incidence rate is unusually high or low.

Chapter 7

7.1 To illustrate the argument, let $p_i = r_i/n_i$, $1 \leqslant i \leqslant k$, in usual notation, and suppose $p_1 > p_2 > p_3 < p_4 < p_5 < ... < p_k$. If we start with the first pair of doses, we form a block with probabilities of response: $\left(\dfrac{r_1 + r_2}{n_1 + n_2} \right)$ at the first two doses that make up the block.

Note that

$$\frac{r_1}{n_1} > \frac{r_2}{n_2}$$

so that

$$r_1 n_2 > r_2 n_1$$
$$(r_1 n_1 + r_1 n_2) > (r_1 n_1 + r_2 n_1)$$

and

$$(r_2 n_2 + r_1 n_2) > (r_2 n_2 + r_2 n_1)$$

Hence,

$$r_1(n_1 + n_2) > n_1(r_1 + r_2)$$

and

$$n_2(r_1 + r_2) > r_2(n_1 + n_2)$$

so that

$$\frac{r_1}{n_1} > \frac{(r_1 + r_2)}{(n_1 + n_2)} > \frac{r_2}{n_2}$$

Hence the block proportion,

$$\left(\frac{r_1 + r_2}{n_1 + n_2}\right) > p_2 > p_3$$

by the original construction.

We therefore combine the first three doses to form a new block, with proportions, $(r_1 + r_2 + r_3)/(n_1 + n_2 + n_3)$ at each dose. Had we started instead with the second and third doses, after the formation of the first block, proportions would be:

$$\frac{r_1}{n_1}, \quad \left(\frac{r_2 + r_3}{n_2 + n_3}\right)$$

and

$$\left(\frac{r_2 + r_3}{n_2 + n_3}\right) < p_2 < p_1$$

by construction and so the same block of size 3 results as before. The generalization of this argument is immediate.

7.4   Expression (7.2) for the Spearman–Kärber estimate is:

$$\hat{\mu} = \frac{1}{2}\left\{\sum_{j=2}^{k-1} \hat{P}_j(d_{j-1} - d_{j+1})\right\} + (d_{k-1} + d_k)$$

For illustration, suppose

$$\frac{r_j}{n_j} > \frac{r_{j+1}}{n_{j+1}}$$

requiring the formation of a block with proportion

$$\left(\frac{r_j + r_{j+1}}{n_j + n_{j+1}}\right)$$

and suppose no further changes are required.

For an equal weight design, for some κ we can set

$$(d_{j+1} - d_{j-1}) = \kappa n_j$$

The relevant terms in $\hat{\mu}$ without pooling are:

$$\frac{r_j}{n_j}(d_{j-1} - d_{j+1}) + \frac{r_{j+1}}{n_{j+1}}(d_j - d_{j+2}) = -\kappa(r_j + r_{j+1})$$

With pooling, the relevant terms are:

$$\left(\frac{r_j + r_{j+1}}{n_j + n_{j+1}}\right)(d_{j-1} - d_{j+1} + d_j - d_{j+2}) = -\kappa(r_j + r_{j+1})$$

7.6
$$\sum_{i=1}^{s} r_i = \sum_{i=s}^{k} (n - r_i)$$

$$\sum_{i=1}^{s} r_i + \sum_{i=s}^{k} r_i = (k - s + 1)n$$

$$\left(\sum_{i=1}^{k} \hat{P}_i\right) + \hat{P}_s = k - s + 1$$

Hence

$$\hat{\theta}_R = d_k - \Delta(k - s) = d_k + \Delta\left(1 - \hat{P}_s - \sum_{i=1}^{k} \hat{P}_i\right).$$

This is of the same form as $\hat{\mu}$ in equation (7.3), but $\frac{1}{2}$ is replaced by $(1 - \hat{P}_n)$.

In many cases we shall have $\hat{P}_n \approx \frac{1}{2}$ (e.g. for a symmetric tolerance distribution). If no such s exists, use interpolation – see Miller (1973) for details.

7.7 $\tilde{P}_{s+1} - \tilde{P}_s \propto \left\{ \left(\sum_{i=1}^{s+1} r_i \right) - \sum_{i=s}^{k} (n - r_i) - \left(\sum_{i=1}^{s} r_i \right) \sum_{i=s+1}^{k} (n - r_i) \right\}$

$\qquad = \left\{ r_{s+1} \sum_{i=s}^{k} (n - r_i) + \left(\sum_{i=1}^{s} r_i \right)(n - r_s) \right\} \geqslant 0$

The constant of proportionality is positive, so the sequence $\{\tilde{P}_s\}$ is non-decreasing.

$$\hat{P}_s = 0.5$$

$$\sum_{i=1}^{s} r_i = \frac{1}{2} \left\{ \sum_{i=1}^{s} r_i + \sum_{i=s}^{k} (n - r_i) \right\}$$

and so

$$\sum_{i=1}^{s} r_i = \sum_{i=s}^{k} (n - r_i)$$

i.e. Dragsted-Behrens and Reed-Muench estimates coincide. If no such s exists, $\hat{\theta}_D$ follows from linear interpolation between the two doses with values $\hat{P}_s$ which span 0.5. In general we can expect $\hat{\theta}_D$ and $\hat{\theta}_R$ to be similar. Asymptotic theory is equivalent.

7.8   Data set 2 has just 4 doses, and so is difficult to discuss in this context. Data set 3 has 9 doses, but only 6 subjects per dose. Deviances for the various fitted models (fitted to doses) are (Goedhart, 1985):

Aranda–Ordaz:	6.02
complementary log-log:	8.46
logit:	6.37
cubic logistic:	6.34
Box–Cox:	5.99

Degrees of freedom $= 7$. We have $\lambda = -0.09$, which suggests taking logarithms before fitting a symmetric model. For fitting to untransformed doses, we would expect an asymmetric model to do better than a symmetric one. This is borne out by the above results. However, differences are small, due to the overall small sample size. The particular skewness of the complementary log–log model is not appropriate for the data.

7.10   For both logit and probit models only the highest age-group satisfies the criterion for trimming. Below we give the results from maximum-likelihood fitting. The effect of omitting the highest age-group is negligible. Cf. discussion of Exercise 1.13. For the probit model the deviance reduction is from 22.89 to 22.85. For the logit model the deviance reduction is from 26.70 to 25.47. Degrees of freedom drop from 23 to 22. In the latter case, for illustration, parameter estimates change from:

$$\begin{cases} \hat{\alpha} = -21.23\,(0.771) \\ \hat{\beta} = 1.632\,(0.059) \end{cases} \quad \text{to:} \quad \begin{cases} \hat{\alpha} = -21.10\,(0.781) \\ \hat{\beta} = 1.622\,(0.060) \end{cases}$$

7.11   We obtain e as the ratio:

$$e = \frac{1}{I(\hat{\mu})}\,\frac{1}{V(\hat{\mu})}$$

using the expressions of (7.5) and (7.6), to give

$$e^{-1} = \left(\int_{-\infty}^{\infty} F(x)\{1 - F(x)\}dx \right)\left(\int_{-\infty}^{\infty} f^2(x)[F(x)\{1 - F(x)\}]^{-1}dx \right)$$

$$= E[F(X)\{1 - F(X)\}/f(X)]E\left[\frac{f(X)}{F(X)\{1 - F(X)\}} \right]$$

where the expectations are taken w.r.t the r.v. X, with probability density function $f(x)$.

Now, by the Cauchy–Schwartz inequality,

$$E\left[\frac{f(X)}{F(X)\{1 - F(X)\}} \right] \geq 1/E\left[\frac{F(X)\{1 - F(X)\}}{f(X)} \right]$$

Hence $e^{-1} \geq 1$. We have equality if and only if

$$\frac{f(x)}{F(x)\{1 - F(x)\}}$$

is constant. Integration gives $F(x)$ as logistic. It is easily verified that if $F(x)$ is logistic, $e = 1$.

7.12 $\displaystyle\int_{0}^{\infty} F(x)\{1 - F(x)\}\,dx$ is finite, but

$$\int_{0}^{\infty} \frac{f^2(x)}{F(x)\{1 - F(x)\}}\,dx \propto \int_{0}^{\infty} \frac{e^{-(x/\mu)}}{1 - e^{-(x/u)}}\,dx \propto \left[\log(1 - e^{-(x/\mu)}) \right]_{0}^{\infty}$$

so that $e = 0$.

7.13   In the normal case we require

$$\int_{-\infty}^{\infty} \frac{\phi^2(x)}{\Phi(x)\{1 - \Phi(x)\}} \, dx \text{ and } \int_{-\infty}^{\infty} \Phi(x)\{1 - \Phi(x)\} \, dx$$

Numerical integration (Govindarajulu, 1988, p. 50) gives

$$\int_{-\infty}^{\infty} \frac{\phi^2(x)}{\{1 - \Phi(x)\}} \, dx = \int_{-\infty}^{\infty} \frac{\phi^2(x)}{\Phi(x)} \, dx = 0.903$$

Then integration by parts provides:

$$\int_{-\infty}^{\infty} \Phi(x)\{1 - \Phi(x)\} \, dx = 2 \int_{-\infty}^{\infty} x\phi(x)\Phi(x) \, dx$$

$$= -2 \int_{-\infty}^{\infty} \Phi(x)\phi'(x) \, dx$$

$$= 2 \int_{-\infty}^{\infty} \phi^2(x) \, dx = 1/\sqrt{\pi}$$

Finally, $e = (2 \times 0.903/\sqrt{\pi})^{-1} = 0.9814$.

7.15   Use the Cauchy–Schwartz inequality.

7.16   The mean square errors for the Cauchy and Slash cases are not as poor as one might have expected from the results of Table 7.4. The LS estimator performs well, dominating the HL estimator. SK5 dominates SK10, as does SK. Conclusions generally parallel those from Table 7.5.

7.17   All models span the full range of responses. With the exception of model V, partial response can be expected to be mainly confined to doses near the ED_{50}.

7.19   For the Spearman–Kärber estimator, $J(u) \equiv 1$, and the result follows. Outliers have the most influence: Spearman–Kärber is not robust.
   For the α-trimmed Spearman–Kärber estimator,

$$J(u) = (1 - 2\alpha)^{-1} \quad \text{for } u \in [\alpha, 1 - \alpha]$$

The α-trimmed Spearman–Kärber estimator is influence curve robust – see James and James (1983) for discussion of violation of regularity conditions.

7.20   The M-estimator based on ψ will be influence curve robust if and only if $\lim\limits_{x \to -\infty} \psi'(x) = \lim\limits_{x \to \infty} \psi'(x) = 0$.

The Tukey biweight estimator has:

$$\psi(x) = \begin{cases} x(1 - x^2)^2 & \text{if } |x| \leqslant 1 \\ 0 & \text{if } |x| > 1 \end{cases}$$

and so is clearly influence curve robust.

7.21   The influence curve given is the limit of sensitivity curves. Since

$$\lim_{d \to -\infty} IC_{L,F}(d, 1) = -1 \text{ and } \lim_{d \to \infty} IC_{L,F}(d, 0) = 1,$$

$\hat{\theta}_L$ is not influence curve robust. The same result for $\hat{\theta}_{SK}$ of Exercise 7.19 is due to the close relationship between $\hat{\theta}_L$ and $\hat{\theta}_{SK}$.

7.23   Here we just consider (i): bell-shaped:
for each pair of values of s and t, for

$$3 \leqslant s \leqslant (k-1); s+1 \leqslant t \leqslant k; \text{ or } s = t = 3, \text{ or } s = t = k,$$

$$P_1 \geqslant 0,$$
$$\Delta P_2 \geqslant 0,$$
$$\Delta^2 P_3 \geqslant 0, \text{ unless } s = 3,$$

then

$$\Delta^2 P_3 \leqslant 0 \text{ if } t = 3, \text{ or}$$
$$\Delta P_3 \geqslant 0 \text{ if } t \geqslant 4,$$
$$\Delta^3 P_i \geqslant 0 \text{ for } 4 \leqslant i \leqslant 3$$
$$\leqslant 0 \text{ for } s+1 \leqslant i \leqslant t$$
$$\geqslant 0 \text{ for } t+1 \leqslant i \leqslant k$$
$$\Delta^2 P_k \leqslant 0, \text{ unless } t = k, \text{ then}$$
$$\Delta^2 P_k \geqslant 0 \text{ if } s = k, \text{ or}$$
$$\Delta P_{k-1} \geqslant 0 \text{ if } s \leqslant k - 1$$
$$\Delta P_k \geqslant 0, \text{ and}$$
$$P_k \leqslant 1$$

Note that if $s = 2$, there are no constraints of the form $\Delta^2 P_i \geqslant 0$, while if $s = k$, there are no constraints of the form $\Delta^2 P_i \leqslant 0$.

7.24 Under the normal model, all observations contribute to the estimator, which is not robust as a result. Stripping away assumptions increases robustness, and reduces the influence of responses to dose levels away from the ED_{50}.

7.25 More information assumed known will reduce the confidence interval width. Glasbey (1987) gives the following illustration.

If the ED_{50} has a normal prior distribution, with mean 1.60 and standard deviation 0.05, then the posterior is normal with mean 1.59 and standard deviation 0.036. A 95% confidence interval for the ED_{50} now has width 0.14. Cf. the non-parametric Bayesian approach of Exercise 3.35.

7.26 $f(x) = \dfrac{dF(x)}{dx}$. Assume $f'(x)$ exists. The point of inflexion, occurring at $x = z$, say, has $f'(z) = 0$. From the definition of sigmoid given in section 7.6.1, for $x < z$, $f'(x) > 0$, and for $x > z$, $f'(x) < 0$. The value $x = z$ therefore corresponds to the unique mode of $f(x)$.

Chapter 8

8.1 Let $P = P(d_1)$ and $Q = P(d_2)$.

$$|\mathbf{V}^{-1}| = \frac{(PQ)^2}{4}(d_1 - d_2)^2 \quad \text{Substitute,}$$

$$d_1 = \frac{1}{\beta}\{\log(P/Q) - \alpha\}$$

$$d_2 = \frac{1}{\beta}\{\log(Q/P) - \alpha\} \quad \text{to give}$$

$$|\mathbf{V}^{-1}| = \left\{\frac{PQ\log(P/Q)}{\beta}\right\}^2$$

$$\frac{d|\mathbf{V}^{-1}|}{dP} = \frac{2PQ\log(P/Q)}{\beta^2}\{1 + (1 - 2P)\log(P/Q)\}$$

(maximizing $|\mathbf{V}^{-1}|$ is equivalent to minimizing $|\mathbf{V}|$).

8.2 $$\mathbf{V} = \frac{2}{PQ(d_1 - d_2)^2}\begin{bmatrix} d_1^2 + d_2^2 & -(d_1 + d_2) \\ -(d_1 + d_2) & 2 \end{bmatrix}$$

$$
= \frac{\beta^2}{2PQ\{\log(P/Q)\}^2}\begin{bmatrix} d_1^2 + d_2^2, & -(d_1 + d_2) \\ -(d_1 + d_2), & 2 \end{bmatrix}
$$

$$
\mathrm{tr}(\mathbf{V}) = \frac{\beta^2}{2PQ\{\log(P/Q)\}^2}[d_1^2 + d_2^2 + 2] = \frac{1}{PQ}\left[1 + \frac{\alpha^2 + \beta^2}{\{\log(P/Q)\}^2}\right]
$$

Setting $d/dP = 0$ gives:

$$
(P - Q) - \frac{(\alpha^2 + \beta^2)\{2 - \log(P/Q)(P - Q)\}}{\{\log(P/Q)\}^3} = 0
$$

8.3   The eigenvalues of $\mathbf{V}$ are given by:

$$
\lambda = \frac{\alpha^2 + \beta^2 + \{\log(P/Q)\}^2 \pm ([\alpha^2 + \beta^2 + \{\log(P/Q)\}^2]^2 - \{2\beta\log(P/Q)\}^2)^{1/2}}{2PQ\{\log(P/Q)\}^2}
$$

Rosenberger and Kalish (1981) used numerical optimization to minimize the larger eigenvalue.

8.4   Application of the delta-method (Appendix A) gives:

$$
\mathrm{V}(\hat{P}(d_0)) = \frac{\left[\left\{\log\left(\frac{P_0 Q}{Q_0 P}\right)\right\}^2 + \left\{\log\left(\frac{P_0 P}{Q_0 Q}\right)\right\}^2\right]P_0^2 Q_0^2}{nPQ\left\{\log\left(\frac{P^2}{Q^2}\right)\right\}^2}
$$

where n is the number of individuals per dose, $P_0 = P(d_0)$, $P_0 + Q_0 = 1$.

Rosenberger and Kalish (1981) found the optimum design to be $P = 0.768$, from using numerical analysis.

As with D-optimality, we are here just dealing with a function of P; (α, β) are not otherwise involved. This is not the case for A and E optimal designs, which are therefore less desirable. Rosenberger and Kalish (1981) define robustness efficiencies for the D and G designs. Neither is ruled out by evaluating the efficiency values. A good compromise design has $P = 0.8$.

8.5   If $\theta_I = \theta_A$, small inaccuracies in β_I still produce high efficiency values. Overestimation of β is less of a problem than underestimation. However, the evidence for the case of a moderate departure from

$\theta_I = \theta_A$, which is perhaps the likely situation, suggests it is better to have $\beta_I < \beta_A$.

8.6   We know, from section 2.2, that

$$-E\left[\frac{\partial^2 l}{\partial \zeta^2}\right] = -\sum_{i=1}^{k} \frac{n_i \left(\frac{\partial P_i}{\partial \zeta}\right)^2}{P_i(1 - P_i)}$$

and

$$-E\left[\frac{\partial^2 l}{\partial \alpha \partial \beta}\right] = -\sum_{i=1}^{k} \frac{n_i \left(\frac{\partial P_i}{\partial \alpha} \frac{\partial P_i}{\partial \beta}\right)}{P_i(1 - P_i)}$$

Also,

$$\frac{\partial P_i}{\partial \alpha} = P_i(1 - P_i); \frac{\partial P_i}{\partial \beta} = d_i P_i(1 - P_i)$$

Hence Fisher (expected) information matrix is:

$$I(\alpha, \beta) = \begin{bmatrix} \sum_i n_i w_i & \sum_i n_i w_i d_i \\ \sum_i n_i w_i d_i & \sum_i n_i w_i d_i^2 \end{bmatrix}$$

leading to:

$$\beta^2 |I(\alpha, \beta)| = \left\{ \sum_{i=1}^{k} n_i w_i \sum_{i=1}^{k} n_i w_i (\beta d_i)^2 - \left(\sum_{i=1}^{k} n_i w_i \beta d_i \right)^2 \right\}$$

If we replace βd_i by $\alpha + \beta d_i$ we introduce the terms:

$$\alpha^2 \sum_i n_i w_i \sum_i n_i w_i + 2\alpha\beta \sum_i n_i w_i \sum_i n_i w_i d_i$$

$$- \alpha^2 \sum_i n_i w_i \sum_i n_i w_i - 2\alpha\beta \sum_i n_i w_i \sum_i n_i w_i d_i = 0$$

Hence we can write

$$\beta^2 |I(\alpha, \beta)| = \left\{ \sum_{i=1}^{m} w_i \sum_{j=1}^{m} w_j \eta_j^2 - \left(\sum_{i=1}^{m} w_i \eta_i \right)^2 \right\}$$

8.7 An exaggerated example is as follows: if the maximum likelihood estimates $(\hat{\theta}, \hat{\beta})$, available after the second stage, turn out so that both $\hat{\beta}(d_1 - \hat{\theta})$ and $\hat{\beta}(d_2 - \hat{\theta})$ are close to 1.5434, the choice of (d_3, d_4) could be made so that $\hat{\beta}(d_3 - \hat{\theta})$, $\hat{\beta}(d_4 - \hat{\theta})$ are close to -1.5434. An appropriate algorithm is given by Minkin (1987).

8.9 Let $w = pq$. From the solution to Exercise 8.6, we can write the Fisher (expected) information matrix as:

$$\mathbf{I}(\alpha, \beta) = \frac{mw}{2} \begin{bmatrix} 2, & d_1 + d_2 \\ d_1 + d_2, & d_1^2 + d_2^2 \end{bmatrix} = \frac{mw}{2} \mathbf{A}, \text{ say.}$$

Note that $d_1 + d_2 = 2\theta$. $|A| = (d_1 - d_2)^2$

$$\mathbf{A}^{-1} = (d_1 - d_2)^{-2} \begin{bmatrix} d_1^2 + d_2^2, & -(d_1 + d_2) \\ -(d_1 + d_2), & 2 \end{bmatrix}$$

From equation (2.9), we estimate $V(\hat{\theta})$ by:

$$\frac{2}{\beta^2 mw(d_1 - d_2)^2} \{ d_1^2 + d_2^2 - 2\theta(d_1 + d_2) + 2\theta^2 \}$$

Substitute $2\theta = (d_1 + d_2)$, to give

$$V(\hat{\theta}) = \frac{1}{\beta^2 mpq}$$

Hence the expression of the exercise is, to order $1/m$, the form given by equation (2.9). It is also the form which results from the direct parameterization, in terms of (β, θ), as seen in Exercise 2.41. Convergence is slow. As $m \to \infty$, the individuals should be placed progressively closer to the (unknown) ED_{50}.

8.10 For $m \leqslant 100$, the D-optimal design performs well for both θ and β separately. As $m \to \infty$, the efficiency shown of the ED_{50} design decreases as it moves the design points together. (Cf. the two-point designs of Exercise 8.9.) The suggested three-point design is close to the D-optimal design of $m \leqslant 100$, and has the advantage of simplicity – see Kalish (1990) for more details.

8.11 Heights for which $r_i/n > 0.5$ are likely to be far higher than the ED_{10}. Kershaw and McRae (1985) recommend setting $0.1 < n_2/n < 0.5$ and $n_1/n < 0.1$. Small n corresponds to larger experiments.

Large n will result in fewer batches to be dropped, and an experiment which is susceptible to a poor start. Kershaw and McRae (1985) justify their choice of $n = 8$ or 10 and $\Delta \approx 1$.

8.12   Introduce weight or log weight as a second covariate.

8.13   The modification will speed up a move away from a poor initial start. Asymptotic results are unchanged. This design was recommended by Davis (1971).

8.14   Only when there is a change of direction (e.g. $d_{i-1} > d_{i-2}$ but $d_i < d_{i-1}$) does the divisor of c increase.

8.16   Let κ be the cost of a non-response. The expected cost is then

$$\sum_{i=1}^{k} (n_i[1 - F\{\beta(d_i - \theta)\}]\kappa + n_i F\{\beta(d_i - \theta)\}(D + \kappa))$$

$$= \sum_{i=1}^{k} n_i[\kappa + D F\{\beta(d_i - \theta)\}]$$

Expression follows if we measure D in units of κ.

   We divide the second term in equation (8.13) by $[1 + DF\{\beta^{(k)}(d_{k+1} - \hat{\theta}^{(k)})\}]$.

8.20   Kalish (1990) describes both fixed and sequential procedures.

8.21   Wu (1985) argues that minimum logit chi-square is inappropriate due to the small number of replications at each dose level, as a result of the sequential generation of doses.

8.22   Kalish (1990) uses a second-order approximation in calculating variances.

8.23   Heuristically, $\pi_i\{1 - F(d_i)\}$ denotes the probability of movement from d_i to d_{i+1}, which, in equilibrium, if it exists, must be balanced by the probability of movement from d_{i+1} to d_i, viz. $\pi_{i+1} F(d_{i+1})$.

8.24   For example, to reach $(d_i, 1)$, two steps before the dose must be d_i, followed by a step down and then a step up. Let π_{ij} denote the equilibrium probability of being in state (d_i, j). We can now express the $\{\pi_{ij}\}$ in terms of the $\{\pi_i\}$. For example:

$$\pi_{i1} = \pi_i F(d_i)\{1 - F(d_{i-1})\}, \text{ etc.}$$

8.25 Let height h_i, when last height was h_{i-1}, be $(i, 0)$.
Let height h_i, when last height was h_{i+1}, be $(i, 1)$.
Let p_i denote the probability that at height h_i the number of positive responses $\leqslant n_2$. Let r_i denote the probability that at height h_i the number of positive responses $\geqslant n_1$. The equilibrium probabilities, $\pi(i, 0)$, $\pi(i, 1)$, etc. then can be seen to satisfy:

$$\pi(i, 0) = \pi(i - 1, 0)p_{i-1} + \pi(i - 1, 1)(1 - r_{i-1})$$
$$\pi(i, 1) = \pi(i + 1, 0)(1 - p_{i+1}) + \pi(i + 1, 1)r_{i+1}$$

See Kershaw and McRae (1985) for comments on solution.

8.27 We see that $\hat{\beta}_2 > \hat{\beta}_1$, suggesting atropine has a larger effect that pralidioxine chloride. Squared terms, with mainly negative coefficients, correspond to toxicity. Pralidioxine chloride may be the more toxic.

8.28 The likelihood, L, is given by:

$$L = \text{constant} + r_{11}\log\{\Phi(c)\} + r_{12}\log\{1 - \Phi(c)\} + r_{21}\log\{\Phi(c - d)\}$$
$$+ r_{22}\log\{1 - \Phi(c - d)\}$$

$$\frac{\partial L}{\partial d} = \frac{\phi(c - d)\{r_{2.}\Phi(c - d) - r_{21}\}}{\Phi(c - d)\{1 - \Phi(c - d)\}}$$

$$\frac{\partial L}{\partial c} = \frac{\phi(c)\{r_{11} - r_{1.}\Phi(c)\}}{\Phi(c)\{1 - \Phi(c)\}} + \frac{\phi(c - d)\{r_{21} - r_{2.}\Phi(c - d)\}}{\Phi(c - d)\{1 - \Phi(c - d)\}}$$

After further differentiation and taking expectations, we obtain the Fisher expected information matrix $\mathbf{I}$, with inverse,

$$\mathbf{I}^{-1} = \begin{bmatrix} \dfrac{1}{r_{1.}w_1} + \dfrac{1}{r_{2.}w_2}, & \dfrac{1}{r_{1.}w_1} \\[2mm] \dfrac{1}{r_{1.}w_1}, & \dfrac{1}{r_{1.}w_1} \end{bmatrix}$$

where

$$w_1 = \frac{\phi^2(c)}{\Phi(c)\{1 - \Phi(c)\}}$$

$$w_2 = \frac{\phi^2(c - d)}{\Phi(c - d)\{1 - \Phi(c - d)\}}$$

(this is from regarding d as the first parameter). The result for $V(\hat{d})$ follows.

8.30   From Exercise 8.9, to order m^{-1},

$$V(\hat{\theta}) = \frac{1}{mpq}$$

Hence the Fisher expected information is $mpq \propto f(d - \theta)$. In general, the expected Fisher information is:

$$I(d.) = mf^2(d - \theta)/[F(d - \theta)\{1 - F(d - \theta)\}]$$

(See, e.g., solutions to Exercise 8.6.)
By symmetry,

$$F(u)\{1 - F(u)\} = \tfrac{1}{4} - \{F(u) - \tfrac{1}{2}\}^2$$

Set $\theta = 0$. For all $u > 0$,

$$I(u) \leqslant I(0) \text{ if and only if } f^2(u) \leqslant 4f^2(0)F(u)\{1 - F(u)\}$$

Now, given

$$\Phi(d) \leqslant \tfrac{1}{2}\{1 + (1 - e^{-d^2})^{1/2}\}$$
$$\Phi(d) - \tfrac{1}{2} \leqslant \tfrac{1}{2}(1 - e^{-d^2})^{1/2}$$
$$\{\Phi(d) - \tfrac{1}{2}\}^2 \leqslant \tfrac{1}{4}(1 - e^{-d^2})$$

But $\phi(u)/\phi(0) = e^{-u^2/2}$ and the result follows.

8.31   Let $P(d) = 1 - e^{-\beta d}$.
Suppose we have n subjects at a single dose d. From the solution to Exercise 8.30, the expected Fisher information for $v = \log \beta$ is:

$$I(v) = n(\beta d)^2/(e^{\beta d} - 1)$$

This is maximized when $\beta d \approx 1.59$.
Sequential approaches are now clear. Details are in Abdelbasit and Plackett (1983).

References

Abbott, W. S. (1925) A method of computing the effectiveness of an insecticide. *J. Econ. Entomol.*, **18**, 265–267.

Abdelbasit, K. M. and Plackett, R. L. (1982) Experimental design for joint action. *Biometrics*, **38**, 1, 171–179.

Abdelbasit, K. M. and Plackett, R. L. (1983) Experimental design for binary data. *J. Amer. Statist. Assoc.*, **78**, 90–98.

Adby, P. R. and Dempster, M. A. H. (1974) *Introduction to Optimization Methods*, Chapman and Hall, London.

Adena, M. A. and Wilson, S. R. (1982) *Generalized Linear Models in Epidemiological Research*, INSTAT, Sydney.

Aeschbacher, H. U., Vuataz, L., Sotek, J. and Stalder, R. (1977) The use of the beta-binomial distribution in dominant–lethal testing for 'weak mutagenic activity' (Part I). *Mutation Research*, **44**, 369–390.

Afifi, A. A. and Clark, V. (1990) *Computer-aided Multivariate Analysis* (2nd edition), Van Nostrand Reinhold Co., New York.

Agha, M. and Ibrahim, M. T. (1984) Maximum likelihood estimation of mixtures of distributions. *Applied Statistics*, **33**, 327–332.

Agresti, A. (1984) *Analysis of Ordinal Categorical Data*, Wiley, New York.

Aitchison, J. and Silvey, S. D. (1957) The generalization of probit analysis to the case of multiple responses. *Biometrika*, **44**, 131–140.

Aitkin, M., Anderson, D., Francis, B. and Hinde, J. (1989) *Statistical Modelling in GLIM*, Clarendon Press, Oxford.

Aitkin, M. and Clayton, D. (1980) The fitting of exponential, Weibull and extreme value distributions to complex censored survival data using GLIM. *Appl. Statist.*, **29**, 156–163.

Aitkin, M. and Rubin, D. B. (1985) Estimation and hypothesis in finite mixture models. *J. R. Statist. Soc. B*, **47**, 67–75.

Albert, A. and Anderson, J. A. (1984) On the existence of maximum likelihood estimates in logistic regression models. *Biometrika*, **71**, 1–10.

Aldrich, J. H. and Nelson, F. D. (1984) *Linear Probability, Logit and Probit Models*, Sage publications, Beverley Hills.

Alling, D. W. (1971) Estimation of hit number. *Biometrics*, **27**, 3, 605–614.

Altham, P. M. E. (1978) Two generalizations of the binomial distribution. *Appl. Statist.*, **27**, 162–167.

Amemiya, J. (1980) The n^{-2}-order mean squared errors of the maximum likelihood and the minimum logit chi-square estimator. *Annals of Statistics*, **8**, 488–505.

Amemiya, J. (1981) Qualitative response models: a survey. *Journal of Economic Literature*, **19**, 1483–1536.

Amemiya, J. (1984) Tobit models: a survey. *J. Econometrics*, **24**, 3–61.

American Public Health Association, American Water Works Association, and Water Pollution Control Federation (1981) Standard Methods for the Examination of Water and Wastewater (15th edition), *APHA*, Washington, DC, 644.

Ames, B. N., McCann, J. and Yamasaki, E. (1975) Methods for detecting carcinogens and mutagens with the Salmonella/microsome mutagenicity test. *Mutation Res.*, **31**, 347–364.

Anbar, D. (1978) A stochastic Newton–Raphson method. *J. Stat. Plan. Inf.*, **2**, 153–163.

Anbar, D. (1984) Cochran's contribution to sequential estimation of median lethal dose (LD_{50}) in *W. G. Cochran's Impact on Statistics* (eds Rao, P. S. R. S. and Sedransk, J.), Wiley, New York, pp. 91–100.

Anderson, D. (1985a) Experiments in wool prickle II. *CSIRO DMS Consulting Report No. VT85/28*.

Anderson, D. (1985b) Behaviour of a sequential estimation scheme for estimating thresholds in wool prickle. *CSIRO DMS Consulting Report No. VT85/9*.

Anderson, D. A. (1988) Some models for overdispersed binomial data. *The Australian Journal of Statistics*, **30**, 2, 125–148.

Anderson, D. A. and Aitkin, M. A. (1985) Variance component models with binary response: interviewer variability. *J. R. Statist. Soc. B*, **47**, 203–210.

Anderson, J. A. and Pemberton, J. D. (1985) The grouped continuous model for multivariate ordered categorised variables and covariate adjustment. *Biometrics*, **41**, 875–885.

Anderson, J. A. and Richardson, S. C. (1979) Logistic discrimination and bias correction in maximum likelihood estimation. *Technometrics*, **21**, 71–78.

Anderson, T. W., McCarthy, P. J. and Tukey, J. W. (1946) 'Staircase' method of sensitivity testing. *Naval Ordnance Report No. 65–46.* Statistical Research Group, Princeton.

Andrews, D. F., Bickel, P. J., Hampel, F. R., Huber, P. J., Rogers, W. H., and Tukey, J. W. (1972) *Robust Estimates of Location-Survey and Advances.* Princeton University Press, Princeton, N.J.

Anscombe, F. J. (1949) Note on a problem in probit analysis. *Ann. Appl. Biol.*, **36**, 203–205.

Anscombe, F. J. (1956) On estimating binomial response relations. *Biometrika*, **43**, 461–464.

Aranda-Ordaz, F. J. (1981) On two families of transformations to additivity for binary response data. *Biometrika*, **68**, 2, 357–364. (Correction: *Biometrika*, **70**, 1, 303)

Aranda-Ordaz, F. J. (1983) An extension of the proportional-hazards model for grouped data. *Biometrics*, **39**, 1, 109–118.

Aranda-Ordaz, F. J. (1985a) Fitting a generalised symmetric model to binary response data. *GLIM Newsletter* No. **9**, 27–29.

Aranda-Ordaz, F. J. (1985b) Asymmetric generalisations of logistic models with GLIM. *GLIM Newsletter* No. **9**, 30–32.

Armitage, P. (1955) Tests for linear trends in proportions and frequencies. *Biometrics*, **11**, 375–386.

Armitage, P. (1959) Host variability in dilution experiments. *Biometrics*, **15**, 1–9.

Armitage, P. (1982) The assessment of low-dose carcinogenicity. Proceedings of: 'Current topics in Biostatistics and Epidemiology', A memorial symposium in honour of Jerome Cornfield. *Biometrics*, **38**, Supplement, 119–129.

Armitage, P., Bennett, B. M. and Finney, D. J. (1976) Point and interval estimation in the combination of bioassay results, *Journal of Hygiene*, **76**, 147–162.

Armitage, P. and Doll, R. (1962) Stochastic models for carcinogenesis. *Proceedings of the 4th Berkeley Symposium*, pp. 19–38.

Ashford, J. R. (1981) General models for the joint action of mixtures of drugs. *Biometrics*, **37**, 3, 457–474.

Ashford, J. R. and Smith, C. S. (1965) An analysis of quantal response data in which the measurement of response is subject to error. *Biometrics*, **21**, 811–825.

Ashford, J. R. and Sowden, R. R. (1970) Multi-variate probit analysis. *Biometrics*, **26**, 3, 535–546.

Ashton, W. D. (1972) *The Logit Transformation With Special Reference To Its Uses In Bioassay*. Griffin, London.

Atkinson, A. C. (1972) A test of the linear logistic and Bradley–Terry models. *Biometrika*, **59**, 1, 37–42.

Atkinson, A. C. (1982) Regression diagnostics, transformations and constructed variables. *J. R. Statist. Soc., B*, **44**, 1, 1–36.

Awa, A. A., Sofuni, T., Honda, T., Itoh, M., Neriishi, S. and Otake, M. (1978) Relationship between the radiation dose and chromosome aberrations in atomic bomb survivors of Hiroshima and Nagasaki. *J. Radiation Res.*, **19**, 126–140.

Ayer, M., Brunk, H. D., Ewing, G. M., Reid, W. T. and Silverman, E. (1955) An empirical distribution function for sampling with incomplete information. *Annals of Mathematical Statistics*, **26**, 641–647.

Bahadur, R. R. (1961) A representation of the joint distribution of responses to n dichotomous items, in *Studies in Item Analysis and Prediction* (ed Solomon, H.), Stanford University Press.

Bailey, N. T. J. (1964) *The Elements of Stochastic Processes With Applications to the Natural Sciences*. Wiley, New York.

Bailey, R. A. and Gower, J. C. (1986) Discussion of paper by Racine *et al. Appl. Statist.*, **35**, 2, 133.

Baker, R. J. (1980) Conversational probit analysis. *GLIM Newsletter*, No. 2, 29–30.

Baker, R. J. and Nelder, J. A. (1978) *The GLIM System Manual, Release 3*. Numerical Algorithms Group Ltd., Oxford.

Baker, R. J., Pierce, C. B. and Pierce, J. M. (1980) Wadley's problem with controls. *GLIM Newsletter*, No. 3, 32–35.

Balakrishnan, N. and Leung, M. Y. (1988) Order statistics from the type I generalized logistic distribution. *Communications in Statistics*, **B17 (1)**, 25–50.

Ball, R. D. (1990) Optimal experimental designs for LD_{50}, LD_{90} and LD_{99} estimation using logit, probit and complementary log-log models. Unpublished report: DSIR Applied Mathematics Division, P. B. Auckland, New Zealand.

Barlow, R. E., Bartholomew, D. J., Bremner, J. M. and Brunk, H. D. (1972) *Statistical Inference Under Order Restrictions: the theory and application of isotonic regression*, Wiley, London.

Barlow, W. E. and Feigl, P. (1985) Analyzing binomial data with a nonzero baseline using GLIM. *Computational Statistics & Data Analysis*, **3**, 201–204.

Barnwal, R. K. and Paul, S. R. (1988) Analysis of one-way layout of count data with negative binomial variation. *Biometrika*, **75**, 2, 215–222.

Beedie, M. A. and Davies, D. M. (1981) History and Epidemiology. Chapter 1 of (ed Davies, D.M.) *Textbook of Adverse Drug Reactions* O.U.P.

Behrens, B. (1929) Zur Auswertung der Digitalisblätter im Froschversuch. *Arch. Exp. Path. Pharmack.*, **140**, 237–256.

Bennett, B. M. (1952) Estimation of LS50 by moving averages. *J. Hygiene*, **50**, 157–164.

Bennett, B. M. (1963) Optimum moving averages for the estimation of medium effective dose in bioassay. *J. Hygiene*, **61**, 401–406.

Bennett, S. (1983) Log-logistic regression models for survival data. *Appl. Statist.*, **32**, 3, 165–171.

Bennett, S. (1986) Distributional properties of non-parametric regression models – a simulation study. Unpublished MS.

Bennett, S. (1989) An extension of Williams' method for overdispersion models. *GLIM Newsletter*, No. **17**, 12–18.

Berenbaum, M. C. (1985) The expected effect of a combination of agents: the general solution. *J. Theor. Biol.*, **114**, 413–431.

Berger, J. O. (1985) *Statistical Decision Theory and Bayesian analysis* (2nd edition). Springer-Verlag, New York.

Bergman, S. W. and Gittins, J. C. (1985) *Statistical Methods for Pharmaceutical Research Planning*. Dekker, New York.

Berkson, J. (1944) Application of the logistic function to bioassay. *J. Amer. Statist. Assoc.*, **39**, 357–365.

Berkson, J. (1953) A statistically precise and relatively simple method of estimating the bioassay with quantal response, based on the logistic function. *J. Amer. Statist. Assoc.*, **48**, 565–599.

Berkson, J. (1955) Maximum likelihood and minimum χ^2 estimates of the logistic function. *J. Amer. Statist. Assoc.*, **50**, 130–162.

Berkson, J. (1980) Minimum chi-square, not maximum likelihood! *Annals of Statistics*, **8**, 447–487.

Bickel, P. J. (1975) One-step Huber estimates in the linear model. *J. Amer. Statist. Assoc.*, **70**, 350, 428–434.

Bickel, P. J. and Doksum, K. A. (1981) An analysis of transformations revisited. *J. Amer. Statist. Assoc.*, **76**, 296–311.

Bishop, Y. M. M., Fienberg, S. E. and Holland, P. W. (1975) *Discrete*

Multivariate Analysis: Theory and Practice. MIT Press, Cambridge, Mass.

Bliss, C. I. (1935) The calculation of the dosage-mortality curve. *Annals of Applied Biology,* **22**, 134–167.

Bliss, C. I. (1938) The determination of dosage-mortality curves from small numbers. *Quart. J. Pharm.,* **11**, 192–216.

Böhning, D. and Lindsay, B. (1988) Monotonicity of quadratic approximation algorithms. *Ann. Inst. Statist. Math.,* **40**, 641–663.

Bonney, G. E. (1987) Logistic regression with dependent binary observations. *Biometrics,* **43**, 4, 951–974.

Boos, D. D. and Brownie, C. (1991) Mixture models for continuous data in dose-response studies when some animals are unaffected by treatment. *Biometrics,* **47**, 4, 1489–1504.

Box, G. E. P. and Cox, D. R. (1982) An analysis of transformations revisited, rebutted. *J. Amer. Statist. Assoc.,* **77**, 377, 209–210.

Box, G. E. P. and Tiao, G. C. (1973) *Bayesian Inference in Statistical Analysis,* Addison-Wesley, Reading, Mass.

Box, G. E. P. and Tidwell, P. W. (1962) Transformations of the independent variables. *Technometrics,* **4**, 531–550.

Boyce, C. B. C. and Williams, D. A. (1967) The influence of exposure time on the susceptibility of *Australorbis gabratus* to *N-tritylmorpholine. Annals of Tropical Medicine & Parasitology,* **61**, 1, 15–20.

Bradley, E. L. (1973) The equivalence of maximum likelihood and weighted least squares estimates in the exponential family. *J. Amer. Statist. Assoc.,* **68**, 199–200.

Bratcher, T. L. (1977) Bayesian analysis of a dose-response experiment with serial sacrifices. *J. Environmental Pathology and Toxicology,* **1**, 287–292.

Bratcher, T. L. and Bell, S. C. (1985) A probability model for teratological experiments, involving binary responses. Unpublished report. Baylor University, Texas, U.S.A.

Breslow, N. E. (1984) Extra-Poisson variation in log-linear models. *Appl. Statist.,* **33**, 1, 38–44.

Breslow, N. E. (1989) Score tests in overdispersed GLMs. Proceedings of GLIM 89 and the Fourth International Workshop on Statistical Modelling (eds Decarli, A., Francis, B. J., Gilchrist, R. and Seeber, G. U. H.), Springer-Verlag, New York, pp. 64–74.

Breslow, N. (1990) Tests of hypotheses in overdispersed Poisson regression and other quasi-likelihood models. *J. Amer. Statist. Assoc.,* **85**, 410, 565–571.

Breslow, N. E. and Day, N. E. (1980) Statistical methods in cancer research. 1: *The analysis of case-control studies*. I. A. R. C., Lyon.

British Toxicology Society (1984) Special report: A new approach to the classification of substances and preparations on the basis of their acute toxicity. *Human Toxicology*, **3**, 85–92.

Brooks, R. J. (1983) Analysing proportions that arise from factorial experiments when the underlying proportions are heterogeneous within groups. *Biometrical J.*, **25**, 99–103.

Brooks, R. J. (1984) Approximate likelihood-ratio tests in the analysis of beta-binomial data. *Appl. Statist.*, **33**, 285–289.

Brooks, R. J., James, W. H. and Gray, E. (1991) Modelling sub-binomial variation in the frequency of sex combinations in litters of pigs. *Biometrics*, **47**, 2, 403–418.

Brown, B. W. (1961) Some properties of the Spearman estimator in bioassay. *Biometrika*, **48**, 293–302.

Brown. B. W. (1966) Planning a quantal assay of potency. *Biometrics*, **22**, 2, 322–329.

Brown, B. W. (1967) Quantal-response assays, in *Statistics in Endocrinology* (eds McArthur, J. W. and Colton, T.), The MIT Press, Cambridge, Massachusetts.

Brown, C. C. (1982) On a goodness of fit test for the logistic model based on score statistics. *Commun. Statist.-Theor. Meth.*, **11**, (10), 1087–1105.

Brownlee, K. A., Hodges, J. L. and Rosenblatt, M. (1953) The up-and-down method with small samples. *J. Amer. Statist. Assoc.*, **48**, 262–277.

Burges, H. D. and Thomson, E. M. (1971) Standardization and assay of microbial insecticides, in *Microbial Control of Insects and Mites* (eds Burges, H. D. and Hussey, N. W.), Academic Press, London, pp. 591–622.

Burn, R. (1983) Fitting a logit model to data with classification errors. *GLIM Newsletter*, **8**, 44–47.

Burn, R. and Thompson, R. (1981) A macro for calculating the covariance matrix of functions of parameter estimates. *GLIM Newsletter*, No. 5, 26–29.

Burr, D. (1988) On errors-in-variables in binary regression – Berkson case. *J. Amer. Statist. Assoc.*, **83**, 739–743.

Burrell, R. J. W., Healy, M. J. R. and Tanner, J. M. (1961) Age at menarche of South African Bantu school girls living in the Transkei reserve. *Human Biology*, **33**, 250–261.

Buse, A. (1982) The likelihood ratio, Wald and Lagrange Multiplier tests: an expository note. *The American Statistician*, **36**, 3, 153–157.

Busvine, J. R. (1938) The toxicity of ethylene oxide to *Calandra oryzae, C. granaria, Tribolium castaneum,* and *Cimex lectularius. Ann. Appl. Bio.,* **25**, 605–632.

Cain, K. C. and Lange, N. T. (1984) Approximate case influence for the proportional hazards regression model with censored data. *Biometrics,* **40**, 2, 493–500.

Cairns, T. (1980) The ED_{01} study: introduction, objectives, and experimental design. *J. Environ. Pathol. Toxicol.,* **3**, 1–7.

Cannon, R. J. C. (1983) Experimental studies on supercooling in two Antarctic micro-orthopods. *J. Insect. Physiol.,* **29**, 8, 617–624.

Carroll, R. J., Spiegelman, C. H., Lan, K. K., Bailey, K. T. and Abbott, R. D. (1984) On errors-in-variables for binary regression models. *Biometrika,* **71**, 19–25.

Carter, E. M. and Hubert, J. J. (1981a) Growth curve models for multivariate quantal bioassays. Department of Mathematics and Statistics, Statistical Series, No. 129. University of Guelph.

Carter, E. M. and Hubert, J. J. (1981b) Covariable adjustment for multivariate growth curve models in quantal bioassays. Department of Mathematics and Statistics Statistical Series, No. 130. University of Guelph.

Carter, E. M. and Hubert, J. J. (1981c) Confidence limits and confidence bands for LC_{50} based on multivariate models in quantal bioassay. Department of Mathematics and Statistics Statistical Series, No. 131. University of Guelph.

Carter, E. M. and Hubert, J. J. (1984) A growth-curve model approach for multivariate quantal bioassay. *Biometrics,* **40**, 699–706.

Carter, E. M. and Hubert, J. J. (1985) Analysis of parallel-line assays with multivariate responses. *Biometrics,* **41**, 703–710.

Carter, E. M. and Hubert, J. J. (1989) Response to paper by Petkau and Sitter. *Biometrics,* **4**, 1307–1308.

Carter, W. H., Jones, D. E. and Carchman, R. A. (1985) Application of response surface methods for evaluating the interactions of soman, atropine and pralidioxime chloride. *Fundamental and Applied Toxicology,* **5**, S232–S241.

Chaloner, K. and Larntz, K. (1989) Optimal Bayesian design applied to logistic regression experiments. *J. Stat. Plan. Inf.*, **21**, 191–208.

Chambers, E. A. and Cox, D. R. (1967) Discrimination between alternative binary response models. *Biometrika*, **54**, 573–578.

Chambers, J. M. (1973) Fitting nonlinear models: numerical techniques. *Biometrika*, **60**, 1, 1–13.

Chanter, D. O. (1975) Modifications of the angular transformation. *Applied Statistics*, **24**, 354–359.

Charnes, A., Frome, E. L. and Yu, P. L. (1976) The equivalence of generalized least squares and maximum likelihood estimates in the exponential family. *J. Amer. Statist. Assoc.*, **71**, 353, 169–171.

Chatfield, Č. and Goodhart, G. J. (1970) The beta-binomial model for consumer purchasing behaviour. *Appl. Statist.*, **19**, 240–250.

Chen, J. J. (1990) Modelling the trinomial responses from reproductive and development toxicity experiments in Contributed Paper Proceedings, XVth International Biometric Conference, p. 264.

Chen, T. S. (1974) An objective criterion for the selection of quantal response models. Contributed paper; 8th Biometric conference, Constanza.

Chernoff, H. (1953) Locally optimum designs for estimating parameters. *Ann. Math. Statist.*, **25**, 573–578.

Chew, R. D. and Hamilton, M. A. (1985) Toxicity curve estimation: fitting a compartment model to median survival times. *Trans. Amer. Fisheries Soc.*, **114**, 403–412.

Chmiel, J. J. (1976) Some properties of Spearman-type estimators of the variance and percentiles in bioassay. *Biometrika*, **63**, 621–626.

Choi, S. C. (1971) An investigation of Wetherill's method of estimation for the up-and-down experiment. *Biometrics*, **27**, 961–970.

Choi, S. C. (1990) Interval estimation of the LD_{50} based on an up-and-down experiment. *Biometrics*, **46**, 2, 485–492.

Chung, K. L. (1954) On a stochastic approximation method. *Ann. Math. Statist.*, **25**, 463–483.

Chung, K. L. (1966) *Markov Chains with stationary transition probabilities* (2nd edition), Springer Verlag, Berlin.

Church, J. D. and Cobb, E. B. (1973) On the equivalence of Spearman-Kärber and maximum likelihood estimates of the mean. *J. Amer. Statist. Assoc.*, **68**, 201–202.

Ciampi, A., Kates, L. and Hogg, S. (1982) Survival analysis by the generalized *F* model. p.19 in *Abstracts of Contributed Papers, XIth International Biometric Conference, Toulouse.*

Clarke, M. R. B. (1982) The Gauss-Jordan sweep operator with detection of collinearity. *Appl. Statist.*, **31**, 2, 166–168.

Clayton, D. G. (1991) A Monte Carlo method for Bayesian inference in frailty models. *Biometrics*, **47**, 2, 467–486.

Clayton, D. G. and Cuzick, J. (1985) The semi-parametric Pareto model for regression analysis of survival times. Paper presented at the Conference of the I.S.I., Amsterdam.

Cobb, E. B. and Church, J. D. (1983) Small-sample quantal response methods for estimating the location parameter for a location-scale family of dose-response curves. *J. Amer. Statist. Assoc.*, **78**, 381, 99–107.

Cochran, W. G. (1943) Analysis of variance for percentages based on unequal numbers. *J. Amer. Statist. Assoc.*, **38**, 287–301.

Cochran, W. G. (1954) Some methods for strengthening the common χ^2 tests. *Biometrics*, **10**, 417–451.

Cochran, W. G. and Davis, M. (1963) Sequential experiments for estimating the median lethal dose. *Le Plan d'Expériences*, pp. 181–194. Centre Nationale de la Recherche Scientifique, Paris.

Cochran, W. G. and Davis, M. (1965) The Robbins-Monro method for estimating the median lethal dose. *J. Roy. Statist. Soc. B*, **27**, 28–44.

Collett, D. and Roger, J. H. (1988) Computation of generalised linear model diagnostics. *GLIM Newsletter*, No. 16, 27–32. (Corrigendum: 1988, *GLIM Newsletter*, No. 17, 3)

Committee on Methods for Toxicity Tests with Aquatic Organisms (1975). Methods for Acute Toxicity Tests with Fish, Macroinvertebrates, and Amphibians. Corvallis, OR. U.S. Environmental Protection Agency Ecological Research Series, EPA-660/3-75-009.

Consul, P. C. and Mittal, S. P. (1975) A new urn model with predetermined strategy. *Biometrical Journal*, **17**, 67–75.

Cook, D. (1986) Assessment of local influence. *J. R. Statist. Soc. B*, **48**, 133–169.

Cook, R. D. and Weisberg, S. (1982) *Residuals and Influence in Regression.* Chapman and Hall, London.

Copas, J. B. (1983) Plotting *p* against *x*. *Appl. Statist.*, **32**, 25–31.

Copas, J. B. (1988) Binary regression models for contaminated data. *J. R. Statist. Soc. B*, **50**, 2, 225–265.

Copenhaver, T. W. and Mielke, P. W. (1977) Quantit analysis: a quantal assay refinement. *Biometrics*, **33**, 175–187.

Cornfield, J. (1977) Carcinogenic risk assessment. *Science*, **198**, 693–699.

Cornfield, J., Carlborg, F. W. and Van Ryzin, J. (1978) Setting tolerances on the basis of mathematical treatment of dose-response data extrapolated to low doses in *Proceedings of the First International Congress on Toxicology* (eds Plaa, G. L. and Duncan, W. A. M.), pp. 143–164, Academic Press, New York.

Cornfield, J. and Mantel, N. (1950) Some new aspects of the application of maximum likelihood to the calculation of the dosage response curve. *J. Amer. Statist. Assoc.*, **45**, 181–120.

Court Brown, W. M. and Doll, R. (1957) Leukaemia and aplastic anaemia in patients irradiated for ankylosing spondylitis. *Special Report Series, Medical Research Council*, No. **295**, HMSO, London.

Cox, C. (1984a) An elementary introduction to maximum likelihood estimation for multinomial models: Birch's theorem and the delta method. *American Statistician*, **38**, 283–287.

Cox, C. (1984b) Generalized linear models – the missing link. *Applied Statistics*, **33**, 1, 18–24.

Cox, C. (1987) Threshold dose-response models in toxicology. *Biometrics*, **43**, 3, 511–524.

Cox, C. (1990) Fieller's Theorem, the likelihood and the delta method. *Biometrics*, **46**, 709–718.

Cox, D. R. (1961) Tests of separate families of hypotheses. *Proc. Fourth Berkeley Symposium, Math. Statist. Probab.*, **1**, 105–123.

Cox, D. R. (1962) Tests of separate families of hypotheses. *J. R. Statist. Soc. B*, **24**, 2, 406–424.

Cox, D. R. (1970) *Analysis of Binary Data*, Chapman and Hall, London.

Cox, D. R. (1972) Regression models and life tables. *Journal of the Royal Statistical Society, Series B*, **34**, 187–220.

Cox, D. R. (1983) Some remarks on over-dispersion. *Biometrika*, **70**, 269–274.

Cox, D. R. (1986) Discussion of paper by Racine et al. *Appl. Statist.*, **35**, 2, 134–135.

Cox, D. R. and Hinkley, D. V. (1974) *Theoretical Statistics*, Chapman and Hall, London.

Cox, D. R. and Reid, N. (1987) Parameter orthogonality and approximate conditional inference. *J. R. Statist. Soc. B*, **49**, 1, 1–39.

Cox, D. R. and Snell, E. J. (1989) *Analysis of Binary Data* (2nd edition), Chapman and Hall, London.

Cramer, E. M. (1964) Some comparisons of methods of fitting the dosage response curve for small samples. *J. Amer. Statist. Assoc.*, **59**, 779–793.

Cramer, J. S. (1991) *The Logit Model.* Edward Arnold, London.

Cran, G. W. (1980) AS149 Amalgamation of means in the case of simple ordering. *Applied Statistics*, **29**, 2, 209–211.

Crowder, M. J. (1978) Beta-binomial ANOVA for proportions. *Appl. Statist.*, **27**, 34–37.

Crowder, M. J. (1979) Inference about the intraclass correlation coefficient in the beta-binomial ANOVA for proportions. *J. R. Statist. Soc. B*, **41**, 2, 230–234.

Crowder, M. (1985) Gaussian estimation for correlated binomial data. *J. Roy. Statist. Soc. B*, **47**, 2, 229–237.

Crowder, M. J. (1987) On linear and quadratic estimating functions. *Biometrika*, **74**, 591–597.

Crump, K. S. (1984) A new method for determining allowable daily intakes. *Fundamental and Applied Toxicology*, **4**, 854–871.

Crump, K. S., Guess, H. A. and Deal, K. L. (1977) Confidence intervals and tests of hypotheses concerning dose response relations inferred from animal carcinogenicity data. *Biometrics*, **33**, 3, 437–452.

Crump, K. S., Hoel, D. G., Langley, C. H. and Peto, R. (1976) Fundamental carcinogenic processes and their implication to low dose risk assessment. *Cancer Research*, **36**, 2973–2979.

Curnow, R. N. and Smith, C. (1975) Multifactorial models for familial diseases in man. *J. R. Statist. Soc. A*, **138**, 131–169.

Daley, D. J. and Maindonald, J. H. (1989) A unified view of models describing the avoidance of superparasitism. *IMA J. of Math., Applied in Med. & Biol.*, **6**, 3, 161–178.

Darby, S. C. and Ellis, M. J. (1976) A test for synergism between two drugs. *Appl. Statist.*, **25**, 296–299.

Darroch, J. N. (1976) No-interaction in contingency tables in *Proceedings of the 9th International Biometric Conference*, **I**, 296–316, Raleigh, North Carolina. Biometric Society.

Davidian, M. and Carroll, R. J. (1987) Variance component estimation. *J. Amer. Statist. Assoc.*, **82**, 1079–1091.

Davidian, M. and Carroll, R. J. (1988) A note on extended quasi-likelihood. *J. R. Statist. Soc. B*, **50**, 1, 74–82.

Davies, D. M. (ed) (1981) *Textbook of Adverse Drug Reactions* (2nd edition), O.U.P.

Davis, M. (1971) Comparison of sequential bioassays in small samples. *J. Roy. Statist. Soc. B*, **33**, 78–87.

Dean, C. and Lawless, J. F. (1989) Tests for detecting overdispersion in Poisson regression models. *J. Amer. Statist. Assoc.*, **84**, 406, 467–472.

Della-Porta, A. J. & Westaway, E. G. (1977) A multi-hit model for the neutralisation of animal viruses. *J. Gen. Virol.*, **38**, 1–19.

Dempster, A. P., Laird, N. M. and Rubin, D. B. (1977) Maximum likelihood estimation from incomplete data via the EM algorithm (with discussion). *J. R. Statist. Soc. B*, **39**, 1–38.

Dietz, E. J. (1985) The rank sum test in the linear logistic model. *The American Statistician*, **39**, 322–325.

Diggle, P. J. and Gratton, R. J. (1984) Monte Carlo methods of inference for implicit statistical models. *J. R. Statist. Soc. B*, **46**, 193–227.

Disch, D. (1981) Bayesian nonparametric inference for effective doses in a quantal-response experiment. *Biometrics*, **37**, 4, 713–722.

Dixon, W. J. (1965) The up-and-down method for small samples. *J. Amer. Statist. Assoc.*, **60**, 967–978.

Dixon, W. J. and Mood, A. M. (1948) A method for obtaining and analyzing sensitivity data. *J. Amer. Statist. Assoc.*, **43**, 109–126.

Dobson, A. J. (1990) *An introduction to generalized linear models*, Chapman and Hall, London.

Dobson, A. J. (1990) *An introduction to generalized linear models*, Chapman & Hall, London.

Douglas, J. B. (1980) *Analysis with Standard Contagious Distributions*, International Co-operative Publishing House, Fairland, Maryland.

Draper, N. R. and Smith, H. (1981) *Applied Regression Analysis* (2nd edition), Wiley, New York.

Dubey, S. D. (1967) Normal and Weibull distributions. *Naval Research Logistics Quarterly*, **14**, 69–79.

Efron, B. (1982) The Jackknife, the Bootstrap and other resampling plans. S.I.A.M., Philadelphia.

Egger, M. J. (1979) Power transformations to achieve symmetry in quantal bioassay. Tech. Report No. 47, Division of Biostatistics, Stanford University, California.

Ekholm, A. and Palmgren, J. (1982) A model for a binary response with misclassifications, in GLIM82: *Proceedings of the International Conference on Generalised Linear Models* (ed Gilchrist, R.), Springer Verlag, New York, pp. 128–143.

Ekholm, A. and Palmgren, J. (1989) Regression models for an ordinal response are best handled as nonlinear models. *GLIM Newsletter*, **18**, 31–35.

Elashoff, J. D. (1981) Repeated-measures bioassay with correlated errors and heterogeneous variances: A Monte Carlo study. *Biometrics*, **37**, 3, 475–482.

Engeman, R. M., Otis, D. L. and Dusenberry, W. E. (1986) Small sample comparison of Thompson's estimator to some common bioassay estimators. *J. Statist. Comput. Simul.*, **25**, 237–250.

Eriksson, U. J. and Styrud, J. (1985) Congenital malformations in diabetic pregnancy: the clinical relevance of experimental animal studies. *Acta Paediatr. Scand.*, Suppl. 320, 72–78.

Everitt, B. S. (1980) *The Analysis of Contingency Tables*, Chapman and Hall, London.

Everitt, B. S. and Hand, D. J. (1981) *Finite Mixture Distributions*, Chapman and Hall, London.

Farewell, V. T. (1982a) A note on regression analysis of ordinal data with variability of classification. *Biometrika*, **69**, 3, 533–538.

Farewell, V. T. (1982b) The use of mixture models for the analysis of survival data with long-term survivors. *Biometrics*, **38**, 1041–1046.

Farewell, V. T. and Prentice, R. L. (1977) A study of distributional shape in life testing. *Technometrics*, **19**, 69–76.

Fazal, S. (1976) A test for the quasi-binomial. *Biometrical Journal*, **18**, 619–622.

FDA Advisory Committee on Protocols for Safety Evaluation (1971) Panel on carcinogenesis report on cancer testing in the safety evaluation of additives and pesticides. *Toxicology and Applied Pharmacology*, **20**, 419–438.

Fedorov, V. V. (1972) *The Theory of Optimum Experiments* (transl. and ed Studden, W. J. and Klimko, E. M.). Academic Press, New York.

Feigin, P. D., Tweedie, R. L. and Belyea, C. (1983) Weighted area techniques for explicit parameter estimation in hierarchical models. *Aust. J. Statist.*, **25**, 1, 1–16.

Fenlon, J. S. (1988) Time to response in insect assay. Unpublished MS.

Fenlon, J. S. (1989) The analysis of dose mortality assay data when dose is measured with error. Unpublished manuscript.

Feuerverger, A. and McDunnough, P. (1981a) On the efficiency of empirical characteristic function procedures. *J. R. Statist. Soc. B,* **43**, 20–27.

Feuerverger, A. and McDunnough, P. (1981b) On some Fourier methods for inference. *J. Amer. Statist. Assoc.,* **76**, 374, 379–387.

Feuerverger, A. and McDunnough, P. (1984) On statistical transform methods and their efficiency. *The Canadian Journal of Statistics,* **12**, 4, 303–317.

Fieller, E. C. (1944) A fundamental formula in the statistics of biological assay, and some applications. *Quart. J. Pharm.,* **17**, 117–123.

Fieller, E. C. (1954) Some problems in interval estimation. *J. R. Statist. Soc. B,* **16**, 175–185.

Fienberg, S. (1981) *The Analysis of Cross-classified Categorical Data* (2nd edition), MIT Press, Cambridge, Mass.

Finkelstein, D. M. (1986) A proportional hazards model for interval-censored failure time data. *Biometrics,* **42**, 4, 845–854.

Finkelstein, D. M. and Ryan, L. M. (1987) Estimating carcinogenic potency from a rodent tumorigenicity experiment. *Appl. Statist.,* **36**, 2, 121–133.

Finney, D. J. (1947) The estimation from individual records of the relationship between dose and quantal response. *Biometrika,* **34**, 320–334.

Finney, D. J. (1950) The estimation of the mean of a normal tolerance distribution. *Sankhyā,* **10**, 341–360.

Finney, D. J. (1953) The estimation of the ED50 for a logistic response curve. *Sankhyā,* **12**, 121–136.

Finney, D. J. (1964) *Statistical Methods in Biological Assay* (2nd edition), Griffin, London.

Finney, D. J. (1947, 1952, 1971) *Probit Analysis* (1st, 2nd & 3rd editions), CUP.

Finney, D. J. (1976) A computer program for parallel line bioassays. *J. Pharm. & Exptl. Therapeutics,* **198**, 497–506.

Finney, D. J. (1978) *Statistical Method in Biological Assay* (3rd edition), Griffin, London.

Finney, D. J. (1980) Review of: Report on the LD50 test. *Biometrics,* **36**, 361–362.

Finney, D. J. (1983) Biological assay: a microcosm of statistical practice. *Bull. Int. Stat. Inst., Proceedings of the 44th Session*, **2**, 790–812.

Finney, D. J. (1985) The median lethal dose and its estimation. *Archives of Toxicology*, **56**, 215–228.

Firth, D. (1987) On the efficiency of quasi-likelihood estimation. *Biometrika*, **74**, 233–245.

Fisher, R. A. (1922) On the mathematical foundations of theoretical statistics. *Phil. Trans. A.*, **222**, 309–368.

Fisher, R. A. (1935) Appendix to Bliss, C. I.: The case of zero survivors. *Ann. Appl. Biol.*, **22**, 164–165.

Fletcher, R. (1970) A new approach to variable metric algorithms. *Computer J.*, **13**, 317–322.

Fletcher, R. and Powell, M. J. D. (1963) A rapidly convergent descent method for minimisation. *Computer Journal*, **5**, 163–168.

Ford, I. (1976) Ph.D. Thesis, University of Glasgow.

Ford, I. and Silvey, S. D. (1980) A sequentially constructed design for estimating a nonlinear parametric function. *Biometrika*, **67**, 381–388.

Fowlkes, E. B. (1987) Some diagnostics for binary logistic regression via smoothing. *Biometrika*, **74**, 3, 503–516.

Freeman, P. R. (1970) Optimal Bayesian sequential estimation of the median effective dose. *Biometrika*, **57**, 79–89.

Friedman, B. (1949) A simple urn model. *Communications on Pure and Applied Mathematics*, **2**, 59–70.

Fryer, J. G. and Pethybridge, R. J. (1975) Maximum likelihood estimation for a special type of grouped data with an application to a dose-response problem. *Biometrics*, **31**, 3, 633–642.

Gart, J. J. (1963) Some stochastic models relating time and dosage in response curves. *Biometrics*, **19**, 583–599.

Gart, J. J. and Tarone, R. E. (1983) The relation between score tests and approximate UMPU tests in exponential models common in biometry. *Biometrics*, **39**, 781–786.

Gart, J. J. and Zweifel, J. R. (1967) On the bias of various estimators of the logit and its variance with application to quantal bioassay. *Biometrika*, **54**, 181–187.

Gaylor, D. W. and Kodell, R. L. (1980) Linear interpolation algorithm for low dose risk assessment of toxic substances. *J. Environ. Path. & Toxicol.*, **4**, 305–312.

Genter, F. C. and Farewell, V. T. (1985) Goodness-of-link testing in ordinal regression models. *Canadian Journal of Statistics*, **13**, 37–44.

Gentleman, W. M. (1973) Least squares computations by Givens transformations without square roots. *J. Inst. Maths. Applics.*, **12**, 329–336.

Gilchrist, R. (1985) Case Control Studies in GLIM. *GLIM Newsletter*, No. 9, 22–26.

Gilmour, A. R., Anderson, R. D. and Rae, A. L. (1985) The analysis of binary data by generalised linear mixed model. *Biometrika* **72**, 593–599.

Giltinan, D. M. Capizzi, T. P. and Malani, H. (1988) Diagnostic tests for similar action of two compounds. *Appl. Statist.*, **37**, 1, 39–50.

Gladen, B. (1979) The use of the jacknife to estimate proportions from toxicological data in the presence of litter effects. *J. Amer. Statist. Assoc.*, **74**, 278–283.

Glasbey, C. A. (1987) Tolerance-distribution-free analyses of quantal dose-response data. *Appl. Statist.*, **36**, 251–259.

Goedhart, P. (1985) Statistical analysis of quantal-assay data. Unpublished M.Sc. dissertation, University of Kent, Canterbury, Kent, England.

Goldstein, H. (1980) Specifying a multivariate logit model using GLIM. *GLIM Newsletter*, No. 1, 23–26.

Good P. I. (1979) Detection of a treatment effect when not all experimental subjects will respond to treatment. *Biometrics*, **35**, 483–489.

Gould, A. and Lawless, J. F. (1988) Estimation efficiency in lifetime regression models when responses are censored or grouped. *Commun. Statist. Simula.*, **17** (3), 689–712.

Govindarajulu, Z. (1988) *Statistical Techniques in Bioassay*, Karger, Basel.

Graubard, B. I. and Korn, E. L. (1987) Choice of column scores for testing independence in ordered $2 \times k$ contingency tables. *Biometrics*, **43**, 2, 471–476.

Green, P. J. (1984) Iteratively reweighted least squares for maximum likelihood estimation, and some robust and resistant alternatives (with discussion). *J. R. Statist. Soc. B*, **46**, 149–192.

Grey, D. R. and Morgan, B. J. T. (1972) Some aspects of ROC curve-fitting: normal and logistic models. *J. Math. Psychol.*, **9**, 1, 128–139.

Griffiths, D. A. (1973) Maximum likelihood estimation for the beta-binomial distribution and an application to the household distribution of the total number of cases of a disease. *Biometrics*, **29**, 637–648.

Griffiths, D. (1977a) Models for avoidance of superparasitism. *J. Anim. Ecol.*, **46**, 59–62.

Griffiths, D. (1977b) Avoidance-modified generalised distributions and their application to studies of superparasitism. *Biometrics*, **33**, 103–112.

Griffiths, W. E., Hill, R. C. and Poper, P. J. (1987) Small sample properties of probit model estimators. *J. Amer. Statist. Ass.*, **82**, 929–937.

Grizzle, J. E. (1971) Multivariate logit analysis. *Biometrics*, **27**, 4, 1057–1061.

Grizzle, J. E. and Allen, D. M. (1969) Analysis of growth and dose response curves. *Biometrics*, **25**, 2, 357–382.

Grizzle, J. E., Starmer, C. F. and Koch, G. G. (1969) Analysis of categorical data by linear models. *Biometrics*, **25**, 489–504.

Guerrero, V. M. and Johnson, R. A. (1982) Use of the Box-Cox transformation with binary response models. *Biometrika*, **69**, 2, 309–314.

Gurland, J., Lee, I. and Dahm, P. A. (1960) Polychotomous quantal response in biological assay. *Biometrics*, **16**, 382–397.

Haberman, S. J. (1974) *The Analysis of Frequency Data*, the University of Chicago Press, Chicago.

Hairston, N. G. (1962) Time-concentration relationships. *World Health Organization, Mol. Inf.*, **4**, 21.

Hamilton, M. A. (1979) Robust estimates of the ED_{50}. *J. Amer. Statist. Assoc.*, **74**, 366, 344–354.

Hamilton, M. A. (1980) Inference about the ED_{50} using the trimmed Spearman-Kärber procedure – a Monte Carlo investigation. *Communications in Statistics*, **B9**, 235–254.

Hamilton, M. A., Russo, P. C. and Thurston, R. V. (1977) Trimmed Spearman-Kärber method for estimating median lethal concentrations in toxicity bioassays. *Environmental Science and Technology*, **11**, 714–719. Correction, 1978, **12**, 417.

Hampel, F. R. (1974) The influence curve and its role in robust estimation. *J. Amer. Statist. Assoc.*, **69**, 383–393.

Harrell, F. E. and Lee, K. L. (1987) Simple tests of global association in the logistic model. *The American Statistician*, **41**, 1, 87.

Harris, C. M. (1983) On finite mixtures of geometric and negative binomial distributions. *Commun. Statist.*, A, **12**, 987–1007.

Hartley, H. O. and Sielken, R. L. Jr. (1977a) Estimation of 'safe doses' in carcinogenic experiments. *Biometrics*, **33**, 1–30.

Hartley, H. O. and Sielken, R. L. Jr. (1977b) Estimation of 'safe doses' in carcinogenic experiments. *J. Environ. Path. & Toxicol.*, **1**, 267–269.

Hartley, H. O. and Sielken, R. L. Jr. (1978) Development of statistical methodology for risk estimation. Technical Report prepared for National Centre for Toxicological Research, Institute of Statistics, Texas A & M University, College Station, Texas.

Haseman, J. K. and Kupper, L. L. (1979) Analysis of dichotomous response data from certain toxicological experiments. *Biometrics*, **35**, 281–293.

Hasemen, J. K. and Soares, E. R. (1976) The distribution of fetal death in control mice and its implications on statistical tests for dominant-lethal effects. *Mutation Research*, **41**, 277–288.

Hasselblad, V., Stead, A. G. and Creason, J. P. (1980) Multiple probit analysis with a nonzero background. *Biometrics*, **36**, 659–663.

Hastie, T. J. and Tibshirani, R. J. (1990) *Generalized Additive Models*, Chapman and Hall, London.

Hastings, C., Mosteller, F., Tukey, J. W. and Winsor, C. P. (1947) Low moments for small samples: a comparative study of order statistics. *Ann. Math. Statist.*, **18**, 413–426.

Hauck, W. W. (1983) A note on confidence bands for the logistic response curve. *American Statistician*, **37**, 158–160.

Hauck, Jr., W. W. and Donner, A. (1977) Wald's tests applied to hypotheses in logit analysis. *J. Amer. Statist. Assoc.*, **72**, (360), 851–853.

Healy, M. J. R. (1986) Discussion of paper by Racine et al. *Appl Statist.*, **35**, 2, 127.

Healy, M. J. R. (1988) *GLIM: An Introduction*, Clarendon Press, Oxford.

Hewlett, P. S. (1974) Time from dosage death in beetles *Tribolium castaneum*, treated with pyrethrins or DDT, and its bearing on dose-mortality relations. *Journal of Stored Product Research*, **10**, 27–41.

Hewlett, P. S. and Plackett, R. L. (1979) *The Interpretation of Quantal Responses in Biology*, Edward Arnold, London.

Hoaglin, D. C. and Welsch, R. (1978) The hat matrix in regression and ANOVA. *Amer. Statistician*, **32**, 17–22.

Hoekstra, J. A. (1987) Acute bioassays with control mortality. *Water, Air and Soil Pollution*, **35**, 311–317.

Hoekstra, J. A. (1989) Estimation of the ED_{50}. *Biometrics*, **45**, 1, 337–338.

Hoekstra, J. A. (1990) Estimators for the LC50. Unpublished MS.

Hoel, D. G. and Yanagawa, D. G. (1986) Incorporating historical controls in testing for a trend in proportions. *J. Amer. Statist. Assoc.*, **81**, 396, 1095–1099.

Hoel, P. G. and Jennrich, R. I. (1979) Optimal designs for dose response experiments in cancer research. *Biometrika*, **66**, 2, 307–316.

Holford, T. R. (1976) Life tables with concomitant information. *Biometrics*, **32**, 3, 587–598.

Hosmer, D. W., Jovanovic, B. and Lemeshow, S. (1989) Best subsets logistic regression. *Biometrics*, **45**, 4, 1265–1270.

Hosmer, D. W. and Lemeshow, S. (1989) *Applied Logistic Regression*, Wiley, New York.

Hsi, B. P. (1969) The multiple-sample up-and-down method in bioassay. *J. Amer. Statist. Assoc.*, **64**, 147–162.

Hsiao, C. (1985) Minimum chi-square. pp. 518–522 in (eds Kotz, S. and Johnson, N. L.) *Encyclopaedia of Statistical Sciences*, Wiley, New York.

Huber, P. J. (1977) Robust Statistical Procedures. *Soc. Ind. Appl. Math.*, Philadelphia.

Hubert, J. J. (1992) *Bioassay* (3rd edition) Kendall/Hunt., Dubuque, Iowa.

Hughes, P. R., Wood, H. A. Burand, J. P. and Granados, R. R. (1984) Quantification of Trichoplusia ni, Heliothis zea, and Spodoptera frugiperda to nuclear polyhedrosis viruses: applicability of an exponential model. *J. Invert. Pathol.*, **51**, 107–114.

Huster, W. J., Brookmeyer, R. and Self, S. G. (1989) Modelling paired survival data with covariates. *Biometrics*, **45**, 1, 145–156.

Ihm, P., Müller, H.-G. and Schmitt, T. (1987) PRODOS: Probit analysis of several qualitative dose-response curves. *The American Statistician*, **41**, 1, 79.

Im, S. and Gianola, D. (1988) Mixed models for binomial data with an application to lamb mortality. *Appl. Statist.*, **37**, 2, 196–204.

Irwin, J. O. (1937) Statistical method applied to biological assays. *J. R. Statist. Soc.*, Supplement, **4**, 1–60.

Izuchi, T. and Hasegawa, A. (1982) Pathogenicity of infectious laryngotracheitis virus as measured by chicken embryo inoculation. *Avian Diseases*, **26**, 18–25.

James, B. R. and James, K. L. (1983) On the influence curve for quantal bioassay. *J. Statist. Pl. Inf.*, **8**, 331–345.

James, B. R., James, K. L. and Westenberger, H. (1984) An efficient *R*-estimator for the ED_{50}. *J. Amer. Statist. Assoc.*, **79**, 385, 164–173.

James, D. A. and Smith, D. M. (1982) Analysis of results from a collaborative study of the dominant lethal assay. *Mutation Research*, **97**, 303–314.

Jansen, J. (1988) Using Genstat to fit regression models to ordinal data. *Genstat Newsletter*, **21**, 28–32.

Jansen, J. (1990) On the statistical analysis of ordinal data when extravariation is present. *Appl. Statist.*, **39**, 1, 75–84.

Jansen, J. (1991) Properties of maximum likelihood estimators for overdispersed binomial data. Submitted for publication.

Jansen, J. and Bowman, A. (1988) Statistical analysis of data involving internal bruising in potato tubers. *J. Agric. Engng. Res.*, **40**, 1–7.

Jarrett, R. G. (1984) A look at the MICE test. CSIRO DMS Consulting Report, No. VT 84/16.

Jarrett, R. G., Morgan, B. J. T. and Liow, S. (1981) The effects of viruses on death and deformity rates in chicken eggs. Consulting report VT 81/37, CSIRO Division of Mathematics and Statistics, Melbourne.

Jarrett, R., Saunders, I. and Wong, R. (1984) Optimal estimation of the ED_{50} in bioassay CSIRO DMS Newsletter 100. Canberra.

Jennings, D. E. (1986) Outliers and residual distributions in logistic regression. *J. Amer. Statist. Assoc.*, **81**, 396, 987–990.

Jennrich, R. I. and Moore, R. H. (1975) Maximum likelihood estimation by means of non-linear least squares, in *Proceedings of the Statistical Computing Section, American Statistical Association*, 57–65.

Jennrich, R. I. and Ralston, M. L. (1978) Fitting nonlinear models to data. *BMDP Technical Report No. 46*. (Also published in *Ann. Rev. Biophys. Bioeng.*, **8**, 1979.)

Johnson, N. L. (1965) Systems of frequency curves generated by methods of translation. *Biometrika*, **36**, 149–176.

Johnson, N. L. and Kotz, S. (1969) *Discrete Distributions*, Houghton Mifflin, New York.

Johnson N. L. and Kotz, S. (1970) *Continuous Univariate Distributions –* 1, Wiley, New York.

Johnson, N. L. and Kotz, S. (1973) Extended and multivariate Tukey lambda distributions. *Biometrika*, **60**, 655–661.

Johnson, W. (1985) Influence measures for logistic regression: another point of view. *Biometrika*, **72**, 1, 59–66.

Joiner, B. L. and Rosenblatt, J. R. (1971) Some properties of the range in samples from Tukey's symmetric Lambda distributions. *J. Amer. Statist. Assoc.*, **66**, 394–399.

Jorgensen, B. (1984) The Delta algorithm and GLIM. *International Statistical Review*, **52**, 283–300.

Jorgensen, M. A. (1990) Influence functions for iteratively defined statistics: applications to iteratively reweighted least squares. Technical Report No: STAT-90-25, Dept. of Statistics and Actuarial Science, University of Waterloo, Ontario, Canada.

Jørgensen, M., Keiding, N. and Skakkeback, N. E. (1991) Estimation of spermarche from longitudinal spermaturia data. *Biometrics*, **47**, 1, 177–194.

Kalbfleisch, J. D., Krewski, D. R. and Van Ryzin, J. (1983) Dose-response models for time-to-response toxicity data. *The Canadian Journal of Statistics*, **11**, 25–49.

Kalbfleisch, J. D., Lawless, J. E. and Vollmer, W. M. (1983) Estimation in Markov models for aggregate data. *Biometrics*, **39**, 907–919.

Kalish, L. A. (1990) Efficient design for estimation of median lethal dose and quantal dose-response curves. *Biometrics*, **46**, 3, 737–748.

Kalish, L. A. and Rosenberger, J. L. (1978) Optimal designs for the estimation of the logistic function. Tech. Rep. 33, Pennsylvania State University.

Kappenman, R. F. (1987) Nonparametric estimation of dose-response curves with application to ED_{50} estimation. *J. Statist. Comput. Simul.*, **28**, 1–13

Kärber, G. (1931) Beitrag zur kollektieven Behandlung pharmakologischer Reihenversuche. *Archiv für Experimentelle Pathologie und Pharmakologie*, **162**, 480–487.

Kay, R. and Little, S. (1986) Assessing the fit of the logistic model:

a case study of children with the haemolytic uraemic syndrome. *Appl. Statist.*, **35**, 1, 16–30.

Kay, R. and Little, S. (1987) Transformations of the explanatory variables in the logistic regression model for binary data. *Biometrika*, **74**, 3, 495–502.

Kendall, M. G. and Stuart, A. (1979) *The Advanced Theory of Statistics* Vol. 2 (4th edition) Griffin, London.

Kennedy, W. J. and Gentle, J. E. (1980) *Statistical Computing*, Marcel Dekker, New York.

Kershaw, C. D. (1983) Sequential estimation for binary response. Ph.D. thesis, University of Edinburgh.

Kershaw, C. D. (1985a) Asymptotic properties of $\bar{w}$, an estimator of the ED_{50} suggested for use in up-and-down experiments in bio-assay. *Annals of Statistics*, **13**, 85–94.

Kershaw, C. D. (1985b) Statistical properties of staircase estimates from two interval forced choice experiments. *Br. J. Math. & Stat. Psychol.*, **38**, 34–43.

Kershaw, C. D. (1987) A comparison of estimators of the ED_{50} in up-and-down experiments. *J. Stat. Comp. and Simul.*, **27**, 175–184.

Kershaw, C. D. and McRae, D. C. (1985) Sequential design for estimating levels with low response rates in potato drop tests and other binary response experiments. *J. Agric. Sci.*, **105**, 51–57.

Kessler, K. T. and Gleason, J. R. (1986) A nonparametric alternative to maximum likelihood estimation in simple logistic regression, in *American Statistical Association Proceedings of the Statistical Computing Section*, 130–134. American Statistical Association, Washington.

Kesten, H. (1958) Accelerated stochastic approximation. *Ann. Math. Statist.*, **29**, 41–49.

Khan, M. K. (1988) Optimal Bayesian estimation of the median effective dose. *J. Statist. Plann. Inference*, **18**, 69–81.

Khan, M. K. and Yazdi, A. A. (1988) On D-optimal designs for binary data. *J. Statist. Plann. Inference*, **18**, 83–91.

Kimball, A. W. and Friedman, L. A. (1986) A Bernoulli-logistic model for analysis of binary sequences. Unpublished MS.

Klaassen, C. D. (1986) Principles of toxicology. Chapter 2 of Cassarett & Doull's *Toxicology, The Basic Science of Poisons*, 3rd Edition, Macmillan.

Klaassen, C. D., Amdur, M. O. and Doull, J. (eds) (1986) Cassarett and Doull's *Toxicology, The Basic Science of Poisons*, 3rd Edition. Macmillan, New York.

Kleinman, J. C. (1973) Proportions with extraneous variance: single and independent samples. *J. Amer. Statist. Assoc.*, **68**, 46–54.

Kleinman, J. C. (1975) Proportions with extraneous variance: two dependent samples. *Biometrics*, **31**, 737–743.

Kolakowski, D. and Bock, R. D. (1981) A multivariate generalization of probit analysis. *Biometrics*, **37**, 3, 541–552.

Kooijman, S. A. L. M. (1981) Parametric analyses of mortality data in bioassays. *Water Research*, **15**, 107–119.

Kooijman, S. A. L. M. (1983) Statistical aspects of the determination of mortality rates in bioassays. *Water Research*, **17**, 749–759.

Korn, E. L. (1982) Confidence bands for isotonic dose-response curves. *Appl. Statist.*, **31**, 1, 59–62.

Krewski, D., Kovar, J. and Bickis, M. (1984) Optimal experimental designs for low-dose extrapolation, in *Topics in Applied Statistics*, (ed Dwivedi, T. W.), Concordia University, Montreal.

Krewski, D. and Van Ryzin, J. (1981) *Dose response models for quantal response toxicity data,* in *Current Topics in Probability and Statistics* (eds Csorgo, M., Dawson, D., Rao, J. N. K. and Saleh, E.), North Holland, New York.

Kupper, L. L. and Haseman, J. K. (1978) The use of a correlated binomial model for the analysis of certain toxicological experiments. *Biometrics*, **35**, 281–293.

Kupper, L. L., Portier, C. and Hogan, M. D. (1988) Response to Williams' reaction, *Biometrics*, **44**, 1, 308–310.

Kupper, L. L., Portier, C., Hogan, M. D. and Yamamoto, E. (1986) The impact of litter effects on dose-response modelling in teratology. *Biometrics*, **42**, 85–98.

Lai, T. L. and Robbins, H. (1979) Adaptive design and stochastic approximation. *Annals of Statistics*, **7**, 1196–1221.

Lai, T. L. and Robbins, H. (1981) Consistency and asymptotic efficiency of slope estimates in stochastic approximation schemes. *Zeitschrift fur Wahrscheinlichkeitstheorie und Verwandte Gebiete*, **56**, 329–360.

Landwehr, J. M., Pregibon, D. and Schoemaker, A. C. (1984) Graphical methods for assessing logistic regression models. *J. Amer. Statist. Assoc.*, **79**, 385, 61–83.

Larsen, R. I., Gardner, D. E. and Coffin, D. L. (1979) An air quality data analysis system for interrelating effects, standards, and needed source reductions: Part 5. NO_2 mortality in mice. *Journal of the Air Pollution Control Association*, **39**, 133–137.

Larson, M. G. and Dinse, G. E. (1985) A mixture model for the regression analysis of competing risks data. *Appl. Statist.*, **34**, 201–211.

Laurence, A. F. and Morgan, B. J. T. (1987) Selection of the transformation variable in the Laplace transform method of estimation. *Austral. J. Statist.*, **29**, 113–127.

Laurence, A. F. and Morgan, B. J. T. (1989) Observations on a stochastic model for quantal assay data. *Biometrics*, **45**, 3, 733–744.

Laurence, A. F. and Morgan, B. J. T. (1990a) Quantal assay data with time-to-response: model-fitting using Laplace transforms. Unpublished MS.

Laurence, A. F. and Morgan, B. J. T. (1990b) Aspects of the multi-hit model. Unpublished MS.

Laurence, A. F., Morgan, B. J. T. and Tweedie, R. L. (1987) Parameter estimation in non-linear models using Laplace transforms. Unpublished MS.

Lawless, J. F. (1987) Negative binomial and mixed Poisson regression. *The Canadian Journal of Statistics*, **15**, 209–225.

Lawless, J. F. and McLeish, D. L. (1984) The information in aggregate data from Markov chains. *Biometrika*, **71**, 419–430.

Lawley, D. N. (1956) A general method for approximating to the distribution of the likelihood ratio criteria. *Biometrika*, **43**, 295–303.

Le, C. T. (1988) Testing for linear trends in proportions using correlated otolaryngology or ophthalmology data. *Biometrics*, **44**, 1, 299–304.

Leach, D. (1981) Re-evaluation of the logistic curve for human populations. *J. R. Statist. Soc. A*, **144**, 1, 94–103.

Lewis, T. (1984) Discussion of Iteratively Reweighted Least Squares by P. J. Green. *J. R. Statist. Soc. B*, **46**(2), 178–179.

Lindley, D. V. (1965) *Introduction to probability and statistics. Part I. Probability.* Cambridge University Press.

Little, R. E. (1968) A note on estimation for quantal response data. *Biometrika*, **55**, 578–579.

Looney, S. W. (1983) The use of the Weibull distribution in bioassay. *American Statistical Association*, Proceedings of the Statistical Computing Section, 272–277.

Luning, K. G., Sheridan, W., Ytterborn, K. H. and Gullberg, U. (1966) The relationship between the number of implantations and the rate of intra-uterine death in mice. *Mutation Research*, **3**, 444–451.

Lwin, T. and Martin, P. J. (1989) Probits of mixtures. *Biometrics*, **45**, 3, 721–732.

McCann, J., Choi, E., Yamasaki, E. and Ames, B. N. (1975) Detection of carcinogens as mutagens in the Salmonella/microsome test. Assay of 300 chemicals. *Proc Nat. Acad. Sci.*, **72**, 5135–5139.

McCaughran, D. A. and Arnold, D. W. (1976) Statistical models for numbers of implantation sites and embryonic deaths in mice. *Toxicology and Applied Pharmacology*, **38**, 325–333.

McCullagh, P. (1980) Regression models for ordinal data. *J. R. Statist. Soc. B*, **42**, 109–142.

McCullagh, P. (1983) Quasi-likelihood functions. *Annals of Statistics*, **11**, 59–67.

McCullagh, P. and Nelder, J. A. (1983) *Generalised Linear Models*, Chapman and Hall, London.

McCullagh, P. and Nelder, J. A. (1989) *Generalised Linear Models*. (2nd edition), Chapman and Hall, London.

McFarlane, N. R., Paterson, G. D. and Dunderdale, M. (1977) Modelling as a management tool in pesticide research, in *Crop Protection Agents* (ed McFarlane, N. R.), Academic Press, London.

McGilchrist, C. A. and Aisbett, C. W. (1991) Regression with frailty in survival analysis. *Biometrics*, **47**, 2, 461–466.

McKee, S. P., Klein, S. and Teller, D. Y. (1985) Statistical properties of forced-choice psychometric functions: implications of probit analysis. *Perception and Psychophysics*, **37**, 4, 268–298.

McKnight, B. and Crowley, J. (1985) Tests for differences in tumor incidence based on animal carcinogenicity experiments. *J. Amer. Statist. Assoc.*, **79**, 639–648.

McLeish, D. L. and Tosh, D. H. (1983) The estimation of extreme quantiles in logit bioassay. *Biometrika*, **70**, 3, 625–632.

McLeish, D. L. and Tosh, D. H. (1990) Sequential designs in bioassay. *Biometrics*, **46**, 1, 103–116.

McNally, R. J. Q. (1990) Maximum likelihood estimation of the parameters of the prior distribution of three variables that strongly influence reproductive performance in cows. *Biometrics*, **46**, 2, 501–514.

McPhee, D. A., Parsonson, I. M., Della-Porta, A. J. and Jarrett, R. G.

(1984) Teratogenicity of Australian Simbu Serogroup and some other Bunyaviridae viruses: The embryonated chicken egg as a model. *Infection and Immunity*, **43**, 1, 413–420.

Magnus, A., Mielke, P. W. and Copenhaver, T. W. (1977) Closed expressions for the sum of an infinite series with applications to Quantal Response Assays. *Biometrics*, **33**, 221–223.

Mantel, N. (1969) Models for complex contingency tables and polychotomous dosage response curves. *Biometrics*, **22**, 83–95.

Mantel, N. (1985) Reader Reaction: Maximum likelihood vs. minimum chi-square. *Biometrics*, **41**, 3, 777–780.

Mantel, N. and Brown, C. (1973) A logistic reanalysis of Ashford and Sowden's data on respiratory symptoms in British coal miners. *Biometrics*, **29**, 4, 649–666.

Mantel, N. and Bryan, W. (1961) 'Safety' testing of carcinogenic agents. *J. Nat. Cancer Inst.*, **27**, 455–470.

Mantel, N. and Myers, M. (1971) Problems of convergence of maximum likelihood iterative procedures in multiparameter situations. *J. Amer. Statist. Assoc.*, **66**, 484–491.

Mantel, N. and Paul, S. R. (1987) Goodness-of-fit issues in toxicological experiments involving litters of varying size, in (eds MacNeill, I. B. and Umphrey, G. J.), *Biostatistics*, pp. 169–176, D. Reidel, Boston.

Margolin, B. H. and Risko, K. J. (1984) The use of historical data in laboratory studies, in Proceedings of the XIIth International Biometric Conference, pp. 21–30. The Biometric Society, Washington, D.C.

Marks, B. L. (1962) Some optimal sequential schemes for estimating the mean of a cumulative normal quantal response curve. *J. R. Statist. Soc. B*, **24**, 393–400.

Marquardt, D. W. (1963) An algorithm for least squares estimation of nonlinear parameters. *S. I. A. M. J. Appl. Math.*, **11**, 431–441.

Marquardt, D. W. (1980) You should standardize the predictor variables in your regression models. *J. Amer. Statist. Assoc.*, **75**, 369, 87–91.

Marriott, M. S. and Richardson, K. (1987) The discovery and mode of action of Fluconazole, in *Recent Trends in the Discovery, Development and Evaluation of Antifungal Agents*, (ed Fromtling, R. A.), J. R. Prous Science Publishers, S. A.

Marsden, J. (1987) An analysis of serological data by the method of spline approximation. Unpublished M. Sc. dissertation, University of Reading.

Martin, J. T. (1942) The problem of the evaluation of rotenone-containing plants. VI. The toxicity of *l*-elliptone and of poisons applied jointly, with further observations on the rotenone equivalent method of assessing the toxicity of derris root. *Ann. Appl. Biol.*, **29**, 69–81.

Martin, P. J., Anderson, N., Lwin, T., Nelson, G. and Morgan, T. E. (1984) The association between frequency of thiabendazole treatment and the development of resistance in field isolates of Ostertagia Spp. of sheep. *Int. J. Parasitology*, **14**, 2, 177–181.

Maruani, P. and Schwartz, D. (1983) Sterility and fecundability estimation. *J. Theor. Biol.*, **105**, 211–219.

Marx, B. D. and Smith, E. P. (1990) Principal component estimation for generalized linear regression. *Biometrika*, **77**, 1, 23–32.

Maynard Smith, J. (1971) *Mathematical Ideas in Biology*, C.U.P.

Milicer, H. (1968) Age at menarche of girls in Wroclaw, Poland, in 1966. *Human Biology*, **40**, 249–259.

Milicer, H. and Szczotka, F. (1966) Age at menarche in Warsaw girls in 1965. *Human Biology*, **38**, 199–203.

Miller, R. G. (1973) Nonparametric estimators of the mean tolerance in bioassay. *Biometrika*, **60**, 3, 535–542.

Miller, R. G. and Halpern, J. W. (1980) Robust estimators for quantal bioassay. *Biometrika*, **67**, 103–110.

Milliken, G. A. (1982) On a confidence interval about a parameter estimated by a ratio of normal random variables. *Communications in Statistics – Theory and Methods*, **11**, 1985–1995.

Minkin, S. (1987) On optimal design for binary data. *J. Amer. Statist. Assoc.*, **82**, 1098–1103.

Moolgavkar, S. H. and Venzon, D. J. (1979) Two-event models for carcinogenesis: incidence curves for childhood and adult tumors. *Mathematical Biosciences*, **47**, 55–77.

Moore, D. F. (1986) Asymptotic properties of moment estimators for overdispersed counts and proportions. *Biometrika*, **73**, 583–588.

Moore, D. F. (1987) Modelling the extraneous variance in the presence of extra-binomial variation. *Appl. Statist.*, **36**, 1, 8–14.

Moore, D. F. and Tsiatis, A. (1991) Robust estimation of the variance in moment methods for extra-binomial and extra-Poisson variation. *Biometrics*, **47**, 2, 383–402.

Moore, R. H. and Zeigler, R. K. (1967) The use of nonlinear regression methods for analyzing sensitivity and quantal response data. *Biometrics*, **23**, 563–567.

Morgan, B. J. T. (1978) A simple comparison of Newton-Raphson and the method of scoring. *Int. J. Math. Educ. Sci. Technol.*, **9**, 3, 343–348.

Morgan, B. J. T. (1982) Modelling Polyspermy. *Biometrics*, **38**, 885–898.

Morgan, B. J. T. (1983) Observations on quantit analysis. *Biometrics*, **39**, 879–886.

Morgan, B. J. T. (1984) *Elements of Simulation*, Chapman and Hall, London.

Morgan, B. J. T. (1985) The cubic logistic model for quantal assay data. *Appl. Statist.*, **34**, 2, 105–113.

Morgan, B. J. T. (1986) Notes on using a mixture model to describe hypersensitivity reactions. Unpublished MS.

Morgan, B. J. T. (1988) Extended models for quantal response data. *Statistica Neerlandica*, **42**, 4, 253–272.

Morgan, B. J. T. and North, P. M. (1985) A model for avian lung ventilation, and the effect of accelerating stimulation in Japanese quail embryos. *Biometrics*, **41**, 1, 215–226.

Morgan, B. J. T., North, P. M. and Pack, S. E. (1987) Collaboration between university and industry, in *The Statistical Consultant in Action* (eds Hand, D. J. and Everitt, B. S.), C.U.P., pp. 134–152.

Morgan, B. J. T. and Pack, S. E. (1986) Modelling a rat screen. Unpublished MS.

Morgan, B. J. T. and Pack, S. E. (1988) QUAD: A program for Analysing QUantal Assay Data. *Statistical Software Newsletter*, **13**, 3, 120.

Morgan, B. J. T., Pack, S. E. and Smith, D. M. (1989) QUAD: A computer package for the analysis of QUantal Assay Data. *Computer Methods and Programs in Biomedicine*, **30**, 265–278.

Morgan, B. J. T. and Smith, D. M. (1990) On Wadley's problem with overdispersion. Unpublished manuscript.

Morgan, B. J. T. and Smith, D. M. (1992) A note on Wadley's problem with overdispersion. *Appl. Statist.* **41**, 2, 349–354.

Morgan, B. J. T. and Watts, S. A. (1980) On modelling microbial infections. *Biometrics*, **36**, 2, 317–321.

Morisita, M. (1971) Measuring of habitat value by environmental density method, in *Statistical Ecology* (eds Patil, G. P. *et al.*) Pennsylvania State University Press, pp. 379–401.

Morris, M. D. (1988) Small-sample confidence limits for parameters under inequality constraints with application to quantal bioassay. *Biometrics*, **44**, 4, 1083–1092.

Morton, R. (1981) Generalized Spearman estimators of relative dose. *Biometrics*, **37**, 2, 223–234.

Morton, R. (1987) A generalized linear model with nested strata of extra-Poisson variation. *Biometrika*, **74**, 2, 247–258.

Morton, R. (1991) Analysis of extra-multinomial data derived from extra-Poisson variables conditional on their total. *Biometrika*, **78**, 1, 1–6.

Mudholkar, G. S. and Phatak, M. V. (1984) Quantile function models for quantal response analysis: an outline, in *Topics in Applied Statistics: Proceedings of Statistics*, **81**, (eds Chaubey, Y. P. and Dwinedi, T. D.), Concordia University Press, Montreal.

Müller, H.-G. and Schmitt, T. (1990) Choice of number of doses for maximum likelihood estimation of the ED_{50} for quantal dose-response data. *Biometrics*, **46**, 1, 117–130.

Nelder, J. A. (1985) Quasi-likelihood and GLIM, in *Generalized Linear Models* (eds Gilchrist, R., Francis, B. and Whittaker, J.), Springer-Verlag, Berlin, pp. 120–127.

Nelder, J. A. (1988) Overdispersion and extended quasi-likelihood. Proceedings, Invited Papers, *XIVth International Biometric Conference*, Namur, 18–23 July, pp. 289–300.

Nelder, J. A. (1990) Two statistical ideas and their algorithms. *NAG Newsletter*, **1**, 3–12.

Nelder, J. A. and Mead, R. (1965) A simplex method for function minimisation. *Computer Journal*, **7**, 308–313.

Nelder, J. A. and Pregibon, D. (1987) An extended quasi-likelihood function. *Biometrika*, **74**, 221–232.

Nelder, J. A. and Wedderburn, R. W. M. (1972) Generalized linear models. *J. R. Statist. Soc. A*, **135**, 370–384.

Neyman, J. (1937) 'Smooth' tests for goodness of fit. *Skand. Aktuar.*, **20**, 149–199.

Neyman, J. (1959) Optimal asymptotic tests of composite statistical hypotheses, in *Probability and Statistics* (ed Grenander, U.), Wiley, New York.

Numerical Algorithms Group (1982) Numerical Algorithms Group Library Manual. Oxford; Numerical Algorithms Group.

Nyquist, H. (1991) Restricted estimation of generalized linear models. *Applied Statistics*, **40**, 1, 133–142.

Ochi, Y. and Prentice, R. L. (1984) Likelihood inference in a correlated probit regression model. *Biometrika*, **71**, 3, 531–543.

Ogilvie, J. C. and Creelman, C. D. (1968) Maximum-likelihood estimation of receiver operating characteristic curve parameters. *J. Math. Psychol.*, **5**, 337–391.

O'Hara Hines, R. J. and Lawless, J. F. (1992) Modelling over-dispersion in toxicological mortality data grouped over time. To appear in *Biometrics*.

O'Hara Hines, R. J., Lawless, J. F. and Carter, E. M. (1990) Diagnostics for a multinomial generalized linear model, with applications to grouped toxicological mortality data. To appear in the *J. Amer. Statist. Assoc.*

Orchard, T. and Woodbury, M. A. (1972) A missing information principle: theory and applications. *Proc. 6th Berkley Symposium on Math. Statist. and Prob.*, **1**, 697–715.

Ostwald, W. and Dernoschek, A. (1910) Über die Beziehungen zwischen Adsorption und Giftigkeit. *Kolloidzschr*, **6**, 297–307.

Otake, M. and Prentice, R. L. (1984) The analysis of chromosomally aberrant cells based on beta-binomial distribution. *Radiation Res.*, **98**, 456–470.

Otto, M. C. and Pollock, K. H. (1990) Size bias in line transect sampling: a field test. *Biometrics.*, **46**, 1, 239–246.

Owen, R. J. (1975) A Bayesian sequential procedure for quantal response in the context of adaptive mental testing. *J. Amer. Statist. Assoc.*, **70**, 351–356.

Pack, S. E. (1985) A comparison of three binomial generalisations for toxicological data. Unpublished MS.

Pack, S. E. (1986a) The analysis of proportions from toxicological experiments. Unpublished Ph.D. thesis, University of Kent, Canterbury.

Pack, S. E. (1986b) Hypothesis testing for proportions with over-dispersion. *Biometrics*, **42**, 4, 967–972.

Pack, S. E. and Morgan, B. J. T. (1990a) A mixture model for interval-censored time-to-response quantal assay data. *Biometrics*, **46**, 3, 749–758.

Pack, S. E. and Morgan, B. J. T. (1990b) A model for fly knock-down. Unpublished manuscript.

Palmgren, J. (1987) Precision of double sampling estimators for comparing two probabilities. *Biometrika*, **74**, 687–694.

Palmgren, J. and Ekholm, A. (1987) Exponential family nonlinear models for categorical data with errors of observation. *Appl. Stochast. Models and Data Anal.*, **3**, 111–124.

Park, C. N. and Snee, R. D. (1983) Quantitative risk assessment: state-of-the-art for carcinogenesis. *The American Statistician* **37**, 427–441.

Parkinson, J. M. and Hutchinson, D. (1972) An investigation into the efficiency of variants on the simplex method, in *Numerical Methods for Non-linear Optimization* (ed Lootsma, F. A.), Academic Press, New York, pp. 115–135.

Pate, I. (1989) Direct use of the likelihood function for ED_{50} estimation. Unpublished manuscript.

Paterson, G. D. (1979) Problems in relating models of insecticidal action to experimental data. *Pesticide Science*, **10**, 271–277.

Patwary, K. M. and Haley, K. D. C. (1967) Analysis of quantal response assays with dosage errors. *Biometrics*, **23**, 747–760.

Paul, S. R. (1979) A clumped beta-binomial model for the analysis of clustered attribute data. *Biometrics*, **35**, 4, 821–824.

Paul, S. R. (1982) Analysis of proportions of affected fetuses in teratological experiments. *Biometrics*, **38**, 361–370.

Paul, S. R. (1985) A three-parameter generalisation of the binomial distribution. *Commun. Statist. – Theor. Meth.*, **14**,(6), 1497–1506.

Paul, S. R., Liang, K. Y. and Self, S. G. (1989) On testing departure from the binomial and multinomial assumptions. *Biometrics*, **45**, 1, 231–236.

Paul, S. R. and Mantel, N. (1988) Model dependent analyses of litter-depletion data. *The Statistician*, **37**, 363–370.

Payne, C. D. (ed) (1986) *The GLIM system release 3.77* (Revision of of 1st edition). Numerical Algorithms Group, Oxford.

Payne, R. W., Lane, P. W., Ainsley, A. E., Bicknell, K. E., Digby, P. G. N., Harding, S. A., Leech, P. K., Simpson, H. R., Todd, A. D., Verrier, P. J., White, R. P., Gower, J. C., Tunnicliffe Wilson, G. and Paterson, L. J. (1987) *GENSTAT 5 Reference Manual*, Clarendon Press, Oxford.

Pearson, E. S. and Hartley, H. O. (1970) *Biometrika Tables for Statisticians*, Vol. I (3rd edition), C.U.P.

Peduzzi, P. W., Hardy, R. J. and Holford, T. R. (1980) A stepwise variable selection procedure for nonlinear regression models. *Biometrics*, **36**, 511–516.

Petkau, A. J. and Sitter, R. R. (1989) Models for quantal response experiments over time. *Biometrics*. **45**, 4, 1299–1306.

Pettitt, A. N. (1984) Proportional odds models for survival data and estimates using ranks. *Appl. Statist.*, **33**, 169–175.

Pettitt, A. N. and Bin Daud, I. (1989) Case-weighted measures of influence for proportional hazards regression. *Appl. Statist.*, **38**, 1, 51–68.

Pierce, D. A. and Sands, B. R. (1975) Extra-binomial variation in binary data. Tech. rep. no. 46, Department of Statstics, Oregon State University.

Pierce, D. A., Stewart, W. H. and Kopecky, K. J. (1979) Distribution-free regression analysis of grouped survival data. *Biometrics*, **35**, 785–793.

Plackett, R., L. and Hewlett, P. S. (1979) *An Introduction to the Interpretation of Quantal Responses in Biology*, Edward Arnold, London.

Portier, C. and Hoel, D. (1983) Low-dose rate extrapolation using the multistage model. *Biometrics*, **39**, 4, 897–906.

Potthoff, R. F. and Roy, S. N. (1974) A generalized multivariate analysis of variance model useful especially for growth curve problems. *Biometrika*, **51**, 313–326.

Pregibon, D. (1980) Goodness of link tests for generalized linear models. *Appl. Statist.*, **29**, 1, 15–24.

Pregibon, D. (1981) Logistic regression diagnostics. *Ann. Statist.*, **9**, 705–724.

Pregibon, D. (1982a) Resistant fits for some commonly used logistic models with medical application. *Biometrics*, **38**, 485–498.

Pregibon, D. (1982b) Score tests in GLIM with applications, in *GLIM 82: Proceedings of the International Conference on Generalised Linear Models* (ed Gilchrist, R.), Springer-Verlag, New York, pp. 87–97.

Pregibon, D. (1982c) Resistant fits for some commonly used logistic models with medical applications. *Biometrics*, **38**, 485–498.

Pregibon, D. (1983) An alternative covariance estimated for generalized linear models. *GLIM Newsletter*, **6**, 51–55.

Pregibon, D. (1985) Link Tests. *Encyclopaedia of Statistical Sciences* **5** (eds Kotz, S. and Johnson, N. L.), Wiley, New York, pp. 82–85.

Preisler, H. K. (1988) Maximum likelihood estimates for binary data with random effects. *Biom. J.*, **3**, 339–350.

Preisler, H. K. (1989a) Fitting dose-response data with non-zero background within generalized linear and generalized additive models. *Comp. Stats. & Data Anal.*, **7**, 279–290.

Preisler, H. K. (1989b) Analysis of a toxicological experiment using a generalized linear model with nested random effects. *Int. Stat. Rev.*, **57**, 2, 145–159.

Preisler, H. K. and Robertson, J. L. (1989) Analysis of time-dose-mortality data. *J. Econ. Entomol.*, **82**, (6), 1534–1542.

Prentice, R. (1976a) Use of the logistic model in retrospective studies. *Biometrics*, **32**, 3, 599–606.

Prentice, R. L. (1976b) A generalization of the probit and logit methods for dose response curves. *Biometrics*, **32**, 761–768.

Prentice, R. L. (1986) Binary regression using an extended beta-binomial distribution, with discussion of correlation induced by covariate measurement error. *J. Amer. Statist. Assoc.*, **81**, 394, 321–327.

Prentice, R. L. (1988) Correlated binary regression with covariates specific to each binary observation. *Biometrics*, **44**, 1033–1048.

Prentice, R. L. and Gloeckler, L. A. (1978) Regression analysis of grouped survival data with applications to breast cancer data. *Biometrics*, **34**, 57–68.

Prentice, R. L. and Pyke, R. (1979) Logistic disease incidence models and case-control studies. *Biometrika*, **66**, 3, 403–412.

Press, S. J. (1972) Estimation in univariate and multivariate stable distributions. *J. Amer. Statist. Assoc.*, **67**, 842–846.

Press, S. J. (1975) Stable distributions: probability inference and applications in finance – a survey, and a review of recent results, in *Statistical Distributions in Scientific Work*, **I** (ed Patil, G. P.), Reidel, Dordrecht.

Puri, P. S. (1975) A linear birth and death process under the influence of another process. *J. Appl. Prob.*, **12**, 1, 1–17.

Puri, P. S. and Senturia, J. (1972) On a mathematical theory of quantal response assays. *Proc. 6th Berkeley Symposium on Math., Stats. & Prob.*, **4**, 231–247. University of California Press, Los Angeles.

Raab, G. M. (1981) Estimation of a variance function, with application to immunoassay. *Appl. Statist.*, **30**, 1, 32–40.

Rabinowitz, P. (1969) New Chebyshev polynomial approximations to Mills' ratio. *J. Amer. Statist. Assoc.*, **64**, 647–654.

Racine, A., Grieve, A. P., Fluhler, H. and Smith, A. F. M. (1986) Bayesian methods in practice: experiences in the pharmaceutical industry. *Appl. Statist.*, **35**, 2, 93–150.

Rai, K. (1987) Response to Williams' reaction. *Biometrics*, **43**, 4, 1016.

Rai, K. and Van Ryzin, J. (1979) Risk assessment of toxic environmental substances using a generalized multihit dose-response model, in *Energy and Health* (ed Breslow, N. E. and Whittemore, A. S.), *SIAM*, Philadelphia, pp. 99–117.

Rai, K. and Van Ryzin, J. (1981) A generalized multihit dose-response model for low-dose extrapolation. *Biometrics*, **37**, 2, 341–352.

Rai, K. and Van Ryzin, J. (1985) A dose-response model for teratological experiments involving quantal experiments. *Biometrics*, **41**, 1–9.

Ramberg, J. S., Tadikamalla, P. R., Dudewicz, E. J. and Mykytka, E. F. (1979) A probability distribution and its uses in fitting data. *Technometrics*, **21**, 2, 201–214.

Ramsey, F. L. (1972) A Bayesian approach to bioassay. *Biometrics*, **28**, 3, 841–858.

Rao, C. R. (1954) Estimation of relative potency from multiple response data. *Biometrics*, **10**, 208–220.

Rayner, J. C. W. and Best, D. J. (1990) Smooth tests of goodness of fit: An overview. *Int. Stat. Rev.*, **58**, 1, 9–18.

Reed, L. J. and Muench, H. (1938) A simple method of estimating fifty percent endpoints. *American J. of Hygiene*, **27**, 493–497.

Reese, A. (1989) A cumulative index for GLIM Newsletter, etc. *GLIM Newsletter*, No. 18, 36–46.

Report on the LD50 test (1979) Home Office, London, 31 pp.

Ridout, M. S. (1990) Non-convergence of Fisher's Method of Scoring – a simple example. *GLIM Newsletter*, No. 20, 8–11.

Ridout, M. S. (1991) Partially sequential experimental designs for dilution series – a seed-testing problem (Abstract). *Proceedings of Joint German British Biometric Conference*, Hamburg, p. 67.

Ridout, M. S. (1992) Fitting digit preference models to fecundability data. *Genstat Newsletter* (in press).

Ridout, M. S. and Cobby, J. M. (1989) A remark on algorithm AS178: The Gauss-Jordan sweep operator with detection of collinearity. *Applied Statistics*, **38**, 420–422.

Ridout, M. S. and Fenlon, J. S. (1990) A generalized one-hit model for bioassays of insect viruses. Unpublished manuscript.

Ridout, M. S. and Fenlon, J. S. (1991) Analysing dose-mortality data when doses are subject to error. *Ann. Appl. Biol.*, **119**, 191–201.

Ridout, M. S. and Morgan, B. J. T. (1991) Modelling digit preference in fecundability studies. *Biometrics*, **47**, 4, 1423–1434.

Ridout, M. S., Morgan, B. J. T. and Pack, S. E. (1992) Fitting mixtures to overdispersed binomial data. Submitted for publication.

Ripley, B. D. (1987) *Stochastic Simulation*, Wiley, New York.

Robbins, H. and Monro, S. (1951) A stochastic approximation method. *Ann. Math. Statist.*, **22**, 400–407.

Roberts, G., Rao, J. N. K. and Kumar, S. (1987) Logistic regression analysis of sample survey data. *Biometrika*, **74**, 1, 1–12.

Robertson, C. C. and Morgan, B. J. T. (1990) Aspects of the design and analysis of signal detection experiments. *Br. J. Math. Statist. Psychol.*, **42**, 7–14.

Robinson, L. D. and Jewell, N. P. (1991) Some surprising results about covariate adjustment in logistic regression models. *Int. Stat. Rev.*, **59**, 2, 227–240.

Roger, J. H. (1985) Nearly linear models using general link functions, in *Generalized Linear Models* (eds Gilchrist, R. Francis, B. and Whittaker, J.), Springer-Verlag, Berlin, pp. 147–159.

Rose, R. M., Teller, D. Y. and Rendelman, P. (1970) Statistical properties of staircase estimates. *Perception and Psychophysics*, **8**, 4, 199–204.

Rosenbaum, P. R. (1991) Sensitivity analysis for matched case-control studies. *Biometrics*, **47**, 1, 87–101.

Rosenberger, J. L. and Kalish, L. A. (1981) Optimal designs for the simple logistic regression model. Unpublished manuscript.

Rosner, B. and Milton, C. (1988) Significance testing for correlated binary outcome data. *Biometrics*, **44**, 2, 505–512.

Ross, G. J. S. (1970) The efficient use of function minimisation in non-linear maximum likelihood estimation. *Appl. Statist.*, **19**, 205–221.

Ross, G. J. S. (1975) Simple non-linear modelling for the general user. *Bull. I. S. I.*, **46**, 2, 585–591.

Ross, G. J. S. (1986) Contribution to the discussion of Racine et al. *Appl. Statist.*, **35**, 2, 129–130.

Ross, G. J. S. (1987) *MLP. Maximum Likelihood Program*, NAG Ltd., Oxford.

Ross, G. J. S. (1990) *Nonlinear Estimation*, Springer-Verlag, New York.

Rudolfer, S. M. (1990) A Markov chain model of extrabinomial variation. *Biometrika*, **77**, 2, 255–264.

Russell, R. M., Robertson, J. L. and Savin, N. E. (1977) POLO: A new computer program for probit analysis. *Bulletin of the Entomological Society of America*, **23**, 209–213.

Ryan, B. F., Joiner, B. L. and Ryan, T. A. (1985) *Minitab Handbook* (2nd edition) Duxbury Press, Boston.

Ryan, L. (1990) Dose response models for developmental toxicity. Unpublished manuscript.

Salsburg, D. S. (1986) *Statistics for Toxicologists*, Marcel Dekker, New York.

Salsburg, D. (1990) Mathematical Modelling and the design of studies in drug research. *Drug Information J.*, **1**, 267–279.

Sampford, M. R. (1952a) The estimation of response-time distribution. I. Fundamental concepts and general methods. *Biometrics*, **8**, 13–32.

Sampford, M. R. (1952b) The estimation of response-time distributions. II: Multi-stimulus distributions. *Biometrics*, **8**, 307–369.

Sampford, M. R. (1954) The estimation of response-time distributions. III. Truncation and survival. *Biometrics*, **10**, 531–561.

Sanathanan, L. P., Gade, E. T. and Shipkowitz, N. L. (1987) Trimmed logit method for estimating the ED_{50} in quantal bioassay. *Biometrics*, **43**, 4, 825–832.

Schaefer, R. L. (1986) Alternative estimators in logistic regression when the data are collinear. *J. Statist. Comput. Simul.*, **25**, 75–91.

Schaefer, R. L., Roi, L. D. and Wolfe, R. A. (1984) A ridge logistic estimator. *Comm. Statist. Theor. Meth.*, **13**, 99–113.

Schell, M. J. and Leysieffer, F. W. (1989) An increasing failure rate approach to low-dose extrapolation. *Biometrics*, **45**, 4, 1117–1124.

Schervish, M. J. (1984) AS195: Multivariate normal probabilities with error bound. *Appl. Statist.*, **33**, 1, 81–94.

Schervish, M. J. (1985) Correction to AS195: Multivariate normal probabilities with error bound. *Appl. Statist.*, **34**, 1, 103–104.

Schmoyer, R. L. (1984) Sigmoidally constrained maximum likelihood estimation in quantal bioassay. *J. Amer. Statist. Ass.*, **79**, 448–453.

Schmoyer, R. L. (1986) Dose-response analysis under unimodality of response-to-dose, in *Advances in order restricted statistical inference* (eds Dykstra, R., Robertson, T. and Wright, F. T.), Springer-Verlag, New York.

Schneiderman, M. A., Mantel, N. and Brown, C. C. (1975) From mouse to man – or how to get from the laboratory to Park Avenue and 59th Street. *Annals of the New York Academy of Sciences*, **246**, 237–246.

Schumacher, M. (1980) Point estimation in quantal response models. *Biomedical Journal*, **22**, 315–334.

Scientific Committee of the Food Safety Council (1978, 1980) Proposed system for food safety assessment. *Food and Cosmetic Toxicology*, **16**, Supplement 2, 1–136. Revised report published June 1980 by the Food Safety Council, Washington, D.C.

Segreti, A. C. and Munson, A. E. (1981) Estimation of the median lethal dose when responses within a litter are correlated. *Biometrics*, **37**, 153–154.

Sheiner, L. B., Beal, S. L. and Sambol, N. C. (1989) Study designs for dose-ranging. *Clin. Pharmacol. Ther.*, **46**, 1, 63–77.

Shenton, L. R. and Bowman, K. O. (1977) *Maximum Likelihood Estimation in Small Samples*, Griffin, London.

Shirley, E. A. C. (1987) Applications of ranking methods to multiple comparison procedures and factorial experiments. *Applied Statistics*, **36**, 2, 205–213.

Shirley, E. A. C. and Hickling, H. (1981) An evaluation of some statistical methods for analysing numbers of abnormalities found amongst litters in teratology studies. *Biometrics*, **37**, 819–829.

Shuster, J. J. and Yang, M. C. K. (1975) A distribution-free approach to quantal response assays. *Canadian Journal of Statistics*, **3**, 57–70.

Silvapulle, M. J. (1981) On the existence of maximum likelihood estimators for the binomial response model. *J. R. Statist. Soc., B*, **43**, 3, 310–313.

Silverstone, H. (1957) Estimating the logistic curve. *J. Amer. Statist. Assoc.*, **52**, 567–577.

Silvey, S. D. (1975) *Statistical Inference*, Chapman and Hall, London.

Silvey, S. D. (1980) *Optimal Design*, Chapman and Hall, London.

Simpson, D. G. and Margolin, B. H. (1986) Recursive nonparametric testing for dose-response relationships subject to downturns at high doses. *Biometrika*, **73**, 3, 589–596.

Singh, M. (1987) A non-normal class of distribution function for dose-binary response curve. *J. Appl. Statist.*, **14**, 1, 91–97.

Sitter, R. R. and Wu, C. F. J. (1991) On the accuracy of Fieller intervals for binary response data. Technical Report No. STAT-91-08, Department of Statistics and Actuarial Science, University of Waterloo.

Slater, M. (1981) A GLIM program for stepwise analysis. *GLIM Newsletter*, **4**, 20–25.

Smith, D. M. (1983) AS189 – Maximum likelihood estimation of the parameters of the beta-binomial distribution. *Appl. Statist.*, **32**, 196–204.

Smith, D. M. (1991) Procedure Wadley, in *GENSTAT 5 Procedure Library, Manual Release 2.2*, Numerical Algorithm Group, Oxford (in press).

Smith, D. M. and James, D. A. (1984) A comparison of alternative distributions of post-implantation death in the dominant lethal assay. *Mutation Research*, **128**, 195–206.

Smith, D. M. and Morgan, B. J. T. (1989) Extended models for Wadley's Problem. *GLIM Newsletter*, **18**, 21–35.

Smith, K. C., Robertson, J. L. and Savin, N. E. (1985) Response to Mantel. *Biometrics*, **41**, 3, 781–784.

Smith, K. C., Savin, N. E. and Robertson, J. L. (1984) A monte carlo comparison of maximum likelihood and minimum chi square sampling distributions in logit analysis. *Biometrics*, **40**, 2, 471–482.

Smith, P. G. (1979) Some problems in assessing the carcinogenic risk to man of exposure to ionizing radiations, in (eds Breslow, N. and Whittermore, A.), *Energy and Health*, Society for Industrial and Applied Mathematics, Philadelphia, pp. 61–80.

Smits, P. H. and Vlak, J. M. (1988) Biological activity of Spodoptera exigua nuclear polyhedrosis virus against S. exigua larvae. *J. Invertebrate Pathology*, **51**, 107–114.

Sowden, R. R. (1971) Bias and accuracy of parameter estimates in a quantal response model. *Biometrika*, **58**, 595–604.

Spearman, C. (1908) The method of 'right and wrong cases' ('constant stimuli') without Gauss's formulae. *Brit. J. Psychol.*, **2**, 227–242.

Spiegelhalter, D. J. (1983) Diagnostic tests of distribution shape. *Biometrika*, **70**, 2, 401–410.

Sprague, J. B. (1970) Measurement of pollutant toxicity to fish. II. Utilizing and applying bioassay results. *Water Research*, **4**, 3–32.

Sprott, D. A. and Kalbfleisch, J. G. (1974) Inferences about hit number in a virological model. *Biometrics*, **30**, 199–208.

Staffa, J. A. and Mehlman, M. (eds) (1979) *Innovations in Cancer Risk Assessment*: (ED_{01} Study). Pathotox Publ., Illinois.

Staniswalis, J. G. and Cooper, V. (1988) Kernel estimates of dose response, *Biometrics*, **44**, 4, 1103–1120.

Stephan, C. E. (1977) Methods for calculating an LC50, in *Aquatic Toxicology and Hazard Evaluation* (eds Mayer, F. L. and Hamelink, J. L.), American Society for Testing and Materials, Philadelphia, pp. 65–84.

Stirling, W. D. (1984) Iteratively reweighted least squares for models with a linear part. *Appl. Statist.*, **33**, 1, 7–17.

Stirling, W. D. (1985) Heteroscedastic models and an application to block designs. *Appl. Statist.*, **34**, 1, 33–41.

Stukel, T. (1985) Implementation of an algorithm for fitting a class of generalized logistic models. *Proceedings of GLIM85*, Springer-Verlag.

Stukel, T. A. (1988) Generalized logistic models. *J. Amer. Statist. Assoc.* **83**, 426–431.

Stukel, T. A. (1989) The similarity of the Weibull, the multihit and the logistic dose-response models for low-dose extrapolation. *Bull. Int. Stat. Inst.*, 2, 363–364.

Stukel, T. A. (1990) A general model for estimating ED_{100p} for binary response dose-response data. *The American Statistician*, **44**, 1, 19–22.

Tamura, R. N. and Young, S. S. (1987) A stabilized moment estimator for the beta-binomial distribution. *Biometrics*, **43**, 4, 813–824.

Tarone, R. E. (1979) Testing the goodness of fit of the binomial distribution. *Biometrika*, **66**, 585–590.

Tarone R. E. (1982a) The use of historical control information in testing for a trend in proportions. *Biometrics*, **38**, 1, 215–220.

Tarone R. E. (1982b) The use of historical control information in testing for a trend in Poisson means. *Biometrics*, **38**, 457–462.

Tarone, R. E. (1985) Score statistics, in (eds Kotz, S. and Johnson, N.L.), *Encyclopaedia of Statistical Sciences*, Wiley, New York pp. 304–308.

Taylor, J. M. G. (1988) The cost of generalizing logistic regression. *J. Amer. Statist. Assoc.*, **83**, 1078–1083.

Thall, P. F. and Simon, R. (1990) Incorporating historical control data in planning Phase II clinical trials. *Statistics in Medicine*, **9**, 215–228.

Thomas, D. C. (1983) Nonparametric estimation and tests of fit for dose-response relations. *Biometrics*, **39**, 1, 263–268.

Thomas, D. G. (1972) Tests of fit for a one-hit vs. two-hit curve. *Appl. Statist.*, **21**, 103–112.

Thompson, R. and Baker, R. J. (1981) Composite link functions in generalised linear models. *Appl. Statist.*, **30**, 2, 125–131.

Thompson, W. A. (1977) On the treatment of grouped observations in life studies. *Biometrics*, **33**, 463–470.

Thompson, W. A. and Funderlic, R. E. (1981) A simple threshold model for the classical bioassay problem, in *Measurement of Risks* (eds Berg, G. G. and Maillie, H. D.), Plenum, New York, pp. 521–533.

Thompson, W. R. (1947) Use of moving averages and interpolation to estimate median-effective dose. I: Fundamental formulas, estimation of error and relation to other methods. *Bacteriological Reviews*, **11**, 115–145.

Tibshirani, R. J. and Ciampi, A. (1983) A family of proportional and additive-hazards models for survival data. *Biometrics*, **39**, 141–148.

Titterington, D. M. and Morgan, B. J. T. (1977) Discussion of paper by Dempster Laird and Rubin. *J. R. Statist. Soc. B*, **39**, 1, 17.

Titterington, D. M., Smith, A. F. M. and Makov, U. E. (1985) *Statistical Analysis of Finite Mixture Distributions*, Wiley, New York.

Tomatis, L., Turusov, V., Day, N. and Charles, R. T. (1972) The effect of long term exposure to DDT in CF-1 mice. *International Journal of Cancer*, **10**, 489–506.

Trajstman, A. C. (1989) Indices for comparing decontaminants when data come from dose-response survival and contamination experiments. *Appl. Statist.*, **38**, 3, 481–494.

Tsiatis, A. A. (1980) A note on a goodnes-of-fit test for the logistic regression model. *Biometrika*, **67**, 250–251.

Tsutakawa, R. K. (1967a) Random walk design in biossay. *J. Amer. Statist. Assoc.*, **62**, 842–856.

Tsutakawa, R. K. (1967b) Asymptotic properties of the block Up-and-Down method in bio-assay. *Ann. Statist.*, **38**, 1822–1828.

Tsutakawa, R. K. (1972) Design of an experiment for bioassay. *J. Amer. Statist. Assoc.*, **67**, 584–590.

Tsutakawa, R. K. (1980) Selection of dose levels for estimating a percentage point of a logistic quantal response curve. *Appl. Statist.*, **29**, 1, 25–33.

Tukey, J. W. (1949) One degree of freedom for non-additivity. *Biometrics*, **5**, 232–242.

Tukey, J. W. (1962) The future of data analysis. *Ann. Math. Statist.*, **33**, 1–67.

Tukey, J. W. (1970) *Exploratory Data Analysis*, Addison-Wesley, Reading, MA.

UKEMS (1983) United Kingdom Environmental Mutagen Society sub-committee report on guidelines for mutagenicity testing, Part I.

Vaeth, M. (1985) Wald's test in exponential families. *Int. Stat. Rev.*, **53**, 199–214.

Van den Heuvel, M. J., Clark, D. G., Fielder, R. J., Koundakjian, P. P., Oliver, G. J. A., Pelling, D., Tomlinson, N. J. and Walker, A. P. (1990) The international validation of a fixed-dose procedure as an alternative to the classical LD_{50} test. *Fd. Chem. Toxic.*, **28**, 469–482.

Vanderhoeft, C. (1985) Macros for calculating the covariance matrix of functions of parameter estimates. *GLIM Newsletter*, No. 11, 21–24.

van Montfort, M. A. J. and Otten, A. (1976) Quantal response analysis: enlargement of the logistic model with a kurtosis parameter. *Biom. Z. Bd.*, **18**, 371–380.

Van Ryzin, J. (1982) Discussion of: P. Armitage, The assessment of low-dose carcinogenicity. *Biometrics*, **38**, *Supplement:* 'Current topics in biostatistics and epidemiology', 130–139.

Van Ryzin, J. and Rai, K. (1987) A dose-response model incorporating nonlinear kinetics. *Biometrics*, **43**, 1, 95–106.

Vince, M. A. (1979) Effects of accelerating stimulation on different indices of development in Japanese quail embryos. *J. Exp. Zool.*, **208**, 2, 201–211.

Vince, M. A. and Chinn, S. (1971) Effect of accelerated hatching on the initiation of standing and walking in the Japanese quail. *Anim. Behav.*, **19**, 62–66.

Viveros, R. and Sprott, D. A. (1986) Maximum likelihood estimation in quantal response bioassay. Tech. Report Series: STAT-85-13; University of Waterloo, Canada.

Volund, A. (1980) Multivariate bioassay. *Biometrics*, **36**, 225–236.

Volund, A. (1982) Combination of multivariate bioassay results. *Biometrics*, **38**, 1, 181–190.

Vuataz, L. and Sotek, J. (1978) Use of the beta-binomial distribution in dominant-lethal testing for 'weak mutgenic activity' (Part 2). *Mutation Research*, **52**, 221–230.

Wadley, F. M. (1949) Dosage-mortality correlation with number treated estimated from a parallel sample. *Ann. Appl. Biol.*, **36**, 196–202.

Walsh, D. (1987) PCPROBIT: A user-friendly probit analysis program for microcomputers. *The American Statistician*, **41**, 1, 78.

Wang, P. C. (1985) Adding a variable in generalized linear models. *Technometrics*, **27**, 3, 273–276.

Watson, A. B. and Pelli, D. G. (1983) QUEST: A Bayesian adaptive psychometric method. *Perception and Psychophysics*, **33**, 2, 113–120.

Watson, G. A. and Elashoff, R. M. (1990) Influence diagnostics for bioassays with failure time data, in *Contributed Paper Proceedings, XVth International Biometric Conference*, 262.

Wedderburn, R. W. M. (1974) Quasi-likelihood functions, generalised linear models and the Gauss-Newton method. *Biometrika*, **61**, 439–447.

Weil, C. S. (1970) Selection of the valid number of sampling units and a consideration of their combination in toxicological studies involving reproduction, teratogenesis or carcinogenesis. *Food Cosmet. Toxicol.*, **8**, 177–182.

Weinberg, C. R. and Gladen, B. C. (1986) The beta-geometric distribution applied to comparative fecundability studies. *Biometrics*, **42**, 547–560.

Wetherill, G. B. (1963) Sequential estimation of quantal response curves. *J. R. Statist. Soc. B*, **25**, 1–48.

Wetherill, G. B. (1981) *Intermediate Statistical Methods*, Chapman and Hall, London.

Wetherill, G. B. and Chen, H. (1965) Sequential estimation of quantal response curves, II. Technical Memo 65-1215-1. Bell Telephone Laboratory, Holmdel, New Jersey.

Wetherill, G. B., Chen, H. and Vasudeva, R. B. (1966) Sequential estimation of quantal response curves: A new method of estimation. *Biometrika*, **53**, 439–454.

Wetherill, G. B. and Glazebrook, K. D. (1986) *Sequential Methods in Statistics* (3rd edition), Chapman and Hall, London.

Wetherill, G. B. and Levitt, H. (1965) Sequential estimation of points on a psychometric function. *Br. J. Math. Statist. Psychol.*, **18**, 1–10.

Wharton, R. M. and Srinivasan, R. (1990) The sequential estimation of quantal response curves for small sample sizes. *Biom. J.*, **32**, 207–212.

White, L. (1975) Ph.D. Thesis, Imperial College, London.

White, R. F. and Graca, J. G. (1958) Multinomially grouped response times for the quantal response bioassay. *Biometrics*, **14**, 462–488.

Whitehead, A. and Curnow, R. N. (1991) Statistical evaluation of the fixed dose procedure (in press).

Whitehead, J. (1983) Fitting stratified case-control models using GLIM-3. *GLIM Newsletter*, No. 8, 37–44.

Whitehouse, G. H. (1985) Screening for breast cancer – a contemporary issue. *Radiology Now*, **8**, 109–114.

Whittemore, A. S. (1983) Tranformations to linearity in binary regression. *SIAM J. Appl. Math.*, **43**, 703–710.

Williams, D. A. (1970) Discussion of Atkinson (1970).

Williams, D. A. (1971) A test for differences between treatment means when several dose levels are compared with a zero dose control. *Biometrics*, **27**, 1, 103–118.

Williams, D. A. (1972) The comparison of several dose levels with a zero dose control. *Biometrics*, **28**, 2, 519–532.

Williams, D. A. (1975) The analysis of binary responses from toxicological experiments involving reproduction and teratogenicity. *Biometrics*, **31**, 949–952.

Williams, D. A. (1982a) Extra-binomial variation in logistic linear models. *Appl. Statist.*, **31**, 144–148.

Williams, D. A. (1982b) GLIM and Hirayama's data. *Royal Statistical Society News and Notes*, **9**, 7.

Williams, D. A. (1983) The use of the deviance to test the goodness of fit of a logistic-linear model to binary data. *GLIM Newsletter*, No. 6, 60–62.

Williams, D. A. (1986a) A note on Shirley's nonparametric test for comparing several dose levels with a zero-dose control. *Biometrics*, **42**, 1, 183–186.

Williams, D. A. (1986b) Interval estimation of the median lethal dose. *Biometrics*, **42**, 3, 641–646.

Williams, D. A. (1986c) A response to recent publications on multivariate bioassay. Invited paper, *XIIIth International Biometrics Conference*, Seattle.

Williams, D. A. (1987a) Generalized linear model diagnostics using the deviance and single case deletions. *Appl. Statist.*, **36**, 2, 181–191.

Williams, D. A. (1987b) Reader Reaction: Dose-response models for teratological experiments. *Biometrics*, **43**, 4, 1013–1016.

Williams, D. A. (1988a) Reader Reaction: Estimation bias using the beta-binomial distribution in teratology. *Biometrics*, **44**, 1, 305–307.

Williams, D. A. (1988b) Extra-binomial variation in toxicology, pp. 301–313 of *Proceedings of the XIVth International Biometric Conference*, Namur. 18–23 July.

Williams, D. A. (1988c) Overdispersion in logistic-linear models, pp. 165–174 of the *Proceedings of The Third International Workshop on Statistical Modelling*, Vienna, 4–8 July.

Williams, D. A. (1988d) Tests for differences between several small proportions. *Applied Statistics*, **37**, 3, 421–434.

Williams, D. A. (1989) Hypothesis tests for overdispersed generalised linear models. *GLIM Newsletter*, **18**, 29–39.

Williams, T. (1965) The basic birth-death model for microbial infections. *J. R. Statist. Soc. B.*, **27**, 338–360.

Wu, C. F. J. (1985) Efficient sequential designs with binary data. *J. Amer. Statist. Assoc.*, **80**, 974–984.

Wu, C. F. J. (1988) Optimal design for percentile estimation of a quantal response curve, in *Optimal Design and Analysis of Experiments* (eds Dodge, Y., Federov, V. V. and Wynn, H. P.), Elsevier, Amsterdam, pp. 213–223.

Wynn, H. P. (1972) Results in the theory and construction of D-optimum experimental designs. *J. Roy. Statist. Soc. B*, **34**, 133–147.

Zacks, S. (1977) Problems and approaches in designs of experiments for estimation and testing in non-linear models, in (ed Krishnaiah, P. R.), *Multivariate Analysis IV*. North-Holland, Amsterdam, pp. 209–223.

Zbinden, G. and Flury-Roversi, M. (1981) Significance of the LD_{50}-test for the toxicological evaluation of chemical substances. *Arch. Toxicol.*, **47**, 77–99.

Zerbe, G. O. (1978) On Fieller's Theorem and the general linear model. *The American Statistician*, **32**, 3, 103–105.

Author index

Subject index

Printed and bound by CPI Group (UK) Ltd, Croydon, CR0 4YY

23/10/2024

01778237-0001